AF615558

Chemical and Molecular Basis of Nerve Activity

Chemical and Molecular Basis of Nerve Activity

DAVID NACHMANSOHN

WITH

SUPPLEMENT I

Properties and Function of the Proteins of the Acetylcholine Cycle in Excitable Membranes

DAVID NACHMANSOHN

Departments of Neurology and Biochemistry, College of Physicians and Surgeons, Columbia University, New York, New York

AND

SUPPLEMENT II

Toward a Molecular Model of Bioelectricity

EBERHARD NEUMANN

Max-Planck-Institut of Biophysical Chemistry Goettingen-Nikolausberg, Germany

1975

ACADEMIC PRESS New York San Francisco London

A Subsidiary of Harcourt Brace Jovanovich, Publishers

ACADEMIC PRESS, INC.
111 Fifth Avenue, New York, New York 10003

United Kingdom Edition published by
ACADEMIC PRESS, INC. (LONDON) LTD.
24/28 Oval Road, London NW1

Library of Congress Cataloging in Publication Data

Nachmansohn, David, (date)
Chemical and molecular basis of nerve activity.

Bibliography: p.
Includes index.
PARTIAL CONTENTS: Supplement 1: Nachmansohn, D. Properties and function of the proteins of the acetylcholine cycle in excitable membranes.– Supplement 2: Neumann, E. Toward a molecular model of bioelectricity.
1. Neurochemistry. 2. Neural transmission. I. Nachmansohn, David (date) Properties and function of the proteins of the acetylcholine cycle in excitable membranes. 1975. II. Neumann, Eberhard. Towards a molecular model of bioelectricity. 1975. III. Title. [DNLM: 1. Neurochemistry. WL104 N122c]
QP356.3.N3 1975 612′.814 75-8750
ISBN 0–12–512757–X

PRINTED IN THE UNITED STATES OF AMERICA

Contents

Preface

Almost sixteen years have elapsed since this monograph was originally published. During this period many dramatic developments in the biological sciences have provided insights into the mechanisms of living cells on the molecular, subcellular, and cellular levels to a degree that only two decades ago seemed beyond the possibility of experimental testing. A few examples, selected at random, follow. The establishment of the tridimensional structure of biopolymers, nucleic acids, and proteins, including enzymes, by the use of X-ray diffraction methods created a new approach to analyzing their mechanisms of action. Electron microscopy, in combination with chemical and physicochemical analyses, provided a basis for the study of cell membranes. Since membranes were recognized as the site of the most vital of cellular functions, biomembranes have become one of the most actively explored fields in present-day biology. Moreover, the study of cellular behavior revealed striking differences among chemical reactions studied in solution and their modification due to membrane structure. Such spectacular progress has been made in genetics, evolution, cell differentiation, immunochemistry, among others, that they have become almost indispensable tools in biochemical studies for the understanding of cellular mechanisms.

Instrumental in all these developments has been the use of new and highly sophisticated physical methods such as fluorospectrometry, nuclear magnetic resonance, circular dichroism, and spin labeling in addition to the already mentioned use of X-ray diffraction and electron microscopy. The development of methods that permit precise measurements of formerly "unmeasurably fast" reactions, presented by Manfred Eigen in his Nobel Prize lecture, have played a special role because of their applicability to an extraordinarily wide range of physical, chemical, and biological problems. Relaxation spectrophotometric techniques have reached a level permitting measurements of reactions covering a time range from seconds to nanoseconds, an achievement thought impossible only fifteen years ago. The rapidity and revolutionary character of all these advances in biological sciences reminds one of the exciting period of the 1920's and 1930's during which atomic physics changed man's grasp of the physical world, and had an impact on the intellectual and philosophical level more profound than the entire preceding century.

As in all fields of biological science, these recent developments obviously have had profound influence on the problems of nerve impulse conduction, nerve excitability, and generation of bioelectricity. The vast number of pertinent and exciting advances in this particular field require a new monograph which is at present in preparation. The question may be raised whether it is useful under these circumstances to reprint a monograph written almost sixteen years ago. For several reasons an affirmative answer seems appropriate. The monograph summarizes the results of the first two decades of biochemical research in a field which until then was predominantly the domain of physiology and pharmacology. The investigations were based on the analysis of the properties and function of the specific proteins and enzymes associated with the action of acetylcholine initiated by the author almost four decades ago. This type of approach was essentially due to the background and training of the author. As outlined in his prefatory chapter Biochemistry as Part of My Life in the *Annual Review of Biochemistry* in 1972, the author had spent his most formative years at the Kaiser–Wilhelm Institutes in Berlin–Dahlem, which at that time comprised the most advanced center for the study of biochemistry. Also at the center, at that time, proteins and enzymes were stressed in research on cellular mechanisms.

When this monograph first appeared, few biologists interested in nerve activity believed that the analysis of proteins was a basic necessity for the understanding of molecular events underlying bioelectricity and nerve excitability. This situation has changed drastically. During the last decade the purely descriptive phenomenology of the events during electrical activity has come under vigorous criticism from many sources. It is now widely recognized that an understanding of nerve excitability requires knowledge of the molecular events within the excitable membrane responsible for this cellular function. Today, literally hundreds of investigators are working on the properties and function of proteins linked to the action of acetylcholine, two of which were already isolated and characterized by the author at the time this book was first published, and another two postulated. This change of attitude has revived the interest in the early phase of the biochemical approach presented in the monograph. There are many facts established in that early period which are still fully valid. Some of the incomplete information has been greatly enhanced by new advances. Some of the data obviously became questionable and some obsolete when tested with the new, more elaborate, and sophisticated techniques. However, much more pertinent is the fact that the fundamental concepts underlying the author's approach are still valid and are the basis of many advances of the last decade. Many proposals and postulates, some of them quite speculative, have been confirmed by recent experimental data. Although the basic concepts remain unchanged, many details have been modified to accommodate the ever-increasing amount of information. Einstein remarked once in a discussion with Werner Heisenberg, reported by the latter in a recent publication: "Whether you can observe a thing or not, depends on the theory you use. It is the theory which decides what can be observed."

It should be of interest to many to read about the origin of postulates confirmed experimentally many years later, such as the existence of an acetylcholine receptor protein, or the assumption that a conformational change of this protein may be induced by reaction with acetylcholine and may trigger the events leading to increased ion permeability, or the proposal of the cyclic nature of the reactions involving acetylcholine (see page 101). In some areas, e.g., those concerning the role of acetylcholine at synaptic junctions, many problems were at that time open to question. However, new data have drastically changed the situation. They have provided the basis for a much better understanding of the differences and similarities between conducting and synaptic parts of the excitable membrane and the specific function of the acetylcholine cycle in both parts.

Copies of this monograph have been unavailable for a decade. New developments have apparently revived interest in the early phase of investigations on the chemical basis of nerve activity. Knowledge of the initial efforts should be informative and useful to many investigators who at present work on the problems discussed in that first publication. Many readers, however, may not be aware of the striking new developments of the last decade supporting many views and postulates presented at that time; others may be interested in the present views of the author as to recent observations pertinent to his original concepts. Therefore two supplements in which recent advances are presented in condensed form have been added. The revised monograph now not only has historical value, but gives the reader an idea of the exciting progress made in this field, and thus acts as a bridge to the new monograph in preparation.

The first supplement concentrates on recent progress in the biochemistry of excitability, while in the second an *attempt* at an integral model of nerve excitability is described. This most exciting recent development was initiated by the late Aharon Katchalsky; a tentative version was published in 1973 (*Proc. Nat. Acad. Sci. U.S.*). The model is an attempt to integrate basic electrophysiological, biochemical, and physicochemical data on nerve excitability. During the last two years Dr. Neumann has greatly extended and broadened the theoretical basis of the model in physicochemical terms. The model, as is the nature of all models, is seen as a working hypothesis detailed to such an extent that it offers many challenges to the researcher.

Dr. Neumann kindly accepted an invitation to write the second supplement in which a presentation of his integral model in its present form is included. The author would like to express to him his thanks for this invaluable addition to the monograph. He also gratefully acknowledges Dr. Neumann's help in the formulation of many ideas which have greatly improved Supplement I.

DAVID NACHMANSOHN

Preface to First Edition

The spectacular progress in biology and biochemistry during the last two decades has revolutionized our understanding of cellular mechanisms and has extended our knowledge of life processes far beyond the limits foreseeable only a generation ago. Many aspects of cellular function are today analyzed on molecular levels. The conduction of nerve impulses, the most important function of the nervous system and one of its most striking and remarkable features, is one of the fields in which much information has been obtained as to the underlying chemical and molecular forces. A comprehensive presentation of the development of this particular problem and an evaluation of its present state appears, therefore, desirable. To provide such a contribution is the aim of this monograph.

In the 1930's there were lively discussions at the meetings of the Physiological Society of England about the question of whether or not acetylcholine was a neurohumoral transmitter across myoneural and synaptic junctions. The interpretation of the role of acetylcholine was based essentially on pharmacological observations and was in contradiction to the conclusions based on electrophysiological evidence. Not the experimental data but interpretations of their meaning were hotly debated; even the most ardent proponents of the hypothesis of neurohumoral transmission admitted that there were many gaps and serious contradictions. The sharp conflict of views was a challenge to initiate an entirely new approach. The development of new chemical and biochemical methods and procedures, the rapid growth of protein and enzyme chemistry, and the notions and principles developed in the study of the chemical basis of other cellular functions, notably that of muscular contraction, seemed to offer great promise of providing a more satisfactory answer to the fascinating problem of nerve activity in general and in particular to that of the role of acetylcholine in this process.

Studies of the author and his associates, initiated about 24 years ago, have made it possible to establish the sequence of energy transformations associated with nerve activity and to integrate the formation and hydrolysis of acetylcholine into the metabolic pathways of the nerve cell. Proteins and enzymes of the acetylcholine system have been isolated, their properties, the molecular forces in their active sites, and their reaction mechanisms have been analyzed. Many relationships between chemical reactions and electrical events, between molecular forces active in the proteins in solutions and manifestations of the intact cell, or even in the intact animal, have been established. The results led very soon to a modification of the original hypothesis of neurohumoral transmission. They have shown that the action of acetylcholine is not an *inter-* but an *intra*-cellular, or rather an intra-membranous process essential for the generation of bioelectric potentials in all conducting membranes throughout the animal kingdom. The acetylcholine system is necessary for controlling the ion movements which form the basis of electrical manifestations in living cells.

As might have been expected, the new concept first proposed in 1940 initially met with much skepticism and frequently vigorous opposition and criticism. This is a natural and healthy reaction in all scientific fields. Some of the objections were helpful in that it became necessary to conduct crucial experiments to obtain an answer to justifiable doubts and questions. During the last two decades a considerable amount of experimental data has accumulated in support of the new concept. The idea has proved to be fruitful and has provoked investigations which have yielded information pertinent to the problem of cellular mechanisms in general.

Acknowledgments

This monograph is dedicated to all those who have stimulated, encouraged, and helped the author in his work. He would like to express his gratitude to Dr. H. Houston Merritt, Professor of Neurology, Dean of the Medical School. Thanks to his efficient and unwavering support over many years it was possible to build up an active group of investigators and all the necessary facilities, thus providing the basis for the progress of the last decade.

Among the scientific colleagues and friends who stimulated his scientific thinking the author would like to mention in the first place Otto Meyerhof and the group of colleagues associated with him. The years spent in the unique atmosphere of the Kaiser Wilhelm Institutes in Berlin-Dahlem in the 1920's, at that time a center with many brilliant scientists, had a decisive impact on the author's scientific formation. The years at the Sorbonne initiated happy and fruitful associations with his French friends, among them René Wurmser, Edgar Lederer, and René Couteaux. The frequent discussion with Sir Frederick Gowland Hopkins, when the author worked in his laboratory, were invaluable.

Dr. John F. Fulton, in whose laboratory at Yale University the author spent three years, was the first neuro-physiologist to fully endorse the new theory and he has forcefully supported the new concept ever since.

The author would like to express his thanks to his many able collaborators who are at present and have been in the past in his laboratory. Without their help the work would never have advanced to the point which it has reached. It was particularly fortunate that Irwin B. Wilson, well trained in modern physical and physical organic chemistry in Columbia's Chemistry Department, joined the laboratory in 1949. The association with him over many years was extremely stimulating and pleasant and, it seems to the author, fruitful for the development of the field.

The generous financial support of several government agencies is gratefully acknowledged, in the first place of the United States Public Health Service, which supplied by far the greatest part of the funds. Essential additional support was received from the National Science Foundation, the Atomic Energy Commission, and the Surgeon General of the Army. Grants from private Foundations have been very helpful, in the earlier phase from the Dazian Foundation for Medical Research and the Josiah Macy, Jr., Foundation, at present from the Rachel Mellon Walton Foundation and from the Muscular Dystrophy Associations of America, Inc.

D. NACHMANSOHN

College of Physicians and Surgeons, New York
August, 1959

CHAPTER I

Physical Events during Nerve Activity

A. Electrical Manifestations

The primary function of nerve cells is that of receiving and carrying messages. They form the communication system between the outer world and the organism and between distant points of the body. The most important and vital functions of our organism are to a great extent controlled and regulated by the nervous system. The brain is the site of the human intellect, of memory, of men's creative forces. It is not surprising, in view of the paramount importance of the nervous system, that Galvani's observations on "animal electricity," published in 1791, and his suggestions about the role of electricity in nerve activity were received with great enthusiasm and passionate interest not only by scientists but by intellectuals all over the world. For the first time there seemed to be a ray of hope of penetrating into one of the great mysteries of the living world. Volta's criticism and his conclusions that Galvani's observations did not really prove his interpretation, were justified. Nevertheless, it turned out that Galvani's ideas were correct and that nerve activity is associated with electric currents. It took, however, half a century until the experiments of Matteucci from Pisa and DuBois-Reymond in Germany firmly established that a flow of current takes place in animal tissue. They discovered, with the aid of the newly developed galvanometers, that current flows through the instrument connected by electrodes with the longitudinal surface of the muscle fiber and the cut end. The former was electropositive, the latter electronegative, and in the external circuit the current flows toward the cut end. It was later recognized that this current flow is due to injury and it is, therefore, referred to as injury or demarcation current. DuBois-Reymond also noticed that the current flow diminished during activity. The observation indicates that the active point becomes less positive in relation to the cut end, i.e., negative to the resting surface. Bernstein later observed that this negativity was propagated like a wave with a velocity equal to that determined by Helmholtz for the rate of propagation of the nerve impulse. Since that time the conduction of the nerve impulse has been identified with the wave of negativity sweeping down the fibers. This flow of current is today usually referred to as action current, and the potential developed, as "spike" potential.

Following the early observations of Matteucci and DuBois-Reymond neurophysiology was for more than a century almost synonymous with electrophysiology. Progress in this field was greatly facilitated by the continuous improvement of highly sensitive recording instruments, culminating in the introduction of the cathode-ray oscillograph for electrophysiological measurements by Erlanger and Gasser more than thirty years ago. Electrophysiology has given us much important information about various properties of neurons in rest and during activity, about such features as facilitation, inhibition, and summation of impulses, about the time relations of the different parameters of the action potentials, and about the refractory period, the absolute and the relative one, during which the nerve excitability is either abolished or greatly decreased. Characteristics of excitability and stimulus strength, differences between axon and cell body, speed of propagation in various types of nerve fibers encountered in the animal kingdom, have been extensively studied. The localization of many specific functions and pathways of many neurons have been elucidated by electrophysiological techniques. Without attempting merely to enumerate all the aspects studied, it may be stated that the knowledge achieved as to the electrical properties of neurons and their electrical manifestations during activity form an integral part of the physiology of the nervous system.

B. The Membrane Theory

Neurophysiologists of the nineteenth century were well aware of the fact that knowledge of the electrical properties and the electrical manifestations of nerve activity are not sufficient for the understanding of the underlying mechanism. Whereas in the copper wire electron movements are responsible for the flow of current, the neuron, being a fluid system, can only be a second-degree conductor. The electric currents must in this case be carried by ions. This was clearly recognized by biologists of the last century. More detailed theories as to the mechanism of the generation of bioelectric currents began to be formulated in the latter part of the nineteenth century, when physicochemical studies revealed the great potential differences which may develop at semipermeable membranes. Notions and ideas proposed by Traube, Ostwald, Nernst, and other physical chemists led to the so-called membrane theory which is still the basis of most modern concepts concerning the mechanism of nerve conduction. The theory is best known through the formulation of Bernstein and Tschermak early in this century (Bernstein, 1902). According to this theory the nerve fiber is surrounded by a semipermeable membrane which has a positive charge on the outside and a negative one on the inside. It is selectively permeable for K^+. When a stimulus reaches a membrane, the permeability at the active site is greatly increased for all ions with a concomitant decrease in resistance. The active part becomes depolarized; thereby small electric currents are generated which stimulate the adjacent points and the same process takes place there. In this way successive parts of the membrane are activated and the impulse propagated along the axon.

It is a remarkable tribute to the ingenuity of the physical chemists of the last century that, in spite of all the great progress of methods and knowledge during the twentieth century, the membrane theory has remained the basis of our present concepts. One major modification has become necessary: at the activated point there is not merely a depolarization but a reversal of charge: the inside becomes positive and the outside becomes negative. This has been shown by Curtis and Cole (1942) and by Hodgkin and Huxley (1945) by insertion of microelectrodes into the interior of an axon. The material used in these experiments was the giant axon of squid (*Loligo pealii*), which has a diameter of about 400–800 μ, depending on the size of the specimen. The technique applied made it possible to measure directly the potential between the inside and the outside electrode. During the passage of the impulse the charge does not merely disappear, as was assumed in the original theory, but is reversed. By this "overshoot" the action potential becomes about twice as great as the resting potential.

An important advance was the experimental evidence by Cole and Curtis (1939) for a breakdown of resistance during activity. According to their measurements, which were carried out on the giant axon of squid, the resistance during activity drops from about 1000 to 40 ohms square centimeter.

C. Ion Movements

One of the characteristics of most living cells is the fact that the concentration of ions in the interior is quite different from that in the extracellular fluid. Sodium (Na) on the outside is usually about ten times as high as inside the cell, the reverse is true for potassium (K). In some types of tissue the differences are even higher. The conducting cell has developed the special ability to make use of the ionic concentration gradients resulting from the unequal distribution of ions for its special function, i.e., for the generation of small electric currents which conduct the impulse. Overton (1902) clearly recognized the specific function of Na for conduction. He found that a muscle kept in Na-free solution became inexcitable. No other ion except lithium was able to replace Na. He also suggested that conduction was associated with an exchange of extracellular Na with intracellular K.

The availability of radioactive ions after the Second World War made it possible to measure quantitatively ion movements in rest and during activity. The first measurements of this type were carried out on the writer's suggestion, and in his laboratory, by Rothenberg in the summers of 1947 and 1948. Rothenberg exposed the giant axon of squid to solutions containing radioactive ions and found that both Na and K move across the membrane in resting condition; the equilibrium is dynamic (Rothenberg and Feld, 1948). During activity a sudden influx of Na takes place. The influx of Na per square centimeter of surface per impulse was estimated to be 4 μμmoles (Rothenberg, 1949, 1950). This figure was later confirmed by experiments of Hodgkin and Huxley and their associates in Cambridge. Keynes and Lewis (1951) found that an equivalent amount of K leaks to the outside. This could have been expected, since the cell interior must keep its electrical neutrality. The analysis of the ionic movements during activity has been extended by the Cambridge group with methods based on principles worked out by Cole (Cole, 1949, 1955; Hodgkin, 1951, 1957; Huxley, 1954). Their investigations have shown that during activity there is at first a specific and rapid, but transitory, increase of permeability to Na. This change permits Na to enter during the rising phase of the action potential. The permeability of the active membrane for this ion species was estimated to be several hundred times as high during the active phase as in resting condition. The movement of Na into the interior makes the inside positive. The current carried by Na^+ reaches its peak very rapidly and falls nearly as fast. The increase in permeability to K is slower and much less pronounced than that to Na. K^+ move during the descending phase to the outside.

The role of calcium (Ca) during activity has also been investigated. The spontaneous activity of nerve fibers in Ca-poor Ringer's solution has been known for a long time. When a squid fiber conducts impulses, a small quantity of Ca enters (Flückiger and Keynes, 1955; Hodgkin and Keynes, 1957): the quantity per square centimeter is only 0.006 μμmoles or 1/700 of that of Na^+. Hodgkin (1957) discusses the possibility that the movement of Ca may have something to do with the change of permeability; the local electric field inside the membrane may be altered by absorption of calcium ions to the membrane without changing the over-all potential difference between external and internal solution.

D. Heat Production

The great number of messages which the nervous system continuously carries throughout the organism makes it nearly imperative to perform this function in an economic way, i.e., with a minimum expenditure of energy. Nature has indeed developed a mechanism which requires an amazingly small amount of energy for the primary event, as indicated by the small amounts of heat production. The development of extremely sensitive recording instruments by A. V. Hill and his associates enabled them to measure amounts of heat of a very small order of magnitude. The results have been well summarized and evaluated by Hill (1932a, b) and by Feng (1936). Only a few figures may be mentioned as an illustration. The greatest rate of heat production in frog sciatic nerve due to steady stimulation (with frequencies below 50/sec.) is of the order of magnitude of 40×10^{-6} cal. per gram nerve per second. At low frequencies the total yield per impulse is about 1 μcal. per gram at 19° C.

As in muscle one may distinguish two phases: the initial heat, closely associated with activity, and the delayed or recovery heat. Whereas in muscle the two phases are of a similar order of magnitude, the initial heat produced during nerve activity is extremely small. In frog nerve at 20° C. the initial heat is about 7×10^{-8} cal. per gram per impulse; at 0° it is about four times as high (26×10^{-8} cal.). The surface per gram nerve in frog sciatic is estimated to be about 1600 cm.^{2}. The initial heat per square centimeter at 20° C. would then be about 4×10^{-11} cal. per impulse. In *Maja* nerve the initial heat at this temperature is about three times as great as that of frog's nerve at 0° C. and at least ten times as great as that at 20° C.

The initial heat appeared to be so small that the question has been raised whether a chemical reaction could be responsible for its origin. It was suggested that the exchange of Na^+ and K^+ between axoplasm and outside fluid may account for it (Hodgkin, 1951). In this view the generation of bioelectric currents does not require a chemical reaction, but may be effected by a purely physical process.

Recently, significant new information about the initial heat has been obtained by A. V. Hill and his associates (Abbott, Hill, and Howarth, 1958) with still more rapid recording equipment than was previously available. The initial heat production of a nonmedullated nerve (*Maja*) has been separated into two successive phases: a positive heat production, averaging in a single impulse at 0° C. about 9×10^{-6} cal. per gram nerve; this phase is rapid and is probably associated with the active phase of the impulse. In the second phase, lasting for about 300 msec., heat is absorbed, averaging about 7×10^{-6} cal. per gram nerve. With the previous methods only the difference between the two processes, the "net heat," i.e., about 2×10^{-6} cal. per gram nerve, was recorded.

Maja nerves contain fibers from 20 to 0.3 μ diameter. About half the heat is probably derived from fibers less than 3 μ. The conduction velocities at 0° C. vary from 1.4 to 0.1 meters per second; impulses reach the recording thermojunctions through a long interval. The observed heat production is the result of positive and negative components in different fibers, and a substantial part of each is masked. The real positive and negative heats, therefore, are substantially greater than those observed. Using the most likely estimate of velocity distribution, the authors arrived at values of 14×10^{-6} cal. and -12×10^{-6} cal. per gram per impulse at 0° C.

For the evaluation of these figures it is important to realize that the process of nerve impulse conduction is generally believed to be a surface phenomenon; the ions are assumed to move across a membrane of about 50 A. thickness. The observed permeability change must take place in this layer, and so should chemical events which may be associated with this change. The structural organization of chemical processes and their significance will be discussed later. The surface in 1 gm. *Maja* nerve is estimated to be 10^4 cm^2. Referring the heat to 1 gm. surface material instead of to 1 gm. nerve, Hill and his associates arrive at the remarkably high value of 2.8×10^{-3} cal. per impulse, which is about the same order of magnitude as the heat produced per gram in a muscle twitch.

What then is the origin of the heat production? In his lecture on "Chemical wave transmission in nerve," A. V. Hill (1932b) considered the possibility of energy liberation in the active membrane in which large changes of permeability are produced and reversed during activity. In their recent paper, the authors (Abbott *et al.*, 1958) write: "It is difficult indeed to imagine an excitable membrane going through a complete cycle involving a several 100-fold increase of permeability to Na ions followed by a similar increase of permeability to K ions, and yet behaving as a conservative system without change of energy. . . . It is hard to believe that so drastic a cycle of physicochemical change could occur in material like that of the excitable membrane without the intervention of work or chemical reaction" (see also A. V. Hill, 1959).

Among some explanations discussed by Hill and his associates is the possibility that the positive heat is derived from the energy released during the rising phase of the action potential, in the discharge of the condenser which exists all over the excitable membrane. The negative heat would then be due to the absorption of energy in recharging the condenser during the falling phase. This "condenser theory" encounters the difficulty that the time relations seem to be wrong, as pointed out by the authors.

The heat of ionic interchange can at present be derived only from very indirect type of information. The quantities involved are much too small for direct measurements in a calorimeter. The actual conditions under which the exchange in nerve takes place are quite different from those of the experimental measurements from which the values have been estimated and extrapolated to the physiological event. Emphasizing these uncertainties, the authors try to evaluate the various available previous data on ion flux and additional new data obtained on *Maja* nerve under conditions as similar as possible to those of their heat experiments. They arrive at a figure which amounts to about half the initial positive heat.

A fact pertinent for overcoming this and other uncertainties in the interpretation of the origin of the heat is the absence of any marked effect of temperature on the net heat. The initial positive and negative heats cannot be evaluated at higher temperatures, since at present instruments are not sufficiently rapid to separate them. The near independence of temperature favors the theory that the net heat is due to chemical reactions involved in the permeability cycle; for if the cycle is the same, though occurring more quickly at a higher temperature, the reactions would probably be the same. For the condenser theory the temperature effect has no bearing, since the net heat cannot be attributed to condenser discharge and recharge. For the theory attributing the net heat to the interchange of sodium and potassium, the small temperature effect is unfavorable, since the amount of interchange observed is much greater at a lower temperature. "For this reason," the authors write, "it seems very unlikely that the net heat could be derived from the ionic interchange alone."

The observations on heat production in *Maja* nerve and the conclusions of A. V. Hill and his associates find strong support from measurements on heat released by the discharge of the electric organ of *Torpedo marmorata* (Abbott, Aubert, and Fessard, 1958). Three successive phases of heat associated with the discharge were observed: the first phase was always a positive heat, produced most likely simultaneously with the discharge; this positive phase is closely followed by a negative one in which heat is absorbed, but which is usually masked by a prolonged third phase in which again heat is produced. This latter heat production may amount within 2 or 3 min. at 20° C. to a value several times as high as that of the initial heat. It is interesting that the total amount of energy released by the organ during and after the discharge increases with the amount of electrical work performed (which may be achieved by varying external resistance). As was recently stated by A. V. Hill (1959), this complex cycle of positive and negative phases must be a sign of chemical changes so extensive that they ought to be susceptible to biochemical analysis. These changes, moreover, must be anaerobic because under the experimental conditions the amount of oxygen available must have been negligible.

The results on the heat production in *Torpedo* are in contradiction with those of Bernstein and Tschermak (1906), who thought to have observed a negative heat associated with the discharge of the fish in cases where external work was performed; they assumed that the energy of the discharge is provided by the diffusion of ions in the direction of their concentration gradients. According to this hypothesis, the work done should be associated with heat absorption. The fallacy of their conclusion that their observations contradict a chemical origin of the generation of bioelectric currents, has been pointed out by Meyerhof (1925) in his "Thermodynamik des Lebensprozesses." The recent observations carried out with greatly improved recording instruments are incompatible with the simple explanation offered by Bernstein and Tschermak and the fundamentally similar views of Hodgkin (1951).

E. Temperature Coefficient

For the question whether purely physical processes may be assumed to be responsible for the action potential, another type of measurement is pertinent: that of the temperature coefficient. There are relatively few data reported in the world literature about the Q_{10} of bioelectric currents. Hodgkin and Katz (1950) did find high temperature coefficients, but this finding was not considered relevant to the problem of the nature of the conducting process. According to Keynes and Martins-Ferreira (1953) the Q_{10} of the action potentials in the electroplax of *Electrophorus electricus* is 1.08 ± 0.04. The authors observed, however, the increased duration of the spike at lower temperature; the Q_{10} of this effect, according to their measurements, was 2.48 ± 0.07. The significance of this high Q_{10} is not discussed.

Recently, Schoffeniels (1958b) has evaluated the Q_{10} and the energy of activation of bioelectric potentials over a wide range of temperature, using the isolated single electroplax of the electric organ of *Electrophorus electricus.* He determined the duration of the action potential of the electroplax, of the postsynaptic potential, and of the latency period of synaptic transmission as a function of temperature. He found that the duration of all three phenomena decreases with rise of temperature, whereas the amplitude of the spike and of the postsynaptic potential remains unchanged. Since during the action potential there is a marked transitory change of permeability, the duration of the spike is a good measure of this change and pertinent to the question whether or not chemical reactions are required for this process. A straight line was obtained when the logarithm of the reciprocal of the half width of the spike was plotted against the reciprocal of the temperature according to the principle of Arrhenius. The Q_{10} of the action potential was found to be around 3.6 and the energy of activation to be 21,000 cal. per mole. The Q_{10} of the latency period and of the postsynaptic potential are close to 2.6 and the energy of activation around 16,000 cal. per mole.

CHAPTER II

Problems of Mechanisms Underlying Nerve Activity

The recordings of the physical events taking place during nerve activity have provided important information. We have briefly referred to the knowledge accumulated about the electrical properties and characteristics of nerve cells and axons. The data on heat production have provided an indication as to the magnitude of energy transformations. The analysis of ion movements has informed us about the amounts and types of ions moving across the membrane and the sequence of changes of Na and K conductance. There is wide agreement that the ionic concentration gradients are the primary source of energy of electromotive force (EMF) and that ions are the carriers of bioelectric currents. But this immediately raises the question: What is the mechanism underlying the generation of bioelectric currents? More precisely, what processes are responsible for the sudden transitory change in permeability to Na and how do they control the ion movements during activity? What is the trigger by which the potential source of energy, i.e., the ionic concentration gradients, becomes suddenly effective? Do we have a clue to the molecular events that take place in this process?

A. Necessity of Correlating Physical Events with Chemical Reactions

Two diametrically opposing views confront each other. Hodgkin and his associates have maintained for many years that the assumption of a chemical reaction during activity is unnecessary. They consider the change of permeability to be a physical process. In his latest summary (1957), Hodgkin, discussing the "nature of the permeability change," does not even refer to the possibility of a chemical reaction being involved. The writer and his colleagues, on the other hand, have always belonged to that group which strongly emphasized the opposite view, i.e., the necessity of a chemical reaction underlying the change in permeability (or conductance) during activity. The character of chemical reactions taking place has been the main subject of investigation of the writer for a period of over twenty years. They form the topic of this monograph. There are many questions still open. But the principle that a permeability change in a living membrane is unthinkable without chemical reactions appears a necessity beyond reasonable doubt. A hundred years ago Justus von Liebig (1846) wrote in his introduction to his

"Thier-Chemie": "Die schoenste und erhabenste Aufgabe des menschlichen Geistes, die Erforschung der Gesetze des Lebens, kann nicht geloest, sie kann nicht gedacht werden, ohne eine genaue Kenntnis der chemischen Kraefte, der Kraefte naemlich, die nicht in Entfernungen wirken, die in einer aehnlichen Weise zur Aeusserung gelangen, wie die letzten Ursachen, von welchen die Lebenserscheinungen bedingt werden, die sich ueberall taetig zeigen, wo sich differente Materien beruehren."* Knowledge of chemical and molecular forces as a prerequisite of a real understanding of any cellular function was repeatedly stressed by leading biologists in this century, as for instance by F. G. Hopkins and by Otto Meyerhof. A few years ago, the French physicist Pierre Auger, in a discussion about the possibility of purely physical changes underlying the electric currents in nerve, made the statement: "It is inconceivable that electricity in a living cell, i.e., in a fluid system, can be generated without a chemical reaction." Perhaps because the writer is primarily a biochemist it appears to him rather surprising that in the second part of the twentieth century neurophysiologists should maintain with such vigor the view that a manifestation of life such as nerve impulse conduction should be possible without a chemical reaction, and that this view is so widely and readily accepted. The smallness of the net heat previously reported was not a strong argument in favor of a pure physical event, considering the extraordinary technical difficulties involved. But the recent brilliant achievement of dissociating the initial heat in its two phases and of demonstrating the surprisingly high positive heat when referred to the active material, combined with the evidence of the high temperature coefficient and the high activation energy, have made untenable the assumption of a purely physical wave propagating the impulse. The physical events form only one aspect of cellular function; they give us essential information, but they cannot explain mechanisms of living cells.

It may be stated at this point that a clear distinction must be made between the events taking place during activity and those in recovery. The flux of Na^+ and K^+ during activity is in the direction of the concentration gradients. The suddenly increased rate of flow of very small amounts of ions for extremely short periods of time (less than 1 msec. according to physical recordings) must obviously require very little energy for effecting the transient change of permeability. This problem

* "The most beautiful and most illustrious assignment of the human mind, that of the exploration of the laws of life, cannot be solved, cannot be even conceived without a precise knowledge of the chemical forces, viz. of those forces that do not act at a distance, that reveal themselves in a way similar to that of the ultimate causes upon which the manifestations of life depend, that are everywhere active where two different types of matter touch each other."

is fundamentally different with respect to ion movements during recovery. The restoration of the original steady state requires extrusion of Na^+ and uptake of K^+ *against* the concentration gradient. The maximum rate at which the ions in the squid giant axon are extruded against the gradients is about 50 μμmoles/cm²/sec., while during activity the movements may reach a rate of about 10,000 μμmoles/cm²/sec. (Hodgkin and Keynes, 1955). The movements against the gradient require relatively large amounts of energy in excess of that necessary for maintaining the resting condition. Most of the extra heat produced is accordingly developed after the electrical changes have taken place, and it is presumably associated to a considerable extent with chemical reactions required for the restoration of the initial electrolyte distribution. The energy for this "active" ion transport clearly must be provided by chemical energy and is probably derived from reactions common to most cells. There is no disagreement on this point. The sudden movements of Na^+ and K^+, on the other hand, carrying bioelectric currents, are a very specific phenomenon of conducting cells. Here one would suspect *a priori* a chemical system specific for conducting cells which effects the sudden permeability changes in the membrane or the special molecular layer responsible for the permeability change.

It was early recognized by Ostwald (1890) that special properties of conducting membranes should provide the clue for the understanding of bioelectric potentials. More specific theories were proposed by Kurt H. Meyer and T. Teorell. Meyer (1937) postulated on the basis of experiments with monomolecular films that a chemical reaction in the active membrane must precede the change in ion permeability. Membranes are formed by chains of protein (or conjugated protein). Meyer assumed that a chemical reaction may result in a rearrangement of acidic and basic groups; positively charged NH_3^+ groups would increase anion permeability, COO^- groups would facilitate the movement of cations. The importance of the role of charges in the membrane has also been stressed for many years by Teorell. In recent publications he has extensively scrutinized the theoretical basis of the permeability problem (Teorell, 1951, 1953).

B. Principles and Difficulties of Approach

The difficulty of identifying and analyzing the chemical reactions generating the electric current is easily recognized if one considers two of its pertinent features: the first one is the small amount of heat produced per gram nerve—an indication of the smallness of the chemical reactions. It is true, as discussed above, that it is misleading to refer the heat to the weight of the whole nerve fiber instead of to that of the active membrane. If it were possible to separate membranes of 100 A. thickness or less and obtain adequate amounts of this material, it might be metabolically quite active, but obviously no such methods are available. The second feature is the high speed of the process. The electrical signs indicate that it must take place within less than a millisecond. Considerable obstacles were encountered even in the recording of the physical events. It was an extraordinary achievement of A. V. Hill and his associates to develop recording instruments of such a high degree of sensitivity and such a high speed, that they were able to analyze the heat production. If it was so difficult for the biophysicist to measure this and other manifestations of the process, obviously the biochemist was faced with still greater problems. However, the magnitude of the task and the difficulties which must be overcome should not be a pretext to avoid or ignore the issue. As in other fields of scientific endeavor, an experimental attack must be tried to find out whether or not the nature of the permeability change can be elucidated.

The rapid advances of physics and chemistry since the turn of the century have provided the biologist with powerful tools and laid the foundation for many dramatic advances in many directions; they are at present going on at an ever increasing rate. At the beginning of the century, knowledge of physical processes associated with cellular function was in an early stage and quite incomplete. But virtually nothing was known about the underlying chemical reactions. The development of highly refined recording instruments, the use of physical and physicochemical methods for chemical analysis, especially for the study of proteins and enzymes, the elaboration of micromethods and isotope techniques, these and other factors have contributed to the rapid progress in biochemistry and form the basis of what is referred to as dynamic biochemistry. Biology has changed from a more or less descriptive science to one in which cellular mechanisms are being studied in terms of physics and chemistry and explanations are being sought at molecular levels.

A classic example in this respect is the function of muscle. The advance in the understanding of the structure and of the physical and chemical events in the elementary process of contraction is one of the fascinating achievements of modern biology. Few problems have attracted the interest of so many outstanding physiologists and biochemists. The challenging question in this case is that of how a cell is able to use the energy of chemical reactions for performing mechanical work. The large amount of energy involved and the convenience of measuring the work performed obviously offered a particularly favorable material for an attempt to explain cellular function in terms of physics and chemistry.

A special role in this development must be attributed to Otto Meyerhof. His philosophical approach, the depth of his thinking, the broadness of his knowledge, the originality of his concepts, and, above all, his ability to integrate a great variety of phenomena, all these factors combined made him a pioneer in this field. At the beginning of this century there was great interest in the question whether the heat produced by the animal body may be accounted for by the energy released through the combustion of foodstuffs. A merely affirmative answer to this problem which seemed obvious to Meyerhof, did not satisfy his searching mind. In a lecture entitled: "Zur Energetik der Zellvorgaenge," delivered in 1913 when he had just become Privatdozent at the University of Kiel, he raised the question how the potential energy of foodstuffs is made available for the special energy requirements of various cellular functions. He pointed out that between the initial energy uptake in the form of foodstuffs and the last step, the dissipation of heat, a series of energy transformations must take place which provide the organism with the requirements necessary for maintaining its function. He recognized that the organism is in a state of dynamic equilibrium. The ideas which Meyerhof expressed in this lecture made a profound impression, and Jacques Loeb invited him to write an enlarged version for his series of "Monographs on Experimental Biology." This essay appeared in 1924 under the title: "The Chemical Dynamics of Life Processes" and many of the ideas expressed there are still very much alive.

Selecting the muscle as material for the analysis of a life process, as envisaged in the article, Meyerhof devoted his life work to the sequence of energy transformations in muscular activity. Among the general principles and notions emphasized and introduced by Meyerhof was the importance of thermodynamics for the study of the sequence of reactions in intermediary metabolism, the cyclic character of these processes, and the necessity to establish relationships between chemical reactions observed in the test tube and physical recordings measured on the intact cell.

The biochemical work of Meyerhof was greatly helped and stimulated, as he himself pointed out in his Nobel Prize lecture and on many other occasions, by the biophysical measurements of A. V. Hill. The close collaboration between these two men, continuing over decades, was a most fortunate event for biology, and the ideas resulting from their work have deeply influenced the thinking of our generation. The way in which relationships were established in this case between metabolism and function has become an integral part of the analysis of cellular function in general; it forms the basis of the approach presented in the following analysis of the elementary processes of the propagation of nerve impulses. The challenging question in this case is: How do chemical reactions generate bioelectric potentials? In spite of the fundamental differences between these two cellular functions, the experience gained in the efforts to explain transformations of chemical into mechanical energy, to correlate metabolism and function in the one type of cell, has been frequently invaluable for the interpretation and evaluation of the information obtained in studies of the other type of cell.

C. Hypothesis of Neurohumoral Transmission

At about the same period in which Du Bois-Reymond's classic work laid the basis of electrophysiology, i.e., in the 1840's, Claude Bernard, in a series of remarkable papers, published his observations on the effect of curare: this poison blocks the transmission of nerve impulses to the muscle, although the conduction in nerve and muscle fibers remains unchanged. These observations marked the beginning of the notion that the junction between nerve and muscle has peculiar properties. This view was supported by a number of developments. Histological examinations showed a structural differentiation of the effector cell at the junction, the motor end plate. The neuron was found to be a cellular unit and it was established that between nerve and muscle and between nerve terminal and nerve cell body, i.e., at the synapse, there is a close contact but no protoplasmic continuity. This notion obviously raised the problem of how nerve impulses were propagated across these junctions.

Physiological observations also seemed to indicate characteristic and special properties of these junctions. Some of the most important may be briefly mentioned: (1) Propagation of nerve impulses from nerve to muscle and across synapses is effected in one direction only. (2) Fatigue of a nerve-muscle preparation often occurs at the junction when stimulation of nerve and muscle fibers does not yet indicate any alteration of their functional ability. (3) A synaptic delay is observed when impulses are propagated from one cell to the other. (4) Junctions are much more sensitive than fibers toward all kinds of physical and chemical agents. Lack of oxygen, for instance, affects transmission across synapses much more readily than conduction along fibers. (5) Various drugs act, like curare, exclusively upon the motor end plate without affecting conduction in the fiber.

It is necessary to keep in mind the special physiological and histological properties of synaptic junctions and of the motor end plate if we want to understand the origin and the background of the hypothesis of neurohumoral transmission of nerve impulses. There was no difference of opinion with respect to the view that nerve impulses along fibers are propagated by electric currents. But in contrast to this process of conduction the transmission of impulses from neuron to neuron or from neuron to the effector cell was assumed to be mediated by chemical substances which are released from the nerve terminal, cross the nonconducting gap between the two conducting cells, and act as a specific chemical stimulant of the next cell. These chemical mediators or neurohumoral transmitters were postulated to substitute for electric currents as propagating agents.

The idea of neurohumoral transmission was first proposed by T. R. Elliott (1905). Impressed by the similarity of the action of adrenaline and the effect of stimulation of sympathetic fibers, he proposed the hypothesis that adrenaline may be the transmitter from sympathetic nerves to their effector organs. A comparable role was later ascribed to acetylcholine in parasympathetic nerve endings. This ester had attracted the interest of physiologists and pharmacologists since Hunt and Taveau (1906) described the extraordinarily strong pharmacological action of this compound. Weiland (1912), in the laboratory of Magnes, observed that various isolated parts of the gastrointestinal tract release a substance into the surrounding fluid which has a stumulating effect upon an isolated intestine. On the basis of these observations he concluded that chemical reactions must be responsible for the automatic movements of the intestine. It was shown in the same laboratory by Le Heux (1919) that at least 75 per cent of the substance is choline. He concluded that choline may be a physiological stimulant, a kind of hormone, for the movement of the intestines. Observations of Rona and Neukirch (1912) that organic acids in very small concentrations may excite the intestine led Le Heux to test whether or not choline esters, especially acetylcholine, may be formed from these acids and choline with the aid of synthesizing enzymes in the intestinal wall. This explanation seemed to him a good possibility in view of the strong pharmacological actions of acetylcholine described by Hunt and Taveau and greatly extended by Dale (1914). In very interesting experiments, Le Heux (1921) succeeded in demonstrating that the action of some organic acids, but especially of acetic acid, must be attributed to the formation of the choline ester. He also demonstrated in these experiments the antagonistic action of atropine toward the choline esters. Otto Loewi (1921) observed that stimulation of the heart vagus of frogs led to the appearance of a substance in the perfusion fluid which, when added to a solution perfusing another frog heart, mimics vagal stimulation. He referred to this substance as "Vagusstoff." Loewi himself considered, in 1926, the possibility that the *Vagusstoff* might be a choline ester (Loewi and Navratil, 1926). This conjecture, supported in fact by the observations of Dale and the school of Magnes, was borne out in the following years by the work of various laboratories, especially by the demonstration of the natural occurrence of acetylcholine in animal tissue (Dale and Dudley, 1929). The various developments, to which only brief allusions have been made, appeared to support the hypothesis of neurohumoral transmission, and acetylcholine was considered to be a chemical mediator at parasympathetic nerve endings analogous to the role proposed by Elliott for adrenaline at sympathetic nerve endings.

It has been known for many years, especially from the work of Boehm (1908), that the frog striated muscle reacts strongly to quaternary ammonium compounds. In 1921 Riesser and Neuschloss, using frog and toad gastrocnemius muscle, first demonstrated that exposure to Ringer's solution containing acetylcholine leads to a contracture. They were impressed by the strength of the effect and its reversible nature. Only the junctional region reacted to acetylcholine; application to either the muscle or nerve fiber was without effect. The authors also described the prominent contracture of the frog rectus abdominis and the antagonism of the effect of acetylcholine by either atropine or curare. In the following decade the behavior of frog and warm-blooded muscle to acetylcholine was investigated in many laboratories and a considerable amount of information accumulated as to the effect of acetylcholine on striated muscle. Dale and his associates, using a technique in which the ester was injected into the artery at close range, obtained twitchlike responses of striated muscle (Brown, Dale, and Feldberg, 1936). Moreover, these investigators were able to demonstrate the appearance of acetylcholine in the perfusion fluid of the neuromuscular junction following stimulation (Dale, Feldberg, and Vogt, 1936). On the basis of these two types of observations they proposed the hypothesis that acetylcholine might be a neurohumoral transmitter across the neuromuscular junction, i.e., that it was released from the nerve terminal and stimulated the muscle. The strong action of acetylcholine on synaptic junctions might have been considered as a purely pharmacological effect, i.e., not necessarily associated with the biological process. The second type of finding, however, indicated that acetylcholine must have some physiological function related to nerve activity, although the type of observation was rather crude and therefore obviously not adequate for an interpretation of the finer cellular mechanism.

Whereas the idea of the neurohumoral transmission appeared acceptable to many physiologists in the case of the autonomous nervous system, the situation changed when the same mechanism was proposed for synapses and motor end plates; in this case the hypothesis encountered strong opposition by many neurophysiologists in this country (see, e.g., Erlanger, 1939; Fulton, 1938, 1939; Lorente de Nó, 1938) and in Europe

(Lapicque, 1936; Monnier, 1936; von Muralt, 1937, 1946, 1954; and others). There were many contradictions, admitted even by the most ardent proponents of the hypothesis, and many difficulties remained unexplained. The time relations between appearance of acetylcholine and the process of transmission were completely unknown (Lorente de Nó, 1938; Fulton, 1938, 1939). In the first experiments the perfusion fluid was collected every 2 min.; later the period of time was reduced to 30 sec. to counter the objection raised. Since, however, even 30 sec. is a period of time about 30,000 times as long as the process of transmission, the problem as to the particular time of action of acetylcholine remained open. The pharmacological action alone did not appear to be sufficient basis for the interpretation of its precise role during the actual process of transmission. One of the greatest difficulties, according to Dale, was the necessity of an extraordinarily high speed of appearance and removal of the ester in a process taking place in a millisecond. At that time, in 1936, no evidence was available to indicate the presence of a mechanism acting with the required speed. Another objection, raised especially by the school of Sherrington (see Fulton, 1938), was the following: the electrical manifestations indicate that the properties of the cell body and of the axon are fundamentally quite similar with respect to excitability. This fact makes it difficult to assume a mechanism of propagation across the synapse fundamentally different from that carrying impulses in the axon. The problem was scrutinized by Erlanger (1939) at a Symposium on the Synapse in Toronto. He showed that many of the so-called peculiarities of the synapse, as for instance latency, one-way transmission, temporal summation and facilitation, transmission across a nonconducting gap, and many other phenomena, may be, under appropriate conditions, demonstrated just as well on the axon as on the synapse. There are many differences as to quantitative aspects, but like Fulton he arrived at the conclusion that the electrical signs do not justify two fundamentally different mechanisms of propagation. With respect to the hypothesis of neurohumoral transmission he raised the pertinent question: "If an inactive zone in the fiber of more than 1 mm length does not prevent the propagation of nerve impulses by electric currents, is it justified to assume that the discontinuity at the synapse would interfere with such a transmission?"

For the reasons outlined and others, the idea of neurohumoral transmission appeared to many biologists unsatisfactory. On the other hand, it was necessary to find a satisfactory explanation for the experimental facts upon which this hypothesis was based and to reconcile them with the conclusions suggested by the electrical manifestations. In this impasse it appeared imperative to approach the problem with new methods. If a chemical compound were to have a physiological role, the new powerful tools available in biochemistry should be used to investigate the chain of chemical events of which the action of acetylcholine could only be a part. The notions and principles applied so successfully to the elementary process of muscular contraction, as mentioned above, appeared to offer the greatest promise of obtaining information as to the elementary processes of nerve conduction and as to the role of acetylcholine in these events.

Investigations carried out over a period of more than twenty years made it indeed possible to establish the sequence of energy transformations during nerve activity, to integrate the formation and the hydrolysis of acetylcholine into the intermediary pathways of the nerve cells, and to correlate a series of chemical processes with electrical manifestations and other events recorded by physical methods. A theory has been proposed, in modification and extension of the original hypothesis of neurohumoral transmission, which attempts to integrate the available physical and chemical data and to eliminate apparently contradictory facts. Previously it was assumed that acetylcholine is released from the nerve terminal and acts upon the effector cell as transmitter. The role of acetylcholine was limited exclusively to the synapse. According to the theory proposed in 1940 and elaborated in subsequent years, the action of acetylcholine is not an *inter*- but an *intra*cellular process taking place within the conducting membrane (Nachmansohn and Bettina Meyerhof, 1941). The action of the ester is necessary for the change of permeability of the membrane. Its action on a receptor is a trigger mechanism by which the ionic concentration gradients, inactive in resting condition, become the effective source of EMF. The action of acetylcholine thus forms an integral part of the elementary processes by which bioelectric potentials are generated in the axon, in the nerve terminal, and in the postsynaptic membrane, but the electric currents are the propagating agent in transmission as well as in conduction.

In the following chapters the most important facts upon which this concept is based will be presented and discussed. They summarize and supplement the accounts given in preceding reviews (Nachmansohn, 1946a, b, 1950a, b, 1951, 1952a, b, c, 1955a–e, 1957, 1959a, b; Nachmansohn and Wilson, 1951, 1955, 1956; Wilson and Nachmansohn, 1954).

CHAPTER III

Physiologically Significant Features of Acetylcholinesterase

In 1932, F. G. Hopkins, in his presidential address to the Royal Society, made the following statement: "It is, I think, difficult to exaggerate the importance to biology, and I venture to say to chemistry no less, of extended studies of enzymes and their action" (Hopkins, 1932). The startling and specific catalytic effects produced by enzymes are not only a unique feature of living cells, but are one of their most amazing and extraordinary characteristics. There is among the huge number of chemical reactions taking place in the living cell hardly one which is not catalyzed by enzymes. This is tantamount to saying that processes of life cannot occur and therefore cannot be understood without knowledge and understanding of enzymes, of their properties, of their mode of action and their relation to a special cellular function. Enzymology is today not only one of the most fertile, but also one of the most extensive, fields of biology, as documented by the vast literature on enzymes, the many handbooks, textbooks, periodicals, etc., devoted to this subject. It seems incredible that only sixty years have passed since the Buchner brothers succeeded in preparing a solution containing the fermentation system of yeast, thereby ending the controversy started by Liebig and Pasteur over the problem of whether enzymes are chemical substances or inseparable parts of a living cell.

The role of acetylcholine in neuromuscular junctions and ganglionic synapses was a frequent topic at the meetings of the Physiological Society of England in the years 1935 and 1936. One of the problems which was so vigorously debated was the question of whether acetylcholine can be removed with the necessary speed compatible with the assumption that it is a neurohumoral transmitter (see, e.g., Eccles, 1937). The two products resulting from the hydrolysis of acetylcholine, acetic acid and choline, have virtually no pharmacological action. The existence of esterases in the animal organism has long been known. It is remarkable that Dale, as early as 1914, ventured the hypothesis that the active ester may be rapidly inactivated by enzymatic action and that this hydrolysis may account for the rapidly passing effect.

Since the writer was trained in enzyme chemistry, it was only natural that his interest should be aroused in the enzyme responsible for the hydrolysis of acetylcholine. Here seemed to be an important opening for the biochemist interested in nerve function, for a new approach which possibly would lead to information capable of overcoming the impasse described above. At that time, in 1936, little was known about the enzyme. Stedman, Stedman, and Easson (1932), had demonstrated that the esterase present in serum seemed to be a cholinesterase: the rate of hydrolysis of the two choline esters acetyl- and butyrylcholine was greater than that of noncholine esters. Many aspects of the enzyme appeared pertinent for evaluation of the role of the ester in nerve activity, such as localization, concentration, occurrence in various tissues, reaction rates, etc. Since then investigations of this enzyme have played an important role in the work of this laboratory, although the aspects studied naturally underwent many changes in direction and emphasis.

Since there are many esterases in the organism, it appeared desirable to refer to the enzyme in more specific terms, especially after it was established that it has certain characteristic features by which it may be distinguished from other esterases; the term acetylcholinesterase was, therefore, proposed (Augustinsson and Nachmansohn, 1949a). In this chapter, a number of physiologically significant features of acetylcholinesterase will be described.

A. Specificity

One question of physiological as well as biochemical interest is whether there is an esterase with a relatively high affinity for acetylcholine. The existence of an enzyme with relative specificity for acetylcholine would obviously suggest that it has the physiological function to inactivate the ester by hydrolysis. Properties of a more or less specific enzyme may give valuable information in many respects.

The question of the existence of an esterase specific for acetylcholine was, as mentioned before, first raised by Stedman, Stedman, and Easson (1932). Their experiments were carried out with horse serum. The choice of the material may appear strange today, but it may be recalled that at that time the view of an intracellular action of acetylcholine was not yet proposed and that the ideas about the localization of the neurohumoral transmitter process were quite vague. The investigators found that butyrylcholine is hydrolyzed by the serum esterase used more rapidly than is acetylcholine. Their partially purified enzyme seemed to split exclusively choline esters. They, therefore, thought that the enzyme may be specific for choline esters and proposed the term cholinesterase for it. Later investigations, however, have shown that serum esterase does not hydrolyze exclusively choline esters. It was found by Vahlquist (1935) that the esterase in human plasma is not specific for choline esters, although these esters are split more rapidly than others. Particularly extensive investigations as to the specificity of horse serum esterase were reported by Glick (1938, 1939, 1941). He showed that quite a few esters were hydrolyzed by serum esterase. He used, however, the material without purification or fractionation.

A significant progress in the characterization of acetylcholine-splitting enzymes resulted from the observations of Alles and Hawes (1940). These investigators found that the esterase of red blood cells differs markedly from that in the serum: the enzyme of red blood cells has a well-defined optimum of substrate concentration. Excess of substrate decreases the activity of the enzyme. This is in sharp contrast to the activity of serum esterase which has no well-defined optimum of substrate concentration but shows the usual Michaelis-Menten type of curve of activity substrate concentration relationship. If a methyl group is attached to the carbon atom next to the ester link, as in acetyl-β-methylcholine, the compound is hydrolyzed by red cell esterase, although at a lower rate. It is not split by serum esterase, as was shown previously by Glick.

The marked differences between these two types of esterases were confirmed by Richter and Croft (1942). They tested the properties of the esterases from the two sources by using specific inhibitors and applying methods of partial separation. Then they found the erythrocyte esterase specific for choline esters and attributed the small amounts of noncholinesters split to the existence of an "ali-esterase." The differentiation by specific inhibitors is, however, not an entirely satisfactory method, since the degree of specificity is usually an open question. Moreover, the substrate concentrations used were quite high, 20 μmoles/ml. This is a concentration several times above the optimum for red cell esterase. Zeller and Bissegger (1943) found the brain esterase to be fundamentally similar to the erythrocyte esterase.

Nachmansohn and Rothenberg (1944, 1945) extended the study of enzyme specificity to many types of conducting tissues. They confirmed the existence of an optimum substrate concentration for the activity of the type of esterase present in these tissues. They found, moreover, an important difference in the behavior of the esterase present in conducting tissue and red blood cells as compared to that in serum. The rate of hydrolysis by serum esterase increases with the length of the acyl chain in the following order: acetyl$<$propionyl$<$butyrylcholine. In contrast, the rate of hydrolysis by the type of esterase present in conducting tissue decreases with increasing length of the acyl chain of the three esters tested. Propionylcholine is usually split at a rate about one-third lower than that of acetylcholine, although there are some minor differences between various tissues. The enzyme prepared from electric tissue of *Electrophorus electricus*, for example, hydrolyzed propionylcholine at the same rate as acetylcholine; the enzyme prepared from electric tissue of *Torpedo,* at a rate which was only one-third of that of acetylcholine. Butyrylcholine was hardly or not at all split under the conditions tested. This, of course, does not indicate that the enzyme is unable to hydrolyze butyrylcholine: the rate of hydrolysis of this ester becomes measurable if the enzyme concentration is increased (Wilson, Bergmann, and Nachmansohn, 1950). A rate of hydrolysis of butyrylcholine approximately equal to that of acetylcholine is obtained when the enzyme concentration is increased by a factor of 140. In this particular case, an enzyme solution prepared from electric tissue of *Torpedo* was used. It may be that the ratio is different with the same type of enzyme, when prepared from other sources. Even if the enzyme concentration is low, noncholine esters, at concentrations optimal for acetylcholine, were hydrolyzed, although at a low rate.

The pattern of different hydrolysis rates obtained with crude extracts of electric organs—and some other types of conducting tissue—was the same as with highly purified enzyme solutions, which showed only one component in the analytical ultracentrifuge and by electrophoretic analysis, and in which 1 mg. of protein was capable of hydrolyzing about 400 mmoles (approximately 70 gm.) of acetylcholine per hour.

On the basis of their data, Nachmansohn and Rothenberg concluded that in conducting tissue and red blood cells there exists a type of esterase which is specific for acetylcholine, although this specificity is relative and not absolute. Since then many investigators have confirmed these data. Adams (1949) also confirmed the experimental findings but criticized the term "specificity." Actually, we had clearly pointed out and emphasized that an absolute specificity could not be expected since the ester linkage is not specific. The essential question raised was not the problem of the degree of specificity, which is partly a question of definition, but that of the existence of an enzyme in conducting tissue with properties different from those of other esterases and possessing a relatively high affinity for acetylcholine. This question had found an affirmative answer.

The specificity for the alcohol group is not fundamentally different between human red cell or electric tissue esterase on the one hand, and serum esterase on the other hand (Adams and Whittaker, 1949, 1950; Wilson, 1954a). Aliphatic esters are more rapidly split by both types of esterases, the more closely they approach the choline configuration. 3,3-Dimethylbutyl acetate was found to be split by both enzymes at a somewhat lower rate (60 per cent) than that at which acetylcholine was split, but at a far greater rate than ethyl acetate. A more detailed discussion of the effect of positive charge in the alcohol group in the interaction between substrate and active surface of the enzyme will be presented later.

For a proper evaluation of the comparative rates of hydrolysis it is necessary to test the effects of substrate concentration not only with acetylcholine, but also with other esters. This aspect was studied by Augustinsson (1948). When working in the writer's laboratory, in 1949, he used enzyme preparations of representative types of conducting tissues, nerve and muscle, which the writer and his associates had found to contain particularly high concentrations of esterase, such as squid head ganglia, nucleus caudatus of ox, *Lebistes* tail muscles, and also the esterase of human erythrocytes. Augustinsson (1949) tested the activity-substrate concentration relationships and found that in these vastly different types of vertebrate and invertebrate conducting tissue the optimal substrate concentration for the hydrolysis of propionyl- and butyrylcholine coincides more or less with that of acetylcholine; other esters which do not contain choline, such as triacetin, show markedly different activity-substrate concentration relationships. At the concentration which is optimal for acetylcholine, triacetin is hydrolyzed at a very low rate. At a high substrate concentration where acetylcholine hydrolysis is already strongly inhibited, triacetin is hydrolyzed at a rate which, in some cases, is higher than that obtained with acetylcholine. Using only these high concentrations of substrate, acetylcholine hydrolysis becomes apparently lower than that of triacetin and one could readily arrive at the conclusion, as has actually been done, that triacetin has a higher affinity to the enzyme than acetylcholine, whereas obviously the opposite is the case.

On the basis of all these investigations, it became apparent that in conducting tissues, nerve and muscle fibers, and in the red blood cell, there exists a type of esterase which may be distinguished by a number of characteristic features from other types of esterases found in many tissues. Since its characteristic feature is the relatively high affinity to acetylcholine, Augustinsson and Nachmansohn (1949a) proposed the term acetylcholinesterase and this term has been widely accepted among enzyme chemists. In addition to the relatively small Michaelis constant, i.e., the relatively high affinity, two other characteristics are: (1) the well-defined optimum substrate concentration resulting in a bell-shaped curve, and (2) the decrease of the rate of hydrolysis if the length of the acyl chain increases from two to four carbon atoms. A few characteristic illustrations are given in Figs. 1–5.

It may be briefly mentioned that Mendel and Rudney (1943) proposed a distinction between "true" and "pseudo" cholinesterase. These terms were apparently appealing to some investigators, especially in clinical and pharmacological fields, but were vigorously objected to by enzymologists (see for instance Glick, 1945). However, the question of terminology appears less important than that of the experimental findings upon which the terms were based. In this respect there were considerable differences between the data of Mendel and his associates

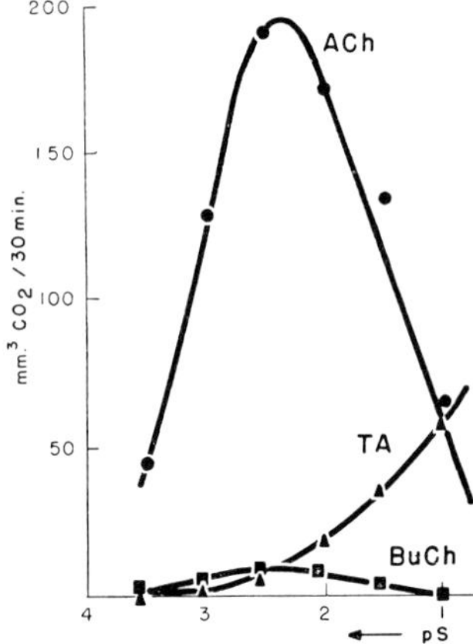

FIG. 1. Acetylcholinesterase prepared from mammalian brain (nucleus caudatus of ox). Activity-pS curves for the enzymatic hydrolysis of various esters.

ACh = acetylcholine, BuCh = butyrylcholine, TA = triacetin.

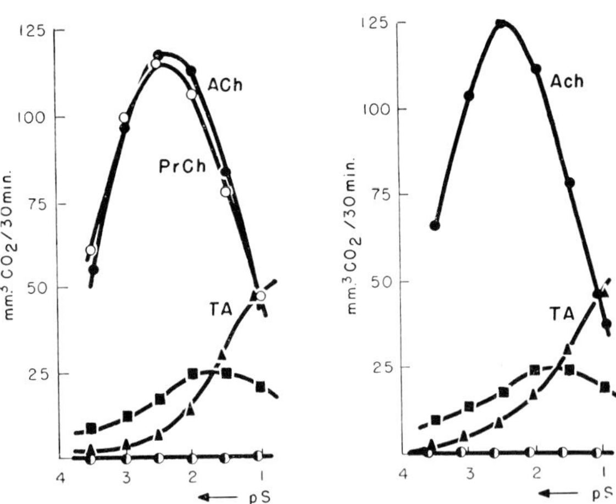

FIG. 2. Acetylcholinesterase prepared from electric tissue of *Electrophorus electricus.*

Left: crude extract, *right*: highly purified preparation (1 mg. of protein hydrolyzed 100 mmoles of ester per hour.) PrCh = propionylcholine, ■ = acetyl-β-methylcholine, ◐ = butyrylcholine; otherwise as in Fig. 1.

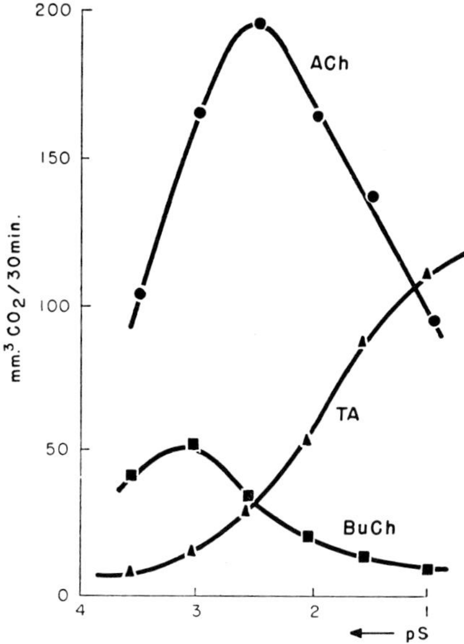

FIG. 3. Acetylcholinesterase prepared from striated muscle (*Lebistes*).

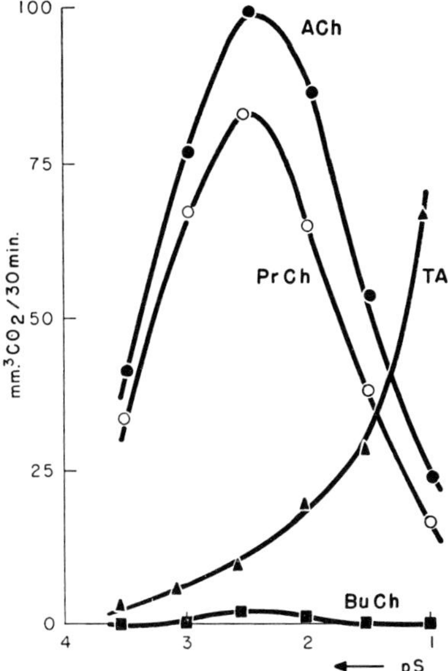

FIG. 4. Acetylcholinesterase prepared from human red blood cells.

and those of other laboratories (see for instance Nachmansohn and Rothenberg, 1945; Agustinsson, 1948). The techniques and experimental conditions used by Mendel and his associates were open to question.

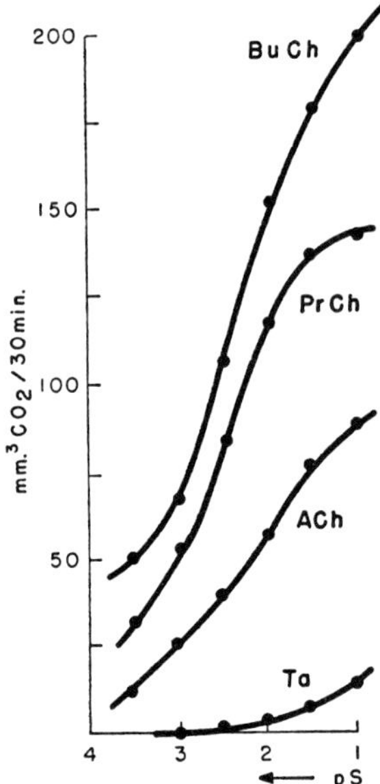

FIG. 5. Human serum esterase.

B. Occurrence in Conducting Tissues and Concentration

If the hydrolysis of acetylcholine is responsible for the inactivation of the ester, it is essential to know where the enzyme occurs. This mechanism of inactivation is, on the basis of all biochemical experience, the only likely one. Unless one speculates about some unknown and never demonstrated type of rapid removal of the ester, the demonstration of the presence of the enzyme is the first, although of course not the only, prerequisite for attributing to the ester an active role in a cellular function. Moreover, in addition to many other criteria to be discussed later, the enzyme must be present in significant concentrations adequate to account for the removal of the amounts of active ester released during activity and in a period of time compatible with the time relationships of the function proposed for the ester.

The presence and the concentration of the enzyme were studied by the writer in extensive investigations during the years 1936 to 1939 with the able technical help of Mlle. Annette Marnay, at that time a student in chemistry (Marnay and Nachmansohn, 1937a–c, 1938; Nachmansohn

1937a, b, 1938b–f, 1939b). The outcome of these studies was the demonstration that an acetylcholine-splitting enzyme is present in remarkably high concentrations in all types of conducting tissue, nerve and muscle, vertebrate and invertebrate, central and peripheral tissues, motor and sensory fibers, sympathetic and parasympathetic, so-called cholinergic and adrenergic fibers. In the course of the years these studies have been still further extended, especially to all kinds of invertebrates. The enzyme was found in such a primitive type of organism as *Tubularia*, a medusa, in which a nerve-muscle tissue is present comparable to that in higher forms. Seaman and Houlihan (1951) demonstrated the existence of the enzyme in a ciliated monocellular organism *Tetrahymena geleii* S. The specific features of the enzyme were not yet worked out at that time; however, later investigations mentioned in the preceding section indicate that the enzyme present in conducting tissue is usually an acetylcholinesterase. In some of these tissues, as might have been expected, other types of esterases are present. Since the figures were obtained by using acetylcholine as the substrate, they should actually be corrected by subtraction of that fraction of the substrate hydrolyzed by the esterase which is not acetylcholinesterase. The ratio between specific enzyme and other esterases is, however, not easy to evaluate quantitatively; in many tissues almost all the esterase is of the specific type, in some others there is definitely some other type of esterase present; usually the percentage is small but may vary somewhat from one type of tissue to the other. But even if half the enzyme, as may be true in some cases, is a different esterase, not specified as yet, the picture as a whole would be little affected by the corrections to be made, namely that the specific enzyme is not only present in all types of conducting tissue, but that it exists there in high concentrations.

Tables I, II, and III summarize the variations of the esterase activity in a variety of representative tissues. The activity coefficient for which the writer proposed the term Q_{AChE} (originally the term Q_{ChE}), expresses the amount of acetylcholine in milligrams hydrolyzed per gram tissue (fresh weight) per hour. Nerve fibers have a Q_{AChE} in the range of 5–15; in a few cases the activity is smaller or greater. At the region where cell bodies and synaptic junctions are located, the concentrations are several times as high as in the corresponding fibers. Usually the concentration increases three to five times, sometimes even more. So, for instance, the Q_{AChE} in preganglionic fibers of the sympathetic cervical ganglion is about 50–70, whereas it is 500–600 in the ganglion itself. In the abdominal chain of lobster, the values are 80–140, but they rise to 180–300 at the regions containing cell bodies.

Myelinated nerve fibers in general have lower concentrations than

TABLE I
RATE OF HYDROLYSIS OF ACETYLCHOLINE IN BRAIN (OR HEAD GANGLIA) OF DIFFERENT SPECIES

Species	Q_{AChE} [a]
Mammals (37° C.)	
Rabbit, rat	80–100
Birds (37° C.)	
Chicken, pigeon	250
Hummingbird, sparrow	300
Amphibia (20° C.)	
Frog	60–70
Turtle	80–100
Fish (20° C.)	
Goldfish	100–150
Invertebrates (20° C.)	
Squid (*Loligo pealei*) head ganglia	3000–4000

[a] Q_{AChE} = milligrams ACh hydrolyzed per gram tissue (fresh weight) per hour.

TABLE II
RATE OF HYDROLYSIS OF ACETYLCHOLINE IN PERIPHERAL NERVE FIBERS[a]

Species and type of tissue	Q_{AChE}
Mammals	
Dog	
Ventral roots	25–30
Dorsal roots	12–15
Sciatic nerve	8–10
Optic nerve	10–15
Preganglionic sympathetic	40–60
Sympathetic ganglion	150–200
Spinal ganglia	30–50
Cat	
Superior cervical sympathetic	60–70
Superior cervical ganglion	400–600
Postganglionic sympathetic ("adrenergic")	25
Amphibia	
Frog	
Splanchnic ("adrenergic")	10
Sciatic nerve	5–10
Fish	
Ray	
Optic nerve	2–4
Invertebrates	
Squid (*Loligo pealei*)	
Fin nerve	4–7
Stellar nerve	2–4

[a] The activity of mammalian fibers was determined at 37° C., of the other fibers, at room temperature.

TABLE III
RATE OF HYDROLYSIS OF ACETYLCHOLINE IN MUSCLE

Species	Muscle	Q_{AChE}
Guinea pig	Gastrocnemius	8–15
Chicken	Leg	4–6
Hummingbird	Breast	80–100
Frog	Gastrocnemius	5–10
Lizard	Leg	30–40
Lebistes	Tail	300–400
Nereis	Body wall	60–65
Lumbricus	Body wall	140–160

the so-called unmyelinated, and white matter contains much less enzyme than gray matter. Great differences were found in the brain of different species and in different centers of the same species (Table IV). In small animals, such as mice, rats, guinea pigs, the concentrations are much

TABLE IV
RATE OF HYDROLYSIS OF ACETYLCHOLINE IN VARIOUS SAMPLES OF BRAIN

	Q_{AChE}			
Sample of brain	Rabbit	Dog	Ox	Human
Cortex	60–80	20–50	20–30	12
White matter (great hemisphere)	—	3	2–3	—
Nucleus caudatus	350	500–600	400	300
Nucleus lentiformis	—	—	680	460
Cerebellum	90–100	120–150	20–40	80
Thalamus opticus	120	60	50	30
Pons	130	70–80	—	60
Corp. quadrigemina ant.	250	140	100–120	60
Corp. quadrigemina post.	130	50	40	30
Retina	—	150	140–200	—

higher than those found in the brain of dog, cat, or ox. The concentrations in human brain are even lower. In the different brain centers, the concentrations may differ by a factor of 10–50. The nucleus caudatus and the nucleus lentiformis have by far the highest concentrations (Nachmansohn, 1937b, 1939b). In contrast to these great variations between different centers and species the enzyme concentrations are remarkably constant for the same species and the same center.

The figures given in the three tables are based exclusively on data obtained in the writer's laboratory, determined mostly in the years 1937 to 1939. The results were later confirmed by many laboratories and additional figures for other tissues have been described. Since, however, methods of determinations and experimental conditions frequently were not the same, the values are not quite comparable and have, therefore, not been included in the tables. This monograph does not aim to give a complete survey of all the data in the field. For the discussion of the essential problem the data presented give an adequate illustration of the variations of acetylcholinesterase in representative types of conducting tissue throughout the animal kingdom.

The existence of an enzyme with a relatively high affinity to acetylcholine in the cell makes possible the assumption that there the ester has a physiological role. One result of general interest which emerged from the work with isotopes is the recognition that enzymes do not lie dormant in cells but are continually active (Schoenheimer and Rittenberg, 1940). This, of course, does not mean that the enzymes are active all the time. A distinction must be made, especially in the case of nerve and muscle, between active and resting state. Moreover, most enzymes are present in excess above the amount required, and this excess may be in some instances quite considerable. But the evidence of the existence of a special type of esterase with a high affinity to acetylcholine in all types of conducting tissue, and mainly there, appears significant.

The importance of these findings was, however, greatly emphasized by the high concentrations of the enzymes. The figures indicate that the amounts of acetylcholine which may be metabolized in milliseconds, i.e., the period of time during which impulses are conducted, are sufficiently high to permit the association of the reaction with the elementary process of conduction (Nachmansohn, 1939b). Ever since these high concentrations of esterase were found in various types of axons, the question inevitably arose whether the role of the ester is really limited to the nerve endings or whether it may have a more general function.

C. Localization

The real significance of the concentrations of the enzyme can be evaluated only in connection with its special localization. When Dale and his associates proposed that acetylcholine may be the neurohumoral transmitter across the myoneural junction and across the ganglionic synapse, they stated that one of the chief difficulties encountered by this hypothesis was the necessity of a removal of the acetylcholine liberated with "flashlike" suddenness, within the refractory period, i.e., within a period of time of the order of 1 msec. It was this special problem, namely the question of whether the rate of hydrolysis of acetylcholine by cholinesterase at myoneural junctions is adequate for assuming an active role of the ester in the transmitter process, which first stimulated the writer's interest in the enzyme.

The rate of hydrolysis of acetylcholine was determined in the pelvic end of frog sartorius, which is free of nerve endings, and compared to that in sections of the muscle rich in nerve endings, a procedure based on the distinction which was used by Keith Lucas (1907a, b) in his physiological studies. The parts rich in nerve endings have an enzyme activity three to six times as high as the pelvic region in which endings are absent. The section richest in nerve endings has an enzyme activity approximately as high as the nerve fibers (Marnay and Nachmansohn, 1937a, 1938). This is a striking fact. It suggested the possibility of a special role of the ester at the level of nerve endings, even if one did not accept the idea of neurohumoral transmission. It was necessary to explain the unequal distribution.

The relationship between the concentration at the junction, and that in the muscle and nerve fibers proper, was at that time incorrectly estimated on the assumption that the enzyme is evenly distributed in the two types of fibers and more concentrated only at the junctions. The volume formed by the nerve endings is small. The high enzyme activity of the sections containing endings seems to suggest that the concentration there is very much higher than in the surrounding muscle tissues. It was only a few years later that the remarkable localization of the enzyme near the surface of the fibers was discovered. It then became apparent that the enzyme is localized in nerve and muscle fibers near their surfaces and that therefore the concentrations in the active zones must also be very high. The increase of the enzyme concentration at the junctions is, therefore, not so extraordinarily high as it appeared to be in the early stage of the studies and may be explained simply in terms of increased surface. These findings made necessary a reevaluation of the ratio of enzyme concentration in muscle and nerve fibers over that in the end plates. But since the number of motor end plates in frog sartorius muscle of *Rana esculenta* was known from the studies of Pézard and May (1937), the determinations permitted a fairly precise evaluation of the enzyme concentration at the level of the neuromuscular junctions of these muscles. At a single motor nerve ending 0.2×10^{-6} μg. ($= 1.6 \times 10^{9}$ molecules) of acetylcholine may be hydrolyzed per millisecond (Marnay and Nachmansohn, 1938; Nachmansohn 1939a).

The unequal distribution of the enzyme and its higher concentration at the neuromuscular junction was subsequently demonstrated with mammalian muscle by Couteaux and Nachmansohn (1940; Couteaux, 1942), using the gastrocnemius muscle of guinea pig. In this muscle all branches including the motor end plates are localized in a well-defined zone. When the muscle is frozen and sections are prepared with the microtome in an appropriate way, most sections contain only a few end plates or none at all. But in a certain zone the sections are full of motor end plates. In these sections the concentration of cholinesterase is four to five times as high as in other sections (Fig. 6). As in the case of frog sartorius muscle, the figures show that the enzyme concentration at motor end plates is several times as high as in the other sections. The evidence for the presence of a high enzyme concentration at synaptic and neuromuscular junctions was extended to a great variety of tissues of

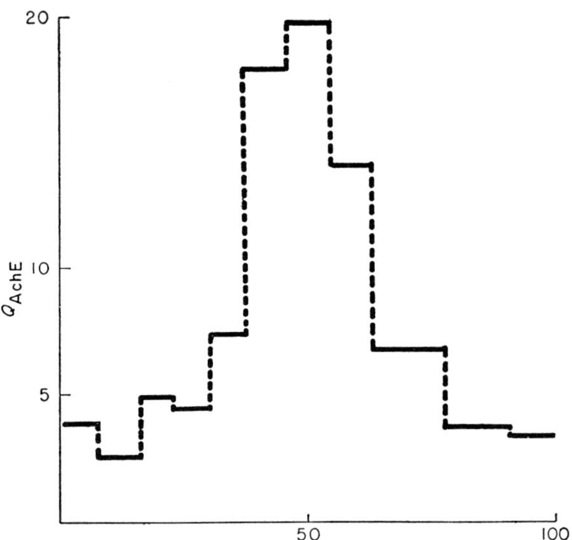

FIG. 6. Distribution of acetylcholinesterase in an interior section of a gastrocnemius (guinea pig) containing most of the nerve fibers and endings.

The middle portion of the section was cut in eleven slices of similar thickness and weight. Each horizontal line corresponds to one slice and indicates its weight in percentage of total weight. Abscissas: Region from which tissue was obtained in terms of order of consecutive slices. Point *50* corresponds to center region where nerve endings are situated. Q_{AChE} = milligrams acetylcholine hydrolyzed per gram fresh tissue per hour.

different types (Nachmansohn, 1939b). However, it soon became apparent, as mentioned before, that the concentration of the enzyme is high in all conducting fibers and that it is only relatively higher at junctions, usually by a factor of three to five times.

An elucidation of the localization and a more quantitative evaluation of the concentration became possible by two further developments. Studying the activity in the superior cervical ganglion of cat following the section of preganglionic fibers, Couteaux and Nachmansohn (1940) found, in 1939, that a rapid decrease takes place during the first week, coinciding with the disappearance of the nerve terminals. The amount of acetylcholine hydrolyzed per gram tissue (wet weight) per hour falls from about 500 to 600 mg. in the normal ganglion to about 250 mg. in ganglia in which the terminals have disappeared. After that period the enzyme concentration remains stable for many weeks. In the preganglionic fibers the enzyme activity amounts to 60–70 mg. of acetylcholine split per gram per hour. The decrease per unit weight coinciding with the disappearance of the nerve terminals is, therefore, much higher than one would expect on the basis of the enzyme concentration in the fibers. This discrepancy suggested that the concentration of cholinesterase in the nerve endings inside the ganglion is several times as high as in the preganglionic fibers. One explanation appeared to be that the enzyme is either exclusively localized, or at least more concentrated, in or near the surface membranes. In that case the increase would be attributable to the extensive endarborization of the presynaptic fibers within the ganglion and consequently to the great increase in axonal surface area per unit volume.

It appeared of interest to test this possibility, since such a localization would obviously have implications of considerable importance. While working in Woods Hole in the following year, in 1940, it appeared to the writer that the giant axon of squid would offer a unique material for an experimental test of the interpretation proposed. The giant axon of squid, *Loligo pealii,* has been described by Young (1936a, b) and has since then become an important tool in electrophysiology. Average-size squids have giant axons of diameters varying from 400 to 800 μ, and fibers 4–5 cm. in length may be readily dissected; Fig. 7 illustrates the huge diameter of these axons if one compares them to the diameters of fibers in rabbit sciatic nerve. These axons permit the separation of the axoplasm from the surrounding envelope by careful extrusion. Hereby it becomes possible to analyze the concentration of an enzyme in the axoplasm and to compare it to that in the envelope. The envelope, of course, does not contain merely the active membrane, but is a composite structure formed mostly by connective tissue to which two membranes are attached, the so-called metatropic and the protoplasmic membrane, the latter possibly being the active membrane (Bear, Schmitt, and Young, 1937). The metatropic membrane is rich in lipids and equivalent to the myelin layers of the myelinated fiber.

Surprisingly it was found that the axoplasm does not contain any measurable esterase activity and that all the enzyme is localized in the envelope (Boell and Nachmansohn, 1940; Nachmansohn and Bettina Meyerhof, 1941). Although the exact distribution of the enzyme within the envelope has not been determined, its exclusive localization in the material which must contain the active membrane and its absence in the axoplasm forming at least 80–90 per cent of the axon, appeared of interest for the problem of the physiological role of the ester, although this finding must be considered in connection with many other factors. But the limitation to a narrow zone of the enzyme present in the axon

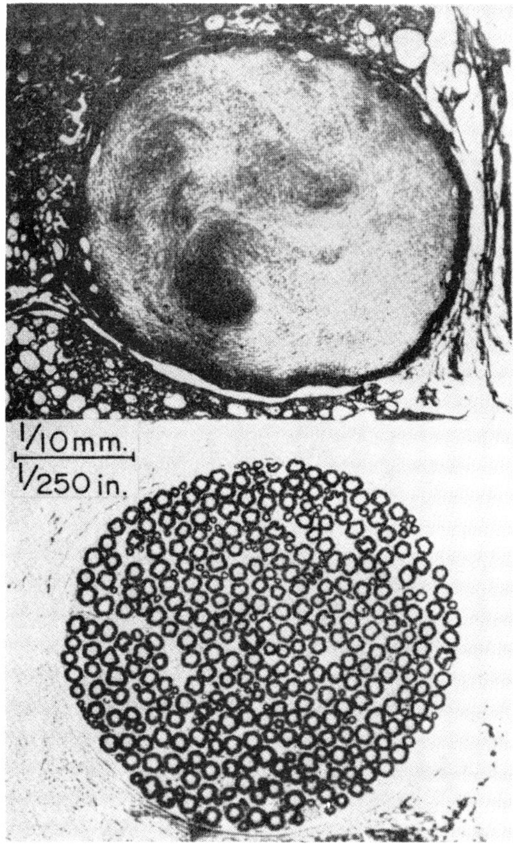

FIG. 7. Cross section of the giant axon of squid (*above*) compared to that of rabbit sciatic nerve (*below*) at the same magnification. (From J. Z. Young, 1952.)

is essential for a proper evaluation of the concentration of cholinesterase in conducting tissues and has thrown new light on its significance. Since the concentrations are all determined per gram fresh weight, the figures do not give a real picture of the extraordinarily high activity, which actually must be higher by several orders of magnitude than the figures per total weight suggest. In the case of heat production, A. V. Hill and his associates in their recent paper referred the heat to a membrane of about 50 A. thickness; this corresponds roughly to the active material to which the process measured must be referred. In evaluating the enzyme activity in similar terms, we must remember that enzymes are present in excess of the minimum requirements, usually three to five times, sometimes even much more. In the giant axon as well as in other axons, the total enzyme activity measured is about five times in excess, since conduction stops at about 20 per cent of the initial value (see following chapter). Where this excess enzyme is located is not known, although it may be assumed to be at least in close vicinity to the active membrane. The minimum amount of acetylcholine metabolized during unimpaired conduction is about 10^{-14} mole per millisecond per square centimeter of surface. We have seen that about 4×10^{-12} mole of Na enter per square centimeter per impulse. The impedance change reaches its maximum in about 100 μsec. Let us assume that the action of the acetylcholine released is responsible for the permeability change observed, i.e., for the increase in Na conductance. We do not know what fraction of the acetylcholine released must be hydrolyzed and within what period of time to permit the return of Na conductance to its normal level. But it seems reasonable to assume that the major part of the effective hydrolytic process takes place within a period of 0.5–1.0 msec. On this basis one would arrive at a ratio of one molecule of acetylcholine metabolized for 500–1000 Na ions entering the fiber. The process may, of course, be slightly faster or slower, which would change the ratio correspondingly. These figures, although only approximations, appear to be of an order of magnitude quite appropriate for a trigger process accelerating the movement of Na by removing charges or opening a barrier in some other way.

In his "Chemical Wave Transmission in Nerve" A. V. Hill (1932b) points out that the molecular change taking place during activity can hardly be expected to cover the whole surface: because of the smallness of the heat observed, the energy per molecule would be several orders of magnitudes smaller than the energy of a quantum of visible light. The chemical reactions would then produce a heat per gram-molecule of reacting substance far below that of any known reaction. It is, therefore, interesting to estimate the surface area which may be covered by the amount of acetylcholine reacting per millisecond, i.e., in the period of time during which the activity takes place. The calculation must be based on the minimum concentration of the enzyme compatible with conduction (20 per cent of the normal value). Let us assume that an acetylcholine molecule covers a surface area of about 30–50 A^2. The amount of surface covered by acetylcholine metabolized in the squid axon per square centimeter surface per millisecond would be of an order of magnitude from 5×10^2 to $5 \times 10^3\ \mu^2$. This is an extremely small fraction of the total surface. The figures are sufficiently low to account for the relatively small amount of heat observed, even assuming the actual heat production to be several times as high as that used in Hill's calculations, in view of the recent data on *Maja* nerve. They are nevertheless impressively high and sufficient to attribute to the action of this molecule the trigger function responsible for the permeability change in the membrane, i.e., the increased Na conductance.

The amount of acetylcholine metabolized per millisecond per gram brain is about 10^{14}–10^{15} molecules. This amount would cover a surface area of about 30–500 millions of square microns. The quantities metabolized per gram brain indicate the total possible amount, since we do not know in this case the excess of enzyme present; but assuming an excess even several times as high as in the axons, the rate is remarkably high and satisfactory in respect to the role proposed for the ester. The special aspects of the enzyme concentration at myoneural and synaptic junctions will be discussed in later chapters.

In two other cases it has been demonstrated that acetylcholinesterase is not evenly distributed in the cell. Brauer and Root (1945) found that in red blood cells the enzyme is not present in the cell interior, but localized exclusively in the stroma. Similarly, Seaman (1951), having found acetylcholinesterase in *Tetrahymena geleii* S, demonstrated that the enzyme is found exclusively in the pellicle, to which the cilia are attached; the cell interior is free of enzyme. It may be mentioned in this connection that the smaller the frogs, the higher is the concentration of the enzyme in the pelvic end of the sartorius muscle which is free of nerve endings; the concentration is about four to eight times as high as in the pelvic end of bullfrog sartorius muscle. These figures also point in the direction of surface localization, since the surface is greater in the small-size specimens than in the large ones. However, these indirect conclusions are obviously not as conclusive as the direct evidence and should be further investigated.

An open question is that of the great variations of the concentration of the enzyme in various types of nerve tissues. The amounts of acetylcholine which may be metabolized per gram tissue per hour range from a few milligrams to a few grams, i.e., by a factor of 1000. The extension of surface area is certainly one important factor. The surface area per gram of fiber varies considerably in the different types of tissue and depends on such factors as fiber diameter, amounts of nonneuronal tissue, etc.; the structures of motor end plates differ greatly even in the same group of animals; the density and the surface area of cell bodies and their dendrites as well as the number of *boutons terminaux* are other features which vary markedly. Exploration of ultrastructures, at present still in the beginning, may provide new information. But it appears very unlikely that the surface area is the only determining factor. The permeability change is a complex phenomenon, in which several factors are involved, as will be discussed later. Physical recordings also show great variations. The speed of propagation, for instance, ranges from 100 meters per second to 1 meter or less. The diameter of the axon is, as we know from the work of Erlanger and Gasser, an important, but not the only, factor. A high speed may be achieved through saltatory conduction in fibers which contain nodes of Ranvier. In view of the complexity of the problem and the many unknowns involved, an answer as to the significance of the differences of enzyme concentration cannot be offered at present.

D. Rate of Hydrolysis of Acetylcholine

A characteristic property of acetylcholinesterase pertinent to the problem of its biological function, is the high rate at which the enzyme hydrolyzes its substrate. The enzyme has not been crystallized as yet, and a final determination of the turnover number has not been achieved. In the most active preparations obtained repeatedly from electric tissue of *Electrophorus electricus* the specific activity has been about 400 mmoles of protein per hour, i.e., 1 mg. of protein was able to hydrolyze per hour about 70 gm. of substrate. These preparations showed only one component in electrophoretic studies and in analytical ultracentrifugation, carried out by Dr. Kurt G. Stern and Dr. D. Moore (Rothenberg and Nachmansohn, 1947). However, there was no sharp peak during ultracentrifugation, the material was polydispersed and a polymerization may well have taken place. The sedimentation rate was very high, the S_{20} was over 40, suggesting a molecular weight of over two million. On this basis one would arrive at a turnover number of 2×10^7 per minute at 25°C. and a turnover time of $3\text{–}4 \times 10^{-6}$ second. In more recent unpublished observations, Dr. Claire Lawler, using different procedures of purification, obtained a sharp peak with a sedimentation rate which would indicate a molecular weight in the range of 300,000. The studies are still in progress and a definite value for the turnover number cannot be given at present. However, the data available have shown that acetylcholinesterase is one of the fastest enzymes known. Only catalase has a much higher turnover number. According to earlier determinations by Britton Chance and his associates, this number was estimated to be 4×10^7 per second at 25°C., i.e., more than 100 times higher than that of the esterase. Later it was found that in these experiments the peroxide concentration had not yet been optimal and therefore the highest velocity rate had not yet been achieved. With a high peroxide concentration an unidentified intermediate becomes rate limiting. However, the lifetime of such an intermediate is very short; at the plateau the turnover time is only $0.693/4 \times 10^7 = 2 \times 10^{-8}$ second at 25°C. (Chance, 1956). Catalase is a metal-containing enzyme and the remarkably high speed of action is less surprising than in a hydrolytic enzyme.

Although the exact turnover number is not yet established, the data obtained indicate that the rate at which a molecule of enzyme hydrolyzes a molecule of acetylcholine is high. Assuming the smaller molecular weight, it would be about 30 μsec. A high rate of hydrolysis in combination with the high concentration of the enzyme appear pertinent in relation to the proposed role of the ester in the elementary process of conduction. A compound associated with the generation of bioelectric potentials should be metabolized at a speed comparable to that of the electrical events. Cole and Curtis (1939) found the peak of the impedance change in the squid giant axon to be reached in about 100 μsec. If the increased Na conductance, responsible for this process, is attributed to the action of acetylcholine, the inactivation of the ester and the return to normal Na conductance should take place within a period of time suggested by the physical recordings. The chemical data show that the activity of acetylcholinesterase satisfies this postulate, a prerequisite of crucial importance for correlating a chemical reaction with the conductance changes.

Other compounds and enzymes have been suggested to be associated with the elementary process of conduction, particularly ATP (adenosine triphosphate) and ATPase; besides many other drawbacks of such an assumption, the speed of the reaction would be inadequate.

E. Coincidence of High Enzyme Activity and Beginning of Function during Growth

If the acetylcholine system is inseparably associated with the elementary process of conduction in nerve and muscle cells, one should expect in the embryo and during growth close relationships between the increase in the concentration of acetylcholinesterase and the period at which certain functions develop. Obviously, the concentration of enzymes rises during growth in general. Only such observations are, therefore, of interest where conditions prevail which permit the establishment of direct relationships between a function and a significant increase in an enzyme concentration. In the case of acetylcholinesterase such studies have been greatly facilitated by the stability of the enzyme, the availability of sensitive and relatively accurate methods of determination, and in addition by the high concentration of the enzyme in conducting tissue.

A few examples will be described as an illustration of how relationships have been established between appearance of function and high enzyme activity. In most cases the enzyme concentration in a tissue rises in an S-shaped curve until a maximum has been reached, and then changes very little. An entirely different behavior was found for the concentrations of acetylcholinesterase in the muscles of the hind leg of chick embryo (Fig. 8) (Nachmansohn, 1938g, 1939a). The concentration rises to high values between the 12th and 14th day of incubation,

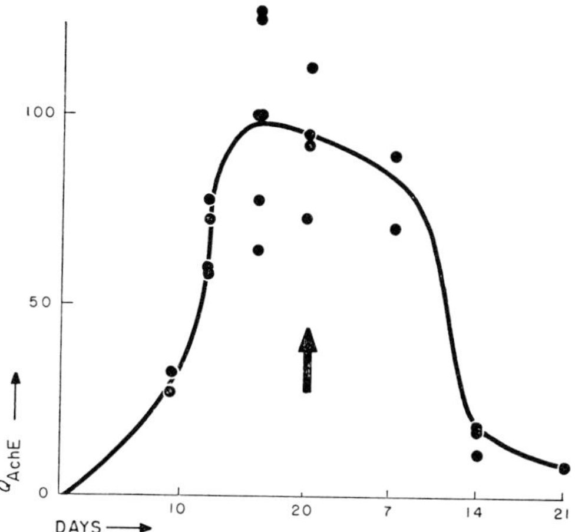

Fig. 8. Changes of the concentration of acetylcholinesterase in muscle of chicken during growth.

Abscissas: days of incubation and after hatching. Arrow marks day of hatching. Q_{AChE} as in Fig. 6.

the period when movements of muscle start. The concentration increases still further, but after hatching it starts to fall and 3 weeks later the values are only 10 per cent of those found on the day of hatching. At hatching the muscle fibers are small; therefore per unit of weight the number of end plates and the total surface of the fibers are high. Since we have seen that the enzyme is localized near the fiber surface, the decrease of concentration after hatching due to the increase of the size of the fiber must indicate that the high enzyme concentration coincides with the time when muscular movements, i.e., conduction along the fibers, starts (about 14th to 15th day). The apparently surprising fall of enzyme concentration is, in fact, not a real decrease of enzyme concentration in the active surface, but an expression of the marked decrease of active surface area per unit weight.

Other observations support the assumption that the high concentration of acetylcholinesterase coincides with the beginning of function. A variety of such relationships has been established with brains of various species and with different centers of the same species. In the brain of the chick embryo, for instance, the Q_{AChE} is about 100 on the 16th day of incubation and rises within the next 4 days to 200 at hatching (Nachmansohn, 1938e, 1939b). During the first few weeks after hatching, the value rises slightly further to about 250; this concentration is approximately that found in the adult chicken. The cerebral functions of the newborn chicken are quite developed. The rapid rise of the enzyme concentration just before hatching is, therefore, remarkable, particularly when compared to the values in young mammals. In the brain of newborn rats and rabbits, which are quite undeveloped, the enzyme concentrations are rather low; they rise rapidly to rather high figures during the first 3 weeks of life, i.e., during that period during which the function of brain develops (see Table V). In contrast, in the brain of newborn guinea pigs, which are fully developed at birth, the concentration of acetylcholinesterase is as high as in the adult animal (Nachmansohn, 1938e, 1939b).

The various centers of the central nervous system do not develop at an equal rate. Extensive investigations of this problem were carried out by Barcroft and Barron(1939), especially in respect to movements and reflexes. They used sheep fetuses, which are particularly suitable for this type of study: in view of the long gestation period, lasting about 150 days, developments are slow and distinctions are easier to make than in embryos with short gestation periods. The investigations show that reflexes of the spinal cord develop at an early stage, whereas the brain centers enter into action at a rather late period. At first local reflexes are observed. Later, the reticulospinal systems acquire a greater significance. Only at a very advanced stage does the brain take over its dominating position.

Studies on the concentration of cholinesterase were carried out in 1939 in Cambridge, in the laboratory of F. G. Hopkins, parallel to those of Barcroft and Barron on function. Interesting relationships and a coincidence between function and high enzyme activity were again found (Nachmansohn, 1940a). In the spinal cord the concentration is high at an early period of gestation; at that time concentrations in various centers of the brain are low; there they begin to rise only in the later period of gestation (Fig. 9). It is noteworthy that in the spinal cord there is in the later stages even an apparent decrease. This fall is, however, as in the muscle of the chick described before, due to the increase in inactive material, such as, for instance, myelin and other supporting

TABLE V
RATE OF HYDROLYSIS OF ACETYLCHOLINE IN A FEW BRAIN CENTERS OF VARIOUS SPECIES DURING GROWTH

Species	Days after birth	Whole brain		Telencephalon		Cerebellum		Corp. quadrigemina	
		W^a	Q^b	W	Q	W	Q	W	Q
	0	796	200	400	195				
Chicken	8	1050	245	541	220	168	208		
	14	1276	243	662	190	210	250		
	0	204	19						
	10	619	36	379	27	36	23		
Rat	21	935	72	527	64	108	46	23	119
	35	1091	95	616	101	136	50	23	135
	110	1195	105	600	110	150	80	29	148
Guinea pig	1	2400	79	1377	73	282	93	55	80
	19	2960	73	1759	64	348	95	68	87
	2	1350	66	778	42	84	54	60	82
Rabbit	10	3183	72	1924	54	297	68	130	133
	20	4419	110	2556	89	546	90	126	206
Cat	1	5299	37	3775	24	300	75		
	21	12261	51	8833	33	1124	108		

[a] W = weight in milligrams.
[b] $Q = Q_{AChE}$.

tissue. Observations on the human fetus led to similar results (Youngstrom, 1941); here, too, significant relationships were found between enzyme activity in various brain centers and the development of motility and function.

A valuable contribution to the problem is the observation of Sawyer (1943) on the changes of enzyme concentration during the larval life of *Amblystoma*. He found significant correlations to exist between the concentration of the enzyme and the functional ability as expressed

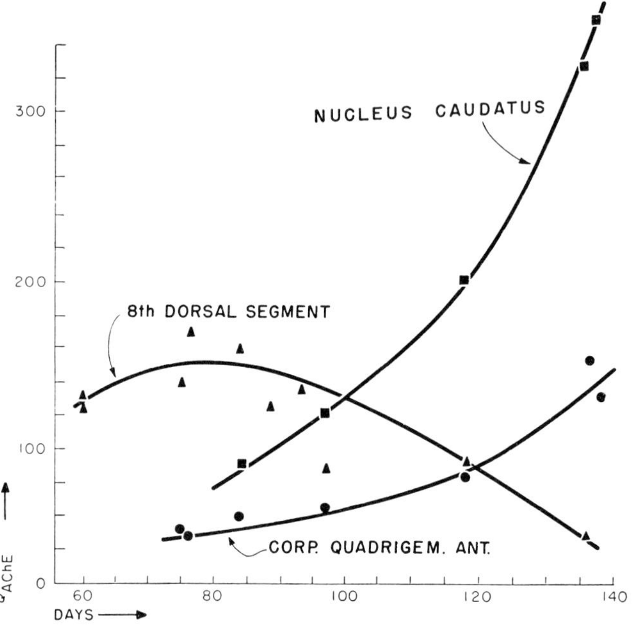

FIG. 9. Concentration of acetylcholinesterase in brain and spinal cord of sheep fetus during growth.

Abscissas: days of gestation time. Q_{AChE} as in Fig. 6.

by behavior manifestations. The enzyme concentration is low during the early stages of development but rises very steeply with the onset of the so-called S-flexure and the rapid movements during the swimming reactions. When the larvae were reared in solutions containing eserine (physostigmine), their functional abilities were strongly impaired, on returning to a normal environment the recovery of motility and enzyme activity were found to parallel each other.

An interesting discussion about the relationships between function and activity of acetylcholinesterase during development took place at a Symposium on the "Biochemistry of the Developing Nervous System" at

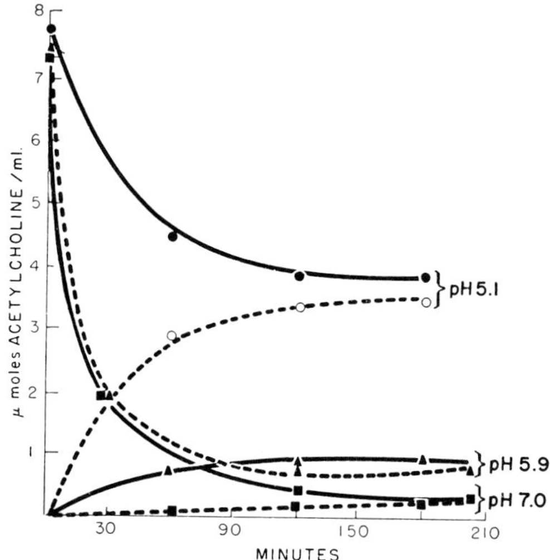

FIG. 10. Synthesis of acetylcholine as a function of pH.

The incubation mixture (total volume 6 ml.) contained either choline chloride and sodium acetate or acetylcholine chloride at the outset. Purified enzyme, prepared from electric tissue, was added in an amount per milliliter which was able to hydrolyze 2 gm. of acetylcholine per hour in optimum conditions. Ester was determined in aliquots. In the absence of either acetate, choline, or esterase no ester formation was observed. $t = 23°$ C.

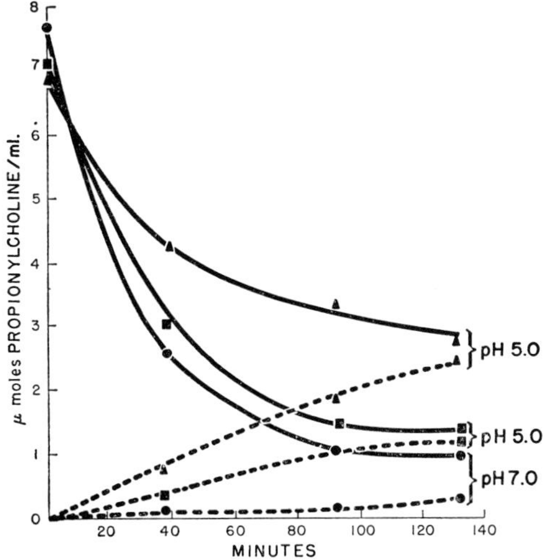

FIG. 11. Synthesis of propionylcholine as a function of pH.

Procedure as described in Fig. 10 except that sodium propionate was used instead of sodium acetate.

Oxford (Waelsch, 1954). The rapid development of various new ultramicrotechniques permits the expectation that more relationships will be established between function and activity of enzymes during growth in general.

F. Free Energy Change of Acetylcholine Hydrolysis

It has been described, a long time ago, that the pharmacological action of solutions containing sodium acetate and choline chloride increases on incubation with esterase (Abderhalden and Paffrath, 1925; Ammon and Kwiatkowski, 1934). Crude tissue preparations were used without any purification. The reaction was enhanced by inhibitors of esterase (Kwiatkowski, 1936). These findings were interpreted as a formation of acetylcholine catalyzed by esterases. The ability of esterases to form esters as well as to hydrolyze them has been frequently investigated.

The availability of a highly active acetylcholinesterase prepared from electric tissue of *Electrophorus electricus* has been very useful for the study of various properties of the enzyme, as will be discussed later in more detail. The specific activity used in most studies has been about 30–60 mmoles of acetylcholine (6–10 gm.) split per milligram of protein per hour. This is not yet an advanced state of purity, since a specific activity of 400 mmoles may be obtained. But it is an adequate degree of purity for many investigations. The preparation has been used by Hestrin (1949b, 1950) for determining the equilibrium

$$K = \frac{[\text{acetylcholine}^+]\ [H_2O]}{[\text{acetic acid}]\ [\text{choline}^+]}$$

The hydrolysis equilibrium constant, K, calculated from molarities was found to have a value of 0.25 at 23° C. The equilibrium position of the hydrolysis of acetylcholine has been studied as a function of pH. An equilibrium at three selected pH values was obtained in each case from both directions (Fig. 10). A corresponding equilibrium was observed in the presence of propionic acid (Fig. 11). A shift to an acid pH, in the range studied, favors the equilibrium in the direction of synthesis.

The standard free energy change of acetylcholine hydrolysis has been calculated from K with the aid of the relationship:

$$\Delta F^\circ = -RT \ln (55.5/K)$$

ΔF° was found to approximate —3200 calories. Although molarities rather than activities were used for the calculation of K, the error from this cause in the value of ΔF° probably does not exceed 10 per cent.

CHAPTER IV

Inseparability of Conduction and Activity of Acetylcholinesterase

A. Competitive Inhibitors

In the introduction to the preceding chapter we have stressed the unique and paramount role of enzymes in all manifestations of life; they are the catalysts which are responsible for virtually all known chemical reactions in a living cell. It is, therefore, obvious that interference with the activity of enzymes may severely interfere with cellular function and may result in death, if either the enzyme which is inhibited is vital or the enzyme forms an essential link in a vital metabolic chain. As an illustration may be mentioned the high toxicity of cyanide which has long been known and is due primarily to the inhibition of the cytochrome-cytochrome oxidase system.

With the rapid development of enzymology during the last few decades there has been a parallel increase of interest in enzyme inhibitors. Quite a few types of inhibitors are known, differing as to the mechanism by which they interfere with an enzymatic process and as to the degree of specificity. Of special interest are the competitive inhibitors, i.e., compounds with a structure similar to that of the substrate or product and competing with them for the active site. Depending on their affinity to the enzyme, which sometimes may be far higher than that of the substrate, they may be weak or very potent inhibitors. Inactivation of enzymatic activity may also occur by a reaction with coenzymes and cofactors, such as metals. The reaction may be readily reversible or may be irreversible. In a reversible inhibition an equilibrium is reached between enzyme and inhibitor. The constant of this equilibrium indicates the affinity between the two components. The criterion for reversible inhibitors is the possibility of removing them by dialysis or of reactivating the enzyme simply by dilution. Enzyme inhibitors which cannot be removed by dialysis have been referred to as "irreversible" inhibitors. This is a somewhat misleading term. As we will see later, in the case of certain organophosphorus compounds which are so-called irreversible inhibitors of acetylcholinesterase, even "irreversibly" inhibited enzymes may be reactivated by a chemical reaction; therefore, the enzyme was actually not irreversibly inhibited. This is a process very different from really irreversible destruction, such as may be obtained by heat, strong acids, or other drastic chemical treatment.

Inhibitors are very useful tools. Enzyme chemists have applied them for a long time to the study of enzyme kinetics and of metabolic pathways. By blocking a certain intermediary step catalyzed by an enzyme, the intermediates accumulate and may be isolated and identified. Such procedures were, for instance, successful in the elucidation of fermentation and glycolysis. A more recent development, especially during the last decade, has been the use of competitive inhibitors for the analysis of the active site. By varying systematically the structure of the inhibitor, much pertinent information may be obtained about the molecular forces of the active protein surface. A particularly illuminating example of the use of the inhibitors has been the analysis of the active surface of acetylcholinesterase which will be described in Chapter VII.

Other fields in which inhibitors may be powerful tools are cellular function and toxicology. As was stressed by Green (1941), many compounds capable of strong action in very low concentration ("trace substances"), such as certain drugs, vitamins, and hormones, produce their effects by reacting with enzymes or enzyme systems. Inhibitors may be used for ascertaining physiological functions of an enzyme or a chemical reaction. A classic example is the demonstration of Lundsgaard (1930) that muscular contraction is still possible after complete suppression of lactic acid formation in a muscle poisoned with monoiodoacetate. His observations profoundly influenced the development of muscle physiology. It was the first time that the inhibition of a specific chemical reaction was used for testing its relationship to a specific cellular function.

In pharmacology and toxicology enzyme inhibitors occasionally have been used, but the implications were frequently not fully recognized. It is indeed often not easy to evaluate the effects; until recently there were hardly any attempts at a systematic application of inhibitors to analysis of the mechanism. However, during the last decade a considerable change has taken place. There is an increase in interest in the correlation of enzyme activity with structure and function; enzymology is penetrating more and more into toxicology as into so many other fields.

Interpretation of the effects of inhibitors on living cells may be difficult and offers many pitfalls. Even in an enzyme solution the interaction between protein and small molecule depends on a variety of factors and their proper analysis is a complicated task. Innumerable other factors must be considered when inhibitors are applied to intact cells or to the whole animal. In view of the complexity of a living cell, conclusions as to the underlying mechanisms cannot be built upon a single fact, but must be corroborated by several types of analysis. In Lundsgaard's experiments it was not merely the absence of lactic acid formation which made his conclusion convincing; it was the correlation between various biochemical and biophysical processes which indicated that lactic acid formation was indeed not essential in the elementary process of muscular contraction as was maintained at that time by Meyerhof. It was the evidence that contraction under the experimental conditions used takes place as long as there is phosphocreatine breakdown and that the chemical energy released by this process is adequate to account for the tension developed.

Before discussing the toxicological effects and the physiological implications of competitive inhibitors of acetylcholinesterase, some chemical features of the most important inhibitors may be briefly mentioned. Two of the strongest and most widely used reversible competitive inhibitors are physostigmine and Prostigmine. Physostigmine, also called eserine, is an alkaloid obtained from the calabar bean, which grows in tropical West Africa. It has been used in medicine, especially in ophthalmology, since the middle of the nineteenth century. Its potentiating effect on the action of acetylcholine was first described by Hunt (1918). Stedman and Barger (1925) elucidated the chemical structure. In the course of the studies of Stedman and his associates on the chemical basis of the inhibition of cholinesterase, Aeschlimann and Reinert (1931) testing systematically a series of substituted phenyl esters of alkyl carbamic esters, synthesized Prostigmine or neostigmine. The structures of these two inhibitors are shown in formulas (I) and (II).

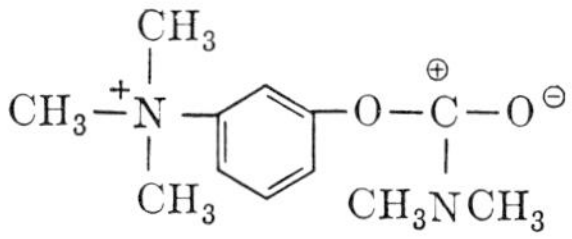

$(CH_3)_3N^+$–C_6H_4–O–C(⊕)(–O⊖)–$N(CH_3)_2$

Prostigmine
(I)

H_2C—C(CH_3)(–C_6H_3–O–C(⊕)(–O⊖)–$NHCH_3$)—CH; H_2C–N(CH_3)–CH–N(CH_3)

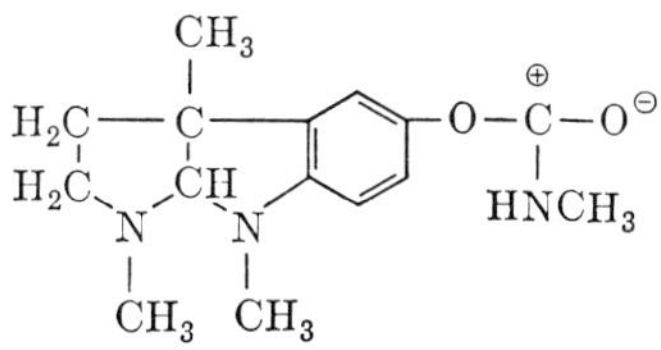

Physostigmine
(eserine)
(II)

The structural similarity with acetylcholine is apparent and suggests the competitive nature of their inhibitory action. The kinetics of these reactions with acetylcholinesterase has been studied with the preparation obtained from electric tissue of *Electrophorus electricus* and purified to a relatively high degree. In the solution used, about 100 mmoles of ester were hydrolyzed per milligram of protein per hour (Augustinsson and Nachmansohn, 1949b). The competitive nature of the inhibition was demonstrated in the following way. If v is the reaction velocity in the absence of the inhibitor I, and v' in its presence, and v/v' is plotted against the concentration of the inhibitor, a straight line is obtained in the case of competitive inhibition at constant concentrations of enzyme according to the equation:

$$\frac{v}{v'} = 1 + [I] \frac{K_S}{K_I([S] + K_S)}$$

$[S]$ and $[I]$ are the concentrations of substrate and inhibitor, K_S and K_I the dissociation constants of the complexes between enzyme and sub-

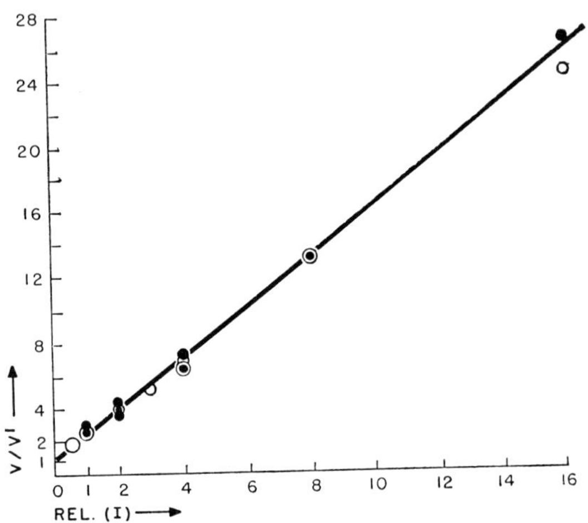

FIG. 12. Inhibition of acetylcholinesterase by eserine as a function of inhibitor concentration.

Purified preparation obtained from electric tissue of *Electrophorus electricus*. v = velocity in absence, v' = in presence of inhibitor, expressed in microliters of CO_2 evolved in 30 min. Relative inhibitor concentration (Rel. $[I]$), $1 = 6 \times 10^{-7}$ *M*. (●) without, (O) with incubation, (⊙) values obtained with different substrate concentrations. $K_I = 6.1 \times 10^{-8}$.

strate and enzyme and inhibitor, respectively. The intercept of the straight line on the ordinate (v/v') is 1. K_S and K_I can be calculated from the slope of the line, that is $K_S/K_I\ ([S] + K_S)$. Figures 12 and 13 show the data obtained in applying this method of analysis to the action of eserine and Prostigmine. They are typical for the competitive nature of the action. The dissociation constants of the enzyme inhibitor complex determined in this way are 1.6×10^{-7} for Prostigmine and 6.1×10^{-8} for eserine. The affinity to eserine of this particular enzyme is thus 2.6 times as high as that of Prostigmine. This is a high degree of affinity indeed, if one considers that some of the tightly bound coenzymes have dissociation constants of the order of 10^{-7} for the enzyme-coenzyme complex.

Competitive inhibition may also be demonstrated in a different way. If substrate and inhibitor compete for the same center, then the inhibitory effect would be expected to become smaller with increasing substrate concentration and at a very high excess of substrate the inhibition may be overcome since the affinity of the inhibitor, although high, may not be high enough to compete with the substrate. The method was used by Augustinsson (1948) in studies on cholinesterase. An illustration

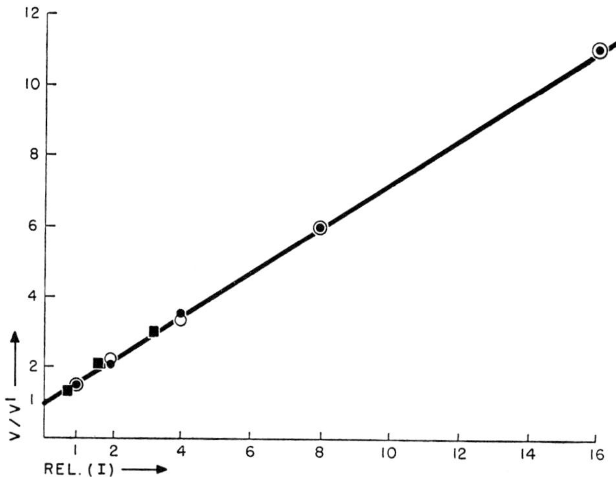

FIG. 13. Inhibibtion of acetylcholinesterase by Prostigmine as a function of inhibitor concentration.

Enzyme preparation and description as in Fig. 12. Relative inhibitor concentration (Rel. [I]), $1 = 6 \times 10^{-7}$ M. (●) without, (O) with incubation, (■) values obtained with different substrate concentration. $K_I = 1.6 \times 10^{-7}$.

for testing the reaction between acetylcholinesterase and Prostigmine is given in Fig. 14 (Augustinsson and Nachmansohn, 1949b).

Since the Second World War a new type of very powerful inhibitors of esterases has become known, a group of organophosphorus compounds. This type of compound has long been known. It was used in the 1930's by Schrader (1952) for developing insecticides. Some of the compounds were found to be volatile and extremely toxic, especially for the nervous system. The German Army took hold of them and developed the famous "nerve gases" as potential chemical warfare agents, the most toxic compounds ever developed. Some of them are the most potent enzyme inhibitors known. The structures (III–V) of a few of the known nerve gases are given as illustrations.

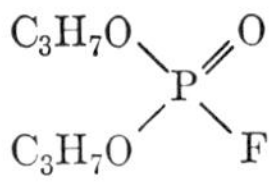

$$(C_3H_7O)_2P(=O)F$$

Diisopropylphosphofluoridate (DFP) (III)

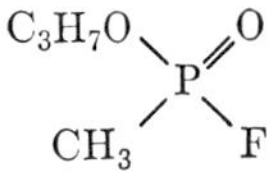

$$C_3H_7O(CH_3)P(=O)F$$

Sarin (IV)

$$(CH_3)_2N(C_2H_5O)P(=O)CN$$

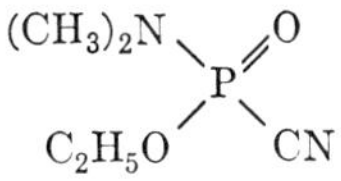

Tabun (V)

Other organophosphorus compounds used as insecticides are tetraethylpyrophosphate (TEPP) (VI) and Paraoxon or Parathion (VII). In the latter an oxygen is substituted by a sulfur atom.

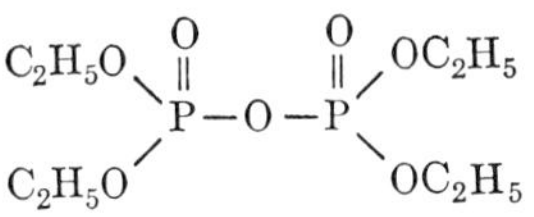

$$(C_2H_5O)_2P(=O)-O-P(=O)(OC_2H_5)_2$$

Tetraethylpyrophosphate (TEPP) (VI)

$$(C_2H_5O)_2P(=O)-O-C_6H_4-NO_2$$

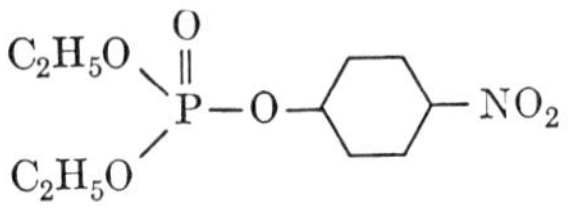

Diethyl *p*-nitrophenyl phosphate (Paraoxon) (VII)

The organophosphorus compounds are specific inhibitors of enzymes capable of splitting ester linkages. They inhibit all types of esterases, lipases, trypsin and chymotrypsin, and thrombin. However, the lethal effect must be attributed, as we will see later, to the inhibition of acetylcholinesterase. This group of compounds has many interesting and unique features which will be discussed in the section on the mechanism of action. They belong to the so-called irreversible inhibitors, i.e., their action is not reversed by simple dialysis, but only by chemical reaction.

The velocity of the enzyme inhibitor combination does not parallel the inhibitory strength. The rate of reaction between diisopropylphosphofluoridate (DFP) and acetylcholinesterase is a rather slow process; TEPP reacts much faster. Velocity rate, inhibitory strength, and other aspects of the reaction depend on the properties of the acid group and on the various alkyl groups attached to the phosphorus atom. Some of these factors will be discussed later. Since there is no equilibrium reached between inhibitor and enzyme, no straight line is obtained if v/v' is plotted against inhibitor concentrations, as is the case with eserine and Prostigmine (Augustinsson and Nachmansohn, 1949b).

Cholinesterase reacts with a very great number of compounds, possibly with more than any other enzyme. Almost all functions of the body are controlled by the nervous system. During the last twenty years a great number of compounds were tested as to their ability to inhibit cholinesterase. A decade ago Augustinsson (1948) quoted in his dissertation something like 900 publications mostly devoted to inhibitors of the enzyme and their pharmacological effects. Only a few inhibitors have been mentioned here, namely those which have been used as tools in the studies on the role of the acetylcholine system in conduction and are thus pertinent to the main topic of this monograph.

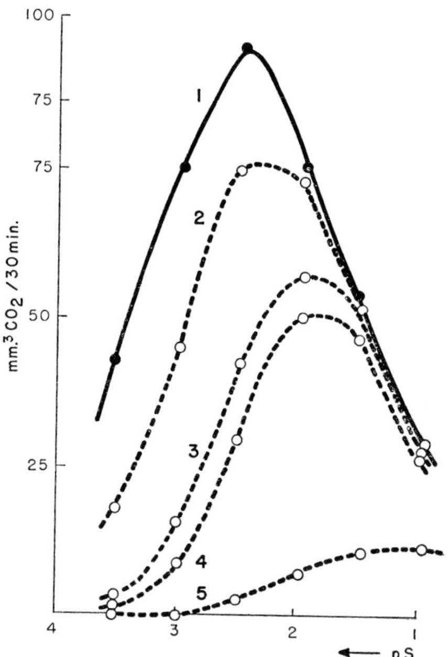

Fig. 14. Activity-pS curves for the enzymatic hydrolysis of acetylcholine by acetylcholinesterase in presence of various concentrations of Prostigmine.

Purified preparation obtained from electric tissue of *Electrophorus electricus*. Curve *1* is the control, curves *2–5* the hydrolysis in presence of 0.4, 1, 2, and 10×10^{-6} *M* Prostigmine.

B. Action of Competitive Inhibitors on Conduction

The various features of acetylcholinesterase, discussed in Chapter III, such as its presence in all types of conducting tissue, the localization, the high concentration and high turnover number, etc., are important prerequisites for the assumption that the action of acetylcholine is essential for conduction; they are suggestive but not conclusive. For a conclusive evidence a direct relationship had to be established between the activity of acetylcholinesterase—implying of course an active role of its substrate—and the function proposed. It was necessary to show that enzyme activity and conduction are inseparable. Here the use of specific inhibitors is an extremely powerful tool if properly applied. A complete inhibition of the enzyme should be incompatible with the generation of bioelectric currents, if the theory of an essential role of the ester in conduction is correct.

In contrast to previously reported negative results (Cantoni and Loewi, 1944), we were able to show, in 1945, that eserine blocks conduction in isolated nerve fibers (Bullock, Nachmansohn, and Rothenberg, 1946). The action of eserine is readily reversible, as might have been expected, since the enzyme inhibition is reversible. The effect was first demonstrated with the squid giant axon. Later, other types of nerve fibers were tested. Figure 15 shows the effect of eserine on the conduction of the bullfrog sciatic nerve in which the blocking effect and its reversal has been repeatedly obtained on the same fiber.

A dramatic development in this field took place in the following year when several observations were presented which seemed to contradict the assumption of an essential role of acetylcholinesterase in conduction. At a Symposium at the New York Academy of Sciences, Gilman reported that it is possible to destroy with DFP all the cholinesterase in the bullfrog sciatic nerve without impairment of conduction (Gilman, 1946; Crescitelli, Koelle, and Gilman, 1946). Several other facts presented at the same Symposium (1946) seemed to be incompatible with the view that acetylcholine is inseparably associated with the elementary processes of nerve activity. One puzzling observation was the evidence that DFP may rapidly produce a reversible block of conduction in contrast to the irreversible destruction of the enzyme.

However, it very soon became apparent that the methods used in the early investigations were not adequate and that many of the experimental conditions were open to criticism. In the following two years the writer and his associates were able to establish conclusively with the aid of DFP and other organophosphorus compounds the essentiality of acetylcholinesterase activity in conduction. A number of quantitative relationships were, moreover, obtained for which reversible inhibitors are not suitable. Under no condition and in no type of nerve fiber has it been possible to separate electrical and enzyme activity.

The alkyl phosphates are, because of the irreversible nature of the inhibition, for various reasons more useful and more versatile tools for studying certain aspects regarding the problem of the function of acetylcholinesterase in nerve activity and, beyond that, of related problems of toxicology of the nervous system. In fact, one can say that the use of these compounds has in the past decade profoundly affected and tremendously helped various fields of biology: enzymology, protein chemistry, physiology, pharmacology, and toxicology. A huge literature has developed, and today many investigators all over the world are using these compounds for a variety of practical and theoretical problems. In describing this development it is more difficult to select properly the

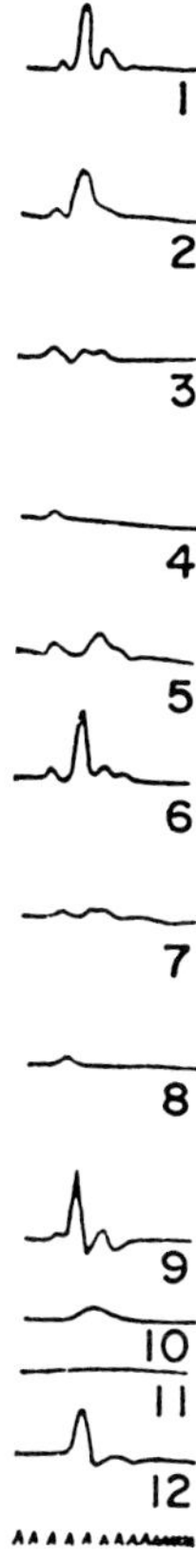

FIG. 15. Effect of eserine on the conduction in bullfrog sciatic nerve, exposed repeatedly to the inhibitor with intervals of recovery in Ringer's solution.

Concentration of eserine: 0.02 *M*. (*1*) control before exposure. First exposure: (*2–4*); 5, 15, and 21 min. (*5–6*), eserine replaced by Ringer's solution; 15 and 48 min. recovery. Second exposure: (*7–8*); 9 and 17 min. (*9*): retested 12 hours later after having been kept in Ringer's solution. Third exposure: (*10–11*); 5 and 73 min. (*12*): 134 min. recovery. $t = 25°$ C. 1000 c.p.s.

facts pertinent to the main topic and to limit the presentation than to discuss some of the fascinating and closely related investigations.

The use of these compounds, especially on the cellular level and *in vivo*, is full of pitfalls. Many factors of chemical, physicochemical, and biological nature have to be considered which, unfortunately, were ignored and neglected in the early observations. Only a few aspects of more general interest will be discussed. Enzymes in a cell are usually present in excess, as mentioned before. This appears to be a necessary security regulation and is known to exist for many biological mechanisms and functions. One can remove one-and-a-half kidneys, or 80 per cent of the liver, and the animals still may survive. Let us assume that acetylcholinesterase is present five times in excess; the function should not be seriously impaired until 80 per cent of the enzyme is inhibited. The remaining amount may be small in absolute values and difficult to detect. The minimal concentration to be expected should be small if the action of acetylcholine is associated with the elementary process, since, as we have seen, the initial heat is very small. This applies particularly to frog sciatic nerve, which was used in the earlier investigations and in which heat production is much smaller than in many unmyelinated invertebrate nerves. A quarter of a century ago, A. V. Hill (1932b) made the statement that physiological laboratories ought to be built near the seashore: "The heat production of crabs' nerves could have been measured twenty years ago: actually fourteen years were wasted before it was measured with frogs." If this is true for physical recordings for which such highly sensitive instruments are available, the warning of Hill should certainly have been heeded in the case of chemical measurements. The choice of the proper biological material is often decisive for finding a correct answer.

One of the greatest difficulties in the measurements of esterase activity after exposure of cells or fibers to organophosphorus compounds is the relatively high concentration required in the outside solution for blocking the enzyme activity inside the cell. Only a very small fraction, as will be discussed below, penetrates the cell interior. Frequently, the enzyme activity was determined after the exposed fibers were ground and a homogenized suspension had been prepared. If one is interested in the activity in the interior of the intact cell, then it is necessary to remove the excess, particularly in the case of DFP, which is more lipid than water soluble, before the actual determination of enzyme activity. Without such removal, if only a small fraction of DFP is retained in the extracellular tissue and released in the homogenized suspension, this free DFP would be more than enough to abolish the enzyme activity still remaining inside the cell. In our earlier experiments we used extensive washing with saline and were able to remove a part of the extracellular DFP. More recently, more efficient removal of the excess DFP was achieved by chloroform treatment (Kewitz, 1957a; Kewitz and Nachmansohn, 1957).

A few days after the symposium in 1946 mentioned above, the relationship between failure of conduction and the level of enzyme activity was checked in lobster nerves after they were exposed to DFP. The material is favorable insofar as it has a high enzyme activity. Moreover, the velocity of the reaction between the enzyme and this particular alkyl phosphate is very low and, therefore, suitable for testing whether the time course of progressive irreversible enzyme inhibition parallels that

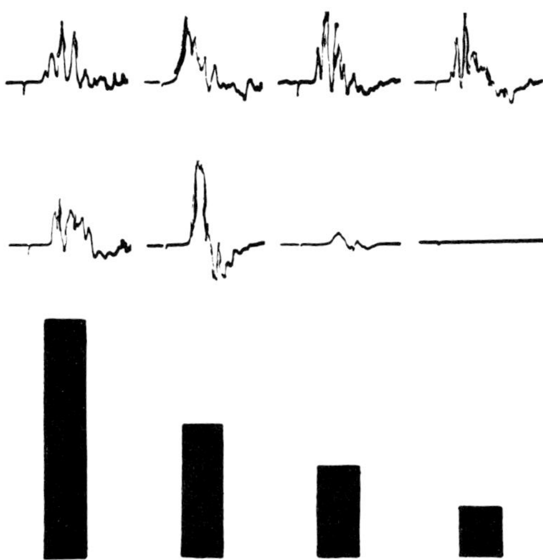

FIG. 16. Irreversible effect of DFP on axonal conduction as a function of time.

The figure shows the parallelism between the progressive irreversible inhibition of conduction and the progressive irreversible inhibition of acetylcholinesterase in the abdominal chain of lobster. Only conduction in the giant fibers is recorded. Four nerve preparations, shown in the four columns, were kept in DFP (0.013 *M*) for 30, 60, 90, and 120 min., respectively and then washed in sea water. Top line of each column shows action currents before exposure. After 30 min. all activity was abolished. Second line shows the degree of recovery of electrical activity after prolonged washing of the nerves. Nerve fibers exposed for 30 min. recovered completely. After 60 min. exposure, recovery is incomplete; after 90 min. exposure, hardly any activity returns. The blocks in the third line indicate the remaining activity of acetylcholinesterase. CO_2 output is 233, 129, 88.5, and 50 mm.^{3} per 100 mg. tissue per hour compared to about 1500 mm.^{3} in the controls.

of decreasing electrical activity. Figure 16 shows the very first experiment. The preparation was exposed for 30 minutes to sea water containing DFP. After this period of time the action potential had disappeared, but it returned after washing and there was still 15–20 per cent of the initial activity. The longer the preparation was exposed to the action of DFP, the stronger was the effect on both electrical and enzyme activity. After 90 minutes the electrical activity had virtually disappeared, but there was still about 8 per cent of enzyme activity left. Even when the last trace of electrical activity had been abolished, after 120 minutes, there was still 5 per cent of the initial enzyme activity.

The observations require several comments. First, the question arises why the action potential disappears in the beginning reversibly, if the enzyme is still active. Since the enzyme inhibition is irreversible, the reversible block cannot be attributed to the enzyme inhibition. Obviously, the action potential does not depend on one single component of the acetylcholine system. This problem will be discussed in connection with the receptor protein (Chapter X). The primary question here was whether conduction is still possible when the enzyme activity is completely abolished, as the early reports had claimed. The experiments clearly contradicted this conclusion. The quantitative relationships in particular experiments offer another problem. The preparation used was the abdominal chain of lobsters. This preparation contains a multitude of fibers and also synaptic junctions. The electrical recordings, under our conditions, show only the activity of the giant fibers, running through the whole preparation; synaptic potentials are not recorded. The enzyme activity at the synaptic junctions is usually higher than in axons, as mentioned before, and is most likely the first to have been abolished, since the junctions are, in contrast to the axons, less protected by lipid barriers. Therefore, the percentage values of enzyme activity in relation to electrical activity are probably too low. For a more quantitative evaluation of the relationship, this preparation, because of its complexity, is less suitable than are simple axons. Moreover, there is still some retention of DFP in the tissue, even after washing, as was experimentally demonstrated. This factor will also contribute to lower the actual value of intracellular enzyme activity still present before the grinding.

The relationships between conduction and enzyme activity on exposure to DFP was then tested in a variety of nerve fibers (Bullock *et al.*, 1947a, b). Figure 17 shows an experiment with the squid giant axon. The electrical activity rapidly disappears on exposure to DFP but reappears on washing. Even after a second short exposure the block of conduction is still reversible. When, however, the exposure is long enough the electrical activity is irreversibly abolished; at that period the cholinesterase activity was found to be about 20 per cent of the initial (ranging from 17 to 25 per cent). Similar results were obtained with the fin nerve of squid.

The essentiality of acetylcholinesterase in conduction can even be demonstrated with bullfrog sciatic nerve. The experiments are, however, technically much more difficult. Not only are the absolute values of enzyme activity small, but there is a huge amount of myelin and lipid surrounding the axon. This situation makes it much more difficult to evaluate the enzyme activity at the time when conduction is blocked. Manometric methods are not suitable; but with much more sensitive biological methods (e.g., assay with frog's rectus abdominis muscle) enzyme activity can still be demonstrated after complete and irreversible block of conduction by DFP. Here too, the block is first reversible and

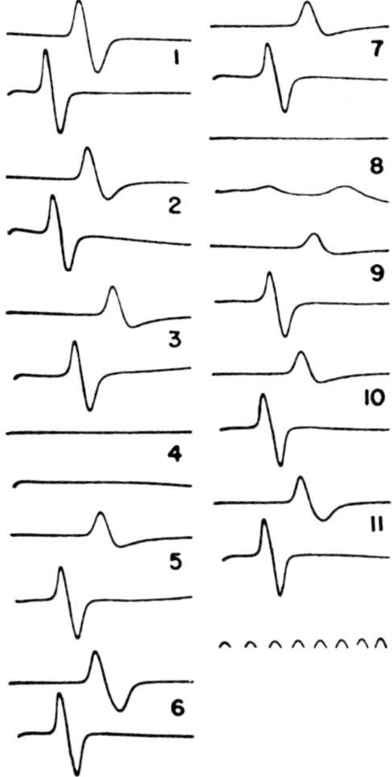

FIG. 17. Effect of DFP on the action currents of the squid giant axon.

Concentration of DFP was 0.005 *M*. Records in pairs, representing potentials from two points on the nerve. After short exposure the currents are reversibly abolished; with longer exposure (not shown in the figure), the effects become irreversible. (*1*) control before exposure. (*2–4*) after 6, 7.5, and 12 min. exposure. The nerve was then washed with sea water. (*5–6*) 4.5 and 105 min. recovery. Second exposure: (*7, 8*); 5 and 8 min. (*9–11*): 2, 7, and 66 min. recovery in sea water. 2000 c.p.s. 23°C.

one has to wait for a certain period of time until it becomes completely irreversible (Fig. 18). But at that period there is still at least 8–10 per cent of the initial enzyme activity present. Here again, the essential point is in the evidence that even under these most unfavorable conditions cholinesterase activity is present and can be detected even after complete and irreversible block of conduction; but the level of enzyme activity is too low to be compatible with electrical activity.

The relationship between irreversible block of conduction and of enzyme activity was tested not only as a function of time but also of temperature. At temperatures of 8–10° C. it takes about 4 hours until conduction in lobster nerves is irreversibly blocked by DFP at concentrations at which the preparation is blocked in 60 min. at room temperature. The process is thus about three to four times slower than in the experiments with a 12–15° C. higher temperature. At 37° C., on the

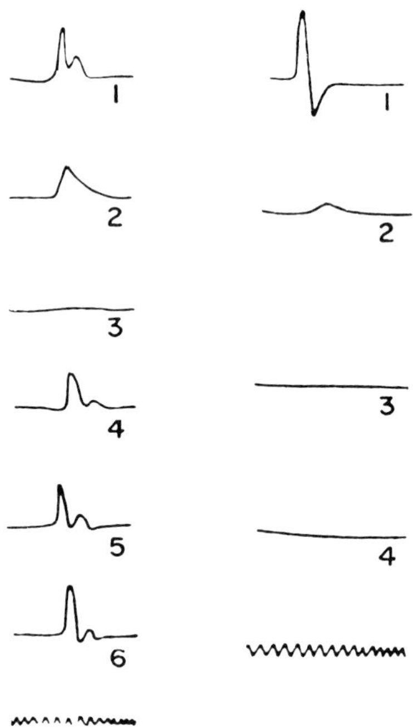

FIG. 18. Irreversible effect of DFP on axonal conduction in bullfrog sciatic nerve.

Concentration of DFP was 0.02 *M*. *Left:* (*1*) before, (*2, 3*) after 11 and 25 min. exposure. (*4–6*) 25, 75 and 135 min. recovery in Ringer's solution. *Right:* (*1*) before, (*2*) after 25 min. exposure. After 120 min. exposure the nerve was returned to Ringer's solution. (*3*) and (*4*) 45 min. and 5 hours later. 1000 c.p.s. Room temperature.

other hand, inactivation is much faster. For experiments at this temperature mammalian nerves, the sympathetic cervical fibers of cat, were selected. The action potential is abolished in 2 min. after exposure to DFP; if returned immediately to normal saline solution, conduction reappears. The esterase activity is high. If, however, the nerves are exposed for an additional 20 min. to DFP, conduction is irreversibly blocked. In control experiments, fibers kept in eserine for the same length of time after disappearance of the action potential, and then returned to saline solution, did not show any impairment of conduction (Fig. 19).

Block of conduction by specific inhibitors of acetylcholinesterase has been demonstrated with all types of nerves, motor and sensory, sympathetic ("adrenergic") and parasympathetic ("cholinergic"), vertebrate

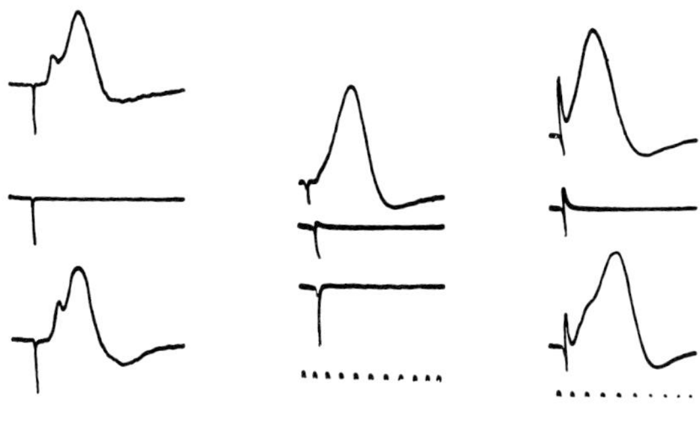

FIG. 19. Irreversible effect of DFP on axonal conduction as a function of temperature.

Superior cervical sympathetic nerves of cat were exposed to DFP (0.02 *M*) at 37° C. *Left* (from top to bottom): before exposure, after 2 min. exposure, and after 11 min. recovery in Ringer's solution. *Middle:* before and after 3 min. exposure. The nerve was then further exposed to DFP for 20 min. No return of activity after 120 min. recovery. *On the right:* the reversible effect of eserine (0.02 *M*) on the same type of preparation. Top: before exposure; middle: after 8 min. exposure. The nerve was then exposed for an additional 20 min. to eserine. Complete recovery after 67 min. in Ringer's solution.

and invertebrate. A few examples may be given as illustration. Figure 20 shows the effect of eserine on the dorsal roots of ox. Two other examples of the action of DFP on purely sensory nerves, the optic nerve and the superficial ophthalmicus of *Raja erinacea* are shown in Fig. 21. Here again, as in previously discussed observations, the block is reversible after short exposure. Only after a more prolonged exposure does it become irreversible. In the same figure is shown the effect of eserine on the splanchnicus of the bullfrog. This nerve fiber contains primarily slow-conducting C-fibers and only very few fibers of a slightly greater diameter. Although the presence of a few preganglionic fibers cannot be excluded, the nerve undoubtedly contains a large number of postganglionic sympathetic, i.e., classic "adrenergic" fibers. As seen in Fig. 21, after 6 min. exposure to eserine, conduction was completely blocked in all fibers, the adrenergic included. When returned to Ringer's solution, a partial recovery was observed after 5 min. and a complete recovery after 70 min.

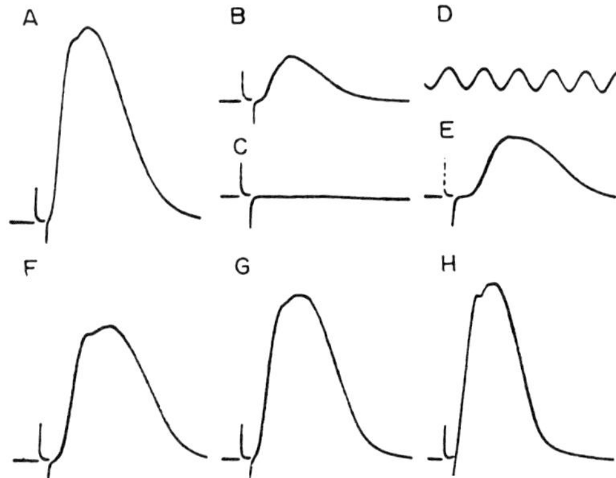

FIG. 20. Effect of eserine upon electrical activity of sensory fibers (dorsal roots of ox).

A: Action current before exposure. *B* and *C:* response following the exposure of the fibers to a Tyrode solution containing 0.013 *M* eserine, 2 and 4 min., respectively. *D:* calibration in 1000 cycles. After 4 min. exposure the fibers were returned to Tyrode solution. *E, F, G,* and *H:* 2, 5, 10, and 15 min. later.

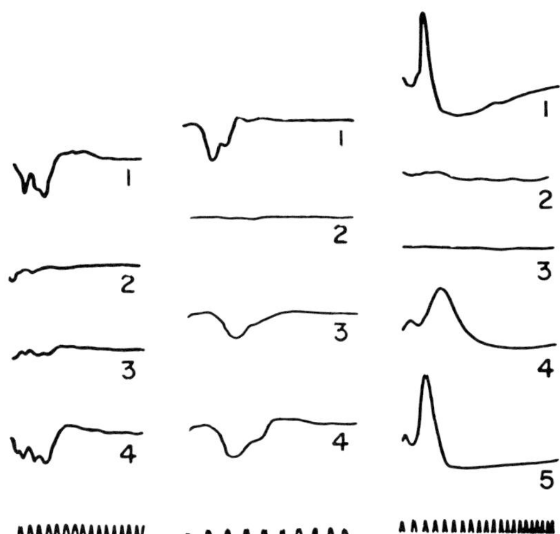

FIG. 21. Effect of cholinesterase inhibition on electrical activity of sensory and adrenergic fibers.

Left: optic nerve of *Raja erinacea,* exposed to 0.015 *M* DFP. (*1*) before, (*2*) after 4 min. exposure. The nerve was then returned to sea water; (*3, 4*): 5 and 16 min. recovery. 1000 c.p.s. 24° C. *Middle:* superficial ophthalmic nerve of *Raja,* exposed to 0.02 *M* DFP. (*1*) before, (*2*) after 6 min. exposure. The nerve was then returned to sea water; (*3, 4*): 1 and 2 hours recovery. 2000 c.p.s. 25° C. *Right:* Splanchnic ("adrenergic") nerve of bull frog, exposed to 0.02 *M* eserine. (*1*) before, (*2, 3*) after 3 and 6 min. exposure. The nerve was then returned to Ringer's solution; (*4, 5*) 4 and 5 min. later 60 c.p.s. 23.5° C.

It is generally believed that the propagation of impulses along muscle fibers is effected by the same mechanism as that along axons. As discussed before, acetylcholinesterase is present in various types of muscle fibers; it is found in parts free of nerve endings and apparently located in the surface region. Conduction in muscle is abolished by eserine and DFP in the same way as in axons. Figure 22 shows the effect of two types of inhibitors on the electrical activity in a frog's sartorius muscle. In order to exclude any effect upon nerve endings, the inhibitors were added only after complete block of the motor end plates by curare and the muscle was stimulated directly.

It may also be mentioned that in the monocellular organism *Tetrahymena geleii* S eserine and DFP were found to block the ciliar movements (Seaman, 1951). The presence of acetylcholinesterase and its exclusive localization in the pellicle to which the cilia are attached, was described in the preceding chapter.

If acetylcholinesterase is essential for conduction in general, an animal should not be able to survive complete inactivation of brain esterase. When a dose of 0.3 mg. of DFP per kilogram is injected into the ear of rabbits, 50 per cent of the animals die within 2 min., the other 50 per cent survive. Very little acetylcholinesterase, or none, was found in the brain of those animals which died; in the surviving animals a concentration of 20–40 per cent of the initial activity as found (Nachmansohn and Feld, 1947). These values were obtained after washing of the slices with saline. With the more efficient method in which the DFP retained in extracellular fluid and tissue is removed by chloroform, possibly even higher values might have been obtained. Respiration, glycolysis, ATPase were, under these conditions, completely unaffected (unpublished data of this laboratory).

The data which indicate the inseparability of acetylcholinesterase activity and conduction, were presented at the International Congress in Oxford, in 1947, and the toxic effect of DFP was attributed to a specific "biochemical lesion" (Nachmansohn, 1947). This viewpoint was vigorously opposed. The complexity of the toxicological picture was considered by many investigators to be evidence of a general toxic effect and the interpretation proposed an oversimplification. One of the main objections was the high concentration of DFP required to block conduction, 1–2 mg. per milliliter. Since the enzyme in solution is completely inactivated in concentrations of less than 1 μg. per milliliter, this discrepancy was considered as incompatible with the assumption that the block of conduction had anything to do with the inactivation of cholinesterase. This objection does not take into account the great difference between applying an inhibitor to an enzyme in solution and to a living cell. The living cell is surrounded by all kinds of structural barriers which either prevent or slow down the entrance of many compounds into the cell interior. Biologists have long been aware of the existence of these barriers, and their properties in living cell were the

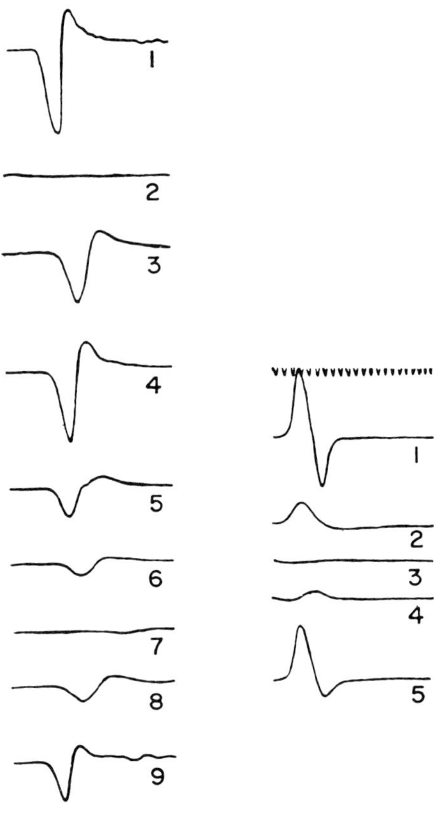

FIG. 22. Effects of esterase inhibitors on conduction of curarized striated muscle.

Frog sartorius muscles were used. Electrical activity was evoked by direct stimulation. Response to indirect (neural) stimulation was completely abolished by keeping the muscles in Ringer's solution containing *d*-tubocurarine (100 μg./ml.). *Left:* exposure to 0.01 *M* DFP, in Ringer's solution containing *d*-tubocurarine. (*1*) before, (*2*) after 9 min. exposure. The muscle was then returned to Ringer's *d*-tubocurarine solution. (*3, 4*) 3 and 10 min. recovery. Second exposure to DFP: (*5–7*) after 2, 5, and 10 min. exposure. Again returned to Ringer's solution. (*8, 9*) 5 and 15 min. recovery. 200 c.p.s. 24° C. *Right:* exposure to 0.01 *M* eserine in Ringer's solution containing *d*-tubocurarine. (*1*) before, (*2, 3*) after 3 and 9 min. exposure. Muscle returned to Ringer's *d*-tubocurarine solution. (*4, 5*) 7 and 10 min. recovery. 1000 c.p.s. 23.5° C.

subject of extensive investigations and many theories and speculations for half a century. During the last decade the use of isotopes and the results of electron microscopy have provided much new information. It has been shown that even inside the cell there exist highly organized structures with barriers which prevent the free movements of many compounds. The objection has been raised that the concentrations of inhibitor required to block conduction are unusually high, implying that this high concentration may indicate an unspecific effect and not be comparable to that with the "conventional" low concentration. This objection is not justified. What is the measure for a "conventional" concentration? As mentioned before, it has been known for a century that most compounds rapidly reach synaptic junctions which are much less protected by permeability barriers than axons. Many compounds affect axons not at all or only in high concentrations. Acetylcholine itself, for instance, which acts in extremely low concentrations at junctions, does not affect the axon even in extremely high concentrations because it does not penetrate into the cell interior. This has been shown experimentally and will be discussed in detail in a later chapter. If an enzyme is blocked at the junction at a certain concentration of an inhibitor, one would *a priori* expect, simply on physicochemical reasoning, that a much higher concentration is required for blocking the same enzyme in the axon. Let us take eserine as an example. The pK_a of the compound is about 8.2. At pH 7.2 only 5 per cent are neutral molecules, and this is the only form likely to penetrate through lipid material. In a frog sciatic nerve the axons are not only surrounded by thick myelin sheaths, but in addition by large amounts of lipid material, connective tissue, etc. Therefore a very small fraction of this inhibitor would be expected to penetrate.

Recently, Dettbarn applied eserine to Ranvier nodes of isolated fibers of frog sciatic nerve. Whereas with intact nerve fibers several milligrams of eserine are required for blocking conduction in 20–30 min., Dettbarn obtained block of conduction in 20–30 seconds with 200 μg. per milliliter of eserine, approximately a thousandfold increase in potency (W. D. Dettbarn, observations presented at the Federation Meetings in Atlantic City, April, 1959). His findings are another striking demonstration of the importance of permeability barriers in evaluating quantitative relationships between the outside concentration of a compound and its intracellular action.

Applying eserine to Ranvier nodes in still lower concentrations, about 2–10 μg. per milliliter, Dettbarn, in unpublished experiments, observed a marked increase of spike height and amplitude and a prolongation of the descending phase. If the action of acetylcholine on the receptor protein is responsible for the increase of Na conductance, and if the removal of the ester by acetylcholinesterase reverses the action, then a decreased speed of enzyme action should produce the effect observed. However, the effect would be expected to take place only at a certain very limited range of enzyme activity, viz. where most of the excess is removed, but the remaining activity is not much below the minimum required for unimpaired function. It is probably quite difficult to find proper conditions for such an effect except in a highly sensitive single fiber preparation, such as that used by Dettbarn. With multifiber preparations the effect may not be detectable for various reasons: it might, for instance, be very transitory and may therefore be masked, because the different components are not simultaneously affected. Since high concentrations of inhibitors are required because of the presence of strong permeability barriers, the transition at the active site from the effective concentration to one which is too low may be too rapid to permit observation of this effect.

The question of penetration also applies to DFP, although for an opposite reason: it is much more lipid than water soluble and it would, therefore, largely accumulate in the lipid material before entering the water phase. Clearly the outside concentration of an inhibitor is irrelevant for its effect on enzyme activity. The effectiveness of a compound depends on the concentration at the site of action. But even inside the cell the distribution of a compound will be determined by its physiochemical properties and the barrier surrounding the microstructures; it is therefore impossible to predict what concentrations are required on the outside for blocking an enzyme inside the cell; nor do the outside concentrations give any indication of the concentration at the site of action.

In order to get some information about the entrance of DFP into the cell interior, squid giant axons were exposed to DFP and the concentration of the compound was determined in the axoplasm extruded just at the time when conduction was blocked. As may be seen from the figures in Table VI, the concentrations of DFP in the interior at this particular period are of the order of 1 μg. per gram, sometimes lower, and in one case 5 μg. per gram. Even this type of experiment does not reveal the concentration of inhibitor at the active site, but it shows that the outside concentration has little meaning and cannot be used as an indication for an "unspecific" effect.

The relationship between electrical and chemical activity has been tested more recently with a method which avoids some of the difficulties encountered previously and which permits the determination of the two activities on the same intact fiber. The method is based on the following principle: whereas acetylcholine is a lipid-insoluble quaternary ammonium salt and does not penetrate into the interior of the axon, its tertiary analog, dimethylaminoethyl acetate, is able to penetrate and is also a fairly good substrate of acetylcholinesterase, although not quite as good as the quaternary ester. The Michaelis-Menten constant for dimethylaminoethyl acetate is about 1×10^{-3} *M*, which is about ten times higher than that for acetylcholine. However, whereas acetylcholine in higher concentrations inhibits the enzyme activity, the tertiary analog follows the usual type of substrate-enzyme activity relationship and does not inhibit at high concentrations. This feature is an additional advantage since high substrate concentrations must be used in testing enzyme activity of intact cells; therefore, one does not have to be concerned about reaching a substrate concentration which is too high and would depress the activity. Using this tertiary amine as substrate, it is possible to determine the rate of acetylcholinesterase activity in an intact fiber. The retention factor is eliminated and the relationship between enzyme and electrical activity in presence of inhibitors can be directly measured at any desired period. However, here too it is important to use the proper material; the fibers should have a high enzyme activity and therefore permit an adequate degree of precision of chemical determinations; they should not offer too great permeability barriers to the substrate. The nerves from the meropodite of the claw and walking legs of the spider crab *Libinia emarginata* (Rathbun) were found to be suitable for such measurements. The fibers are unmyelinated and have a very small diameter; the enzyme activity is very high: about 50–100 mg. of acetylcholine are hydrolyzed per gram (fresh weight) per hour. Figure 23 shows the relationship between chemical and electrical activity in these fibers on exposure to DFP (Wilson and Cohen, 1953). The electrical activity ceases when the enzyme activity falls to 20 per cent

TABLE VI

CONCENTRATION OF DFP IN THE AXOPLASM OF SQUID GIANT AXON AFTER COMPLETE BLOCK OF CONDUCTION

Outside concentration of DFP (μmoles/ml.)	Exposure (min.)	Inside concentration of DFP		
		μmoles × 10^{-3} per ml.	μg./gm.	Per cent of outside concentration
2.2	44	2.0	*0.36*	0.09
2.2	38	8.4	*1.5*	0.38
5.5	11.5	30.0	*5.4*	0.54
5.5	4	2.5	*0.45*	0.05
5.5	27	1.3	*0.23*	0.02

of the initial. The outside concentration of DFP affects only the rapidity with which the critical concentration is reached but not its level. This shows how irrelevant the outside concentration is except for the time factor, i.e., for the rate of penetration. Moreover, using different types of inhibitors such as eserine or the tertiary analog of Prostigmine, one obtains block of electrical activity at the same critical concentration of the enzyme although the two compounds differ in many respects, as to affinity, interaction with the proteins etc. These findings have clearly disposed of the objection to the unusually high concentration of inhibitor required and of the implication of an unspecific effect.

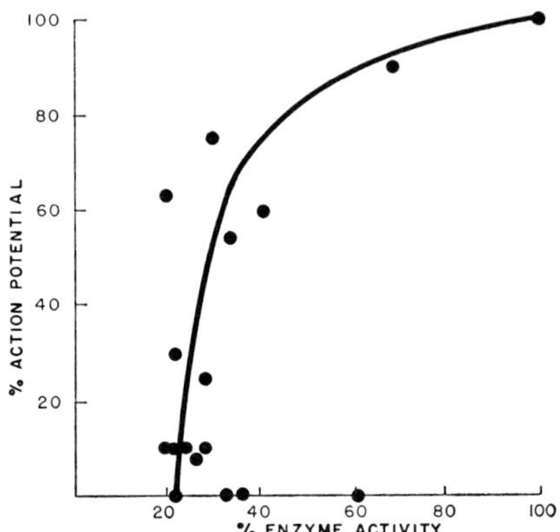

FIG. 23. Inseparability of electrical and esterase activity in axons.

Crab nerve fibers were exposed to DFP (0.02–0.04 *M*). Enzyme activity was determined on intact fibers. No electrical activity can be evoked when enzyme activity falls to about 20 per cent of the initial.

The procedure described is the only one which permits determination of the level of cholinesterase activity in relation to the ability to conduct impulses with the reversible type of inhibitor. A method using homogenized suspensions for determining the esterase activity in tissues exposed to reversible inhibitors is not feasible. A relatively high outside concentration of inhibitor for the inactivation of the enzyme in the intact fiber is required for reasons explained above. If, therefore, eserine is used and the fibers are washed to remove the excess, all or most of the enzyme activity may return, since the enzyme inhibitor complex is readily reversible. If the eserine is not removed, a small fraction of the eserine applied will be sufficient to inactivate all the enzyme when the fibers are homogenized and all the structural barriers have been removed. This explains the failure of Cantoni and Loewi (1944) to find the relationship between cholinesterase and conduction. Applying 10 μg.

per millilter of eserine to the frog sciatic nerve, they found no effect on conduction. This outside concentration is much too low to affect the internal enzyme activity in view of the strong permeability barriers of this preparation. On grinding, however, there was a huge excess of inhibitor in the suspension, since 0.1 μg. per milliliter would block all the enzyme present. It is quite surprising to find a similar argument in 1953, after all the extensive investigations and discussions in regard to this problem and all the information available in this field. The esterase of electric tissue, ground after exposure to high concentrations of eserine (without washing), was found to be inhibited, and this is quoted as demonstration that cholinesterase is unnecessary for electrical activity because the electrical activity was still present before the grinding (Keynes and Martins-Ferreira, 1953). This is a sad illustration of the permeability barriers existing at present between various fields of biology.

For a few years there was a vigorous controversy about the question whether electrical and esterase activity are separable. This question seems to the writer to be definitely settled. Therefore, many details of experimental techniques and many of the pitfalls encountered in these studies have not been discussed. The reader interested in this particular aspect is referred to the original articles which appeared between 1946 and 1948.

One more question may be briefly mentioned in this connection, namely, the possible role of acetylcholinesterase in red blood cells. Here too the enzyme is localized exclusively in the stroma, and not in the interior. The lipids found in the cell membrane are similar to those in the membranes surrounding the nerve fiber. The potassium concentration in most red blood cells is high. The question has been raised whether the acetylcholine system has some function in the ion movements of these cells. Observations of Greig and Holland (1949, Holland and Greig, 1950) on this question were carried out with methods which are open to criticism. The effect of cholinesterase inhibitors on the ion movements were measured with radioactive species by Taylor, Weller, and Hastings (1952). They found that exposure to eserine and DFP leads to a loss of K, primarily because the rate of entrance of K into the cell is decreased. These observations would support the assumption of some role of the acetylcholine system in the regulation of ion movements. On the other hand, the same investigators did not find a relationship between the strength of enzyme inhibition and the effect on ion movements. The enzyme was inhibited at such low concentrations of inhibitor that there was no effect on the K movements. The observations were, however, not made under strictly comparable experimental conditions. The ion movements were measured on intact cells; the enzyme activity was determined in a suspension. In the light of the experience obtained with nerve fibers, a reevaluation of this problem would be desirable.

CHAPTER V

Sequence of Energy Transformations

Nearly simultaneously with the demonstration by A. V. Hill and his associates that nerve activity is associated with heat production, Fenn (1927) and independently and at about the same time Gerard and Meyerhof (1927) found extra oxygen uptake in frog sciatic nerve due to activity. The amount calculated in the two laboratories was in good agreement: 2.5 μμmoles of oxygen were found to be taken up per gram nerve per impulse. Moreover, the figures fit well with the total energy used during activity when estimated on the basis of the heat production. In more recent years the results as to oxygen uptake have been confirmed and extended by Bronk and his associates with refined techniques, using the oxygen electrode, which is a method of great precision and sensitivity (Brink, Bronk, Carlson, and Connelly, 1952).

The data on heat production and oxygen uptake have provided information as to the over-all utilization of energy. The data do not indicate which specific chemical reactions are directly associated with either activity or recovery. The major part of the chemical energy is certainly used for the restoration of the unequal ionic distribution prevailing in the resting state. Since the recent values of Hill and his associates regarding the initial heat made untenable the view that conduction can be explained purely with ion movements, this development added new emphasis to the necessity of elucidating the underlying molecular, i.e., chemical events.

Important information required for correlating a cellular function with chemical reactions is that concerning the sequence of energy transformations taking place between the primary event and oxidation, the ultimate source of energy. In spite of much progress achieved in the knowledge of intermediary metabolism of muscle, that associated with nerve conduction remained uncertain and unsatisfactory, as was pointed out by Feng (1936). The methods available at that time were not adequate for the smallness of the metabolic reactions associated with nerve activity. This situation changed when investigations were started, in 1937, on the electric organ of electric fish with the aim to obtain information about the chemical reactions underlying the generation of bioelectric currents and specifically the role of acetylcholine. The choice of this material proved indeed to be of considerable value for the further development of the field and has contributed to our understanding of intermediary pathways in general. It thereby became possible to integrate acetylcholine into the metabolic pathways of the nerve cell.

A. Electric Fish

Bioelectric potentials were described many years before Galvani. Michel Adanson (1757) compared the sensation felt in touching the *Malopterurus* to that by a discharge of a Leyden flask. John Walsh (1773) demonstrated that the shock of *Torpedo,* known to Greeks and Romans, is an electric discharge, and at about the same time Williamson (1775) made corresponding observations on *Electrophorus electricus.* After Galvani's sensational observations on nerve-muscle preparations, biologists became interested in these fish. Galvani himself, in the last two years of his life worked with *Torpedo.* During the last century many physiologists analyzed different aspects of the electric discharge and other physiological and structural properties of the fish. For the physiologist the most important feature of these organs is the fact that their strong bioelectric potentials are generated in the same way as those of nerve and muscle. This was recognized by physiologists of the last century and particularly stressed by Du Bois-Reymond (1877), who devoted many years of his life to the study of electric fish. He expressed his conviction that analysis of the electric discharge of these fish would eventually lead to a better understanding of the electrical manifestations in nerve and muscle: "The unusual interest which electric fish have always aroused, has been still further increased at present by the knowledge that the ability which made these animals for so long an object of inquisitive curiosity, is not a unique feature but appears to be a special application of a common property frequently encountered in the animal kingdom. Since we know that all nerves and muscles of all animals are capable of electrical activity, electric fish lost somewhat their miraculous nature; but this loss is largely compensated by the hope attached to their investigation, namely that their study will promote the solution of the great problems of the general physiology of nerve and muscle."

The electric organ is formed by compartments—each containing an electric plate, the electroplax—which are arranged in columns. The action potential developed by a single electroplax is about 0.14 volt, which is of the same order of magnitude as that found in ordinary nerve and muscle fibers. It is only the arrangement of these plates in series as in a Voltaic pile which enables these organs to develop the high voltage. Volta recognized the analogy. Describing his pile in a memorandum read before the Royal Society in London, in 1800, he wrote that he wished to give to the pile the name: "an artificial electric organ." Only one face of the electroplax is innervated. Half a century ago Bernstein suggested that the arrangement in series may be explained by the change of potential only at the innervated face. In conformance with his idea on conduction he assumed a simple depolarization. Recent work with intracellular electrodes on electroplax of *Electrophorus electricus* has shown that Bernstein's hypothesis was essentially correct; however, the innervated face is not merely depolarized but actually the charge there is reversed (Altamirano *et al.*, 1953; Keynes and Martins-Ferreira, 1953). Summation of the voltages developed by the individual cells would be impossible if both faces were to reverse their potentials. Figure 24 is a schematic illustration of the arrangement in series as proposed by Bernstein and the modification which has become necessary.

There are several species of fish provided with electric organs. The great differences of the discharge in various species do not depend on the units which show relatively small variations, but on the shape and

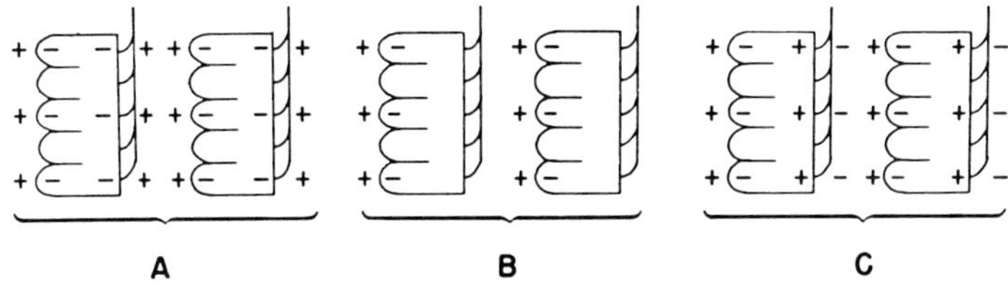

FIG. 24. Schematic presentation of the arrangement in series of the cellular units in the electric organ of *Electrophorus electricus*.

In *A* the two electroplax are in resting condition. Only the right side is innervated. The interior of the cell is negative, the outside positive. *B* shows the active state of the two electroplax according to the hypothesis of Bernstein: During the discharge the innervated side is depolarized, the charge of the noninnervated side remains unchanged. *C*. Actual event measured with microelectrodes inserted into the interior of the electroplax: reverse of charge at the innervated side during the discharge; that at the noninnervated side remains essentially unchanged.

dimensions of the organs. In the species with the most powerful electric organ known, *Electrophorus electricus* (Linnaeus), about 5000 to 6000 electroplax are arranged in series from the cephalic to the caudal end of the organ. The voltage of the discharge is on the average about 600 volts. In *Torpedo marmorata* the number of elements in series does not exceed 400 to 500. The discharge here is on the average 40–60 volts.

Electric organs have evolved phylogenetically from striated muscle. In the strong electric organs of muscular origin the contractile elements have completely disappeared; they exist as rudiments in the plates of the weak electric organs of rays and have been found in the embryonic tissue of *Torpedo*. The large literature on anatomical data and electric characteristics of electric fish accumulated by intensive investigations during the last century has been summarized in an excellent review by Rosenberg (1928). Pertinent recent observations on electrical characteristics carried out mostly by Fessard and Chagas and their associates and by Coates and Cox may be found in papers by Fessard (1946), Albe-Fessard (1950), Albe-Fessard, Chagas, and Martins-Ferreira (1951), Abbe-Fessard *et al.* (1959) and by Cox *et al.* (1945, 1946).

Studies with isolated rows of electroplax of *Torpedo* were initiated by Auger and Fessard (1939) in the late thirties (see also Fessard, 1946). The electrical characteristics of the electric plates of this species appear to resemble those observed at motor end plates. This is appar-

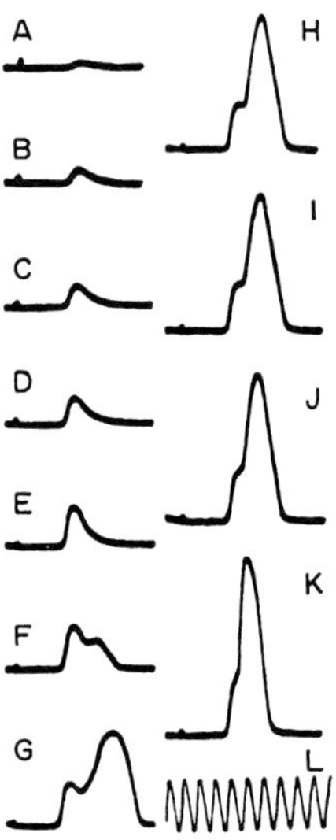

FIG. 25. The two components of the response of the electroplax of *Electrophorus electricus* to neural (indirect) stimulation.

A graded response is evoked by weak stimuli to the nerve and increases with increasing strength of the stimulus (*A–E*). This response is the postsynaptic potential. With still further increase of the stimulus strength the second component appears. The latency decreases and the potential increases until reaching a maximum (*F–K*). The second component of the potential is the spike response of the electroplax.

ently also the case with the electroplax of rays (Brock *et al.*, 1953; Schoffeniels, 1958a). More recently, isolated rows of electroplax of *Electrophorus electricus* were used in the writer's laboratory and in that of Chagas in Rio. The electric response of the electroplax of this species is analogous to that of a muscle fiber rather than to that of an end plate. There are two distinct types of electric response: a graded one, which resembles the end plate potential, and an all-or-nothing spike which appears at a critical size of the prepotential and has all the characteristics of a propagated spike (Fig. 25) (Altamirano *et al.*, 1953; Keynes and Martins-Ferreira, 1953). This confirms and extends the previous observations of direct excitability of the electroplax of this species (Albe-Fessard *et al.*, 1951).

If one believes in the biochemical unity of life, a notion so greatly cherished by Pasteur and Meyerhof and which was so useful in the development of dynamic biochemistry, these most powerful bioelectric generators which nature has created should be a favorable material for the analysis of the basic mechanism by which bioelectric potentials in general are generated. The underlying chemical processes should be in a range more suitable for analysis and identification than those of ordinary nerves, in which metabolic reactions are so small that the methods available at that time were not adequate. Moreover, since the electric organs are highly specialized in their function, the hope appeared justified of finding those special features of intermediary metabolism which are connected with the primary event, i.e., with the ion movements during activity. Thereby, the relationships of metabolism and function could be more readily elucidated.

B. The Concentration of Acetylcholinesterase in Electric Organs

When the remarkable distribution of cholinesterase in muscle and its special concentration at motor end plates had been demonstrated, in 1937, the writer, reading some literature about motor end plates, came across Lindhard's (1931) article in which he states that electric organs are kinds of modified motor end plates. It was this remark that suggested to the writer the idea of testing the concentration of cholinesterase in electric tissue.* The results of the first test were striking: 2–3 gm. of acetylcholine were split per gram of tissue (wet weight) per hour (Marnay, 1937). This was an extraordinarily high activity compared to that of other conducting tissues, as may be seen in Table VII. The high concentration of enzyme appeared all the more significant in view of the high water and low protein content of electric organs: 92 per cent of the organ is water and only 2 per cent is protein. The ability of the organs to hydrolyze amounts of acetylcholine several times their own weight in 1 hour immediately suggests the possibility of a close relationship with their highly specialized function, i.e., the generation of bioelectric potentials. In the following year, in 1938, a specimen of *Electrophorus electricus* died in the Institut Océanographique in Paris and, thanks to the efforts of Fessard, there was an opportunity to test the

* In that year the International World's Fair was held in Paris and there were a few specimens of *Torpedo marmorata* shown at the scientific exhibit. Thanks to the efforts of Dr. Catherine Veil from the same laboratory, the writer obtained two specimens, and his assistant, Mlle. Annette Marnay, made a few determinations of cholinesterase in the electic tissue.

enzyme activity in the electric tissue of this species. Two samples of the main electric organ were used. Here again the activities were extraordinarily high: 900 and 1400 mg. of acetylcholine were split per gram tissue (fresh weight) per hour (Nachmansohn, 1940b).

TABLE VII

ACETYLCHOLINESTERASE CONCENTRATION IN THE MAIN ELECTRIC ORGAN OF *Electrophorus Electricus* COMPARED WITH SOME OTHER TISSUES

Tissue	Acetylcholine hydrolyzed (mg./gm./hr.)
Mammalian (guinea pigs) at 37° C.	
Muscle fibers	8–15
Nerve fibers	10–15
Brain	80–100
Frog at 23° C.	
Muscle fibers	3–6
Nerve fibers	5–10
Brain	40–80
Electric organ, *E. electricus,* at 23° C.	2000–4000
Mammalian kidney	0
Mammalian liver	0

As Selig Hecht once remarked, such an extraordinary activity of a special enzyme in a highly specialized organ could not be just a joke of nature. A depolarizing substance which could appear and disappear in milliseconds might well be a trigger in the elementary process of bioelectric currents in general. Although this idea appeared very appealing, it was not until the end of 1940 that the association of acetylcholine metabolism with bioelectric activity was proposed. The theory that the mechanism of synaptic transmission differs fundamentally from that of axonal conduction appeared less satisfactory than the assumption of a basic similarity, the view maintained by Fulton, Erlanger and Gasser, Lorente de Nó, and several European investigators. The biochemical findings, such as the high enzyme activity in all conducting tissues, in axons as well as at synaptic junctions, the exclusive localization in the envelope surrounding the axon, various findings with electric tissue, the evidence for a depolarizing action of acetylcholine (see below), all these facts combined seemed to indicate a more general role of acetylcholine than that of a neurohumoral transmitter. The concept proposed was aimed at bridging the gap between the conflicting opinions and seemed best to integrate the then available data. The emphasis on a more general role of acetylcholine was expressed in the following statement: "Thus, the battle cries of the last years: 'electrical transmission' and 'chemical transmission' lose their significance, because in view of the results reported here it is difficult to maintain the concept that acetylcholine has a 'transmitter' function specifically limited to nerve endings. The experiments suggest—in modification of the original theories of Loewi and Dale—that acetylcholine metabolism is intrinsically connected with the electrical changes of nerve activity which occur at the surface of nerve cells" (Nachmansohn and Bettina Meyerhof, 1941). The viewpoint was fully endorsed by Fulton (Fulton and Nachmansohn, 1943). At that time it was difficult to foresee the vigorous opposition which this view would at first encounter among many investigators, but, in retrospect, the concept proved fruitful and has been substantiated by a great variety of data.

The assumption of an intimate relation between electrical and enzyme activity appeared to be further supported by the striking parallelism between the concentration of the enzyme and the voltage and the number of plates per centimeter in *Electrophorus electricus.* The number of plates and the voltage per centimeter vary considerably in this species with the size of the specimen and decrease, moreover, markedly from the anterior to the posterior end of the organ (see Fig. 26). Since the same voltage is developed by each plate, the voltage per centimeter differs markedly. The question arose, whether under these conditions electrical and chemical activity would show a relationship. Determinations carried out on a great number of specimens of various sizes, covering a range of action potential from 0.5 to 22 volts per centimeter, have shown a direct proportionality between voltage per centimeter and enzyme concentration (see Fig. 27) (Nachmansohn *et al.*, 1941, 1942, 1946a). In contrast to this remarkable relationship, other enzymes tested, respiratory and glycolytic enzymes, ATPase, etc., do not show any parallelism with the voltage developed. Their activity is more or less uniform in the main electric organ. In the relatively weak electric organ of Sachs where the extracellular space is extremely large and forms more than 95 per cent of the compartment, the absolute values are smaller. Most of the energy-yielding enzymes, being required for a great variety of functions, must be present at many places all over the cell and the density of plates per centimeter should therefore not change the activity of these enzymes expressed per gram of tissue.

In view of the evidence for the localization of cholinesterase in or near the surface membranes, one might have suspected that the enzyme would be localized also in the surface of the electroplax. However, in view of certain structural and functional features of the surface of the individual electroplax of this species, the direct proportionality cannot be explained simply in terms of surface localization. At the anterior end

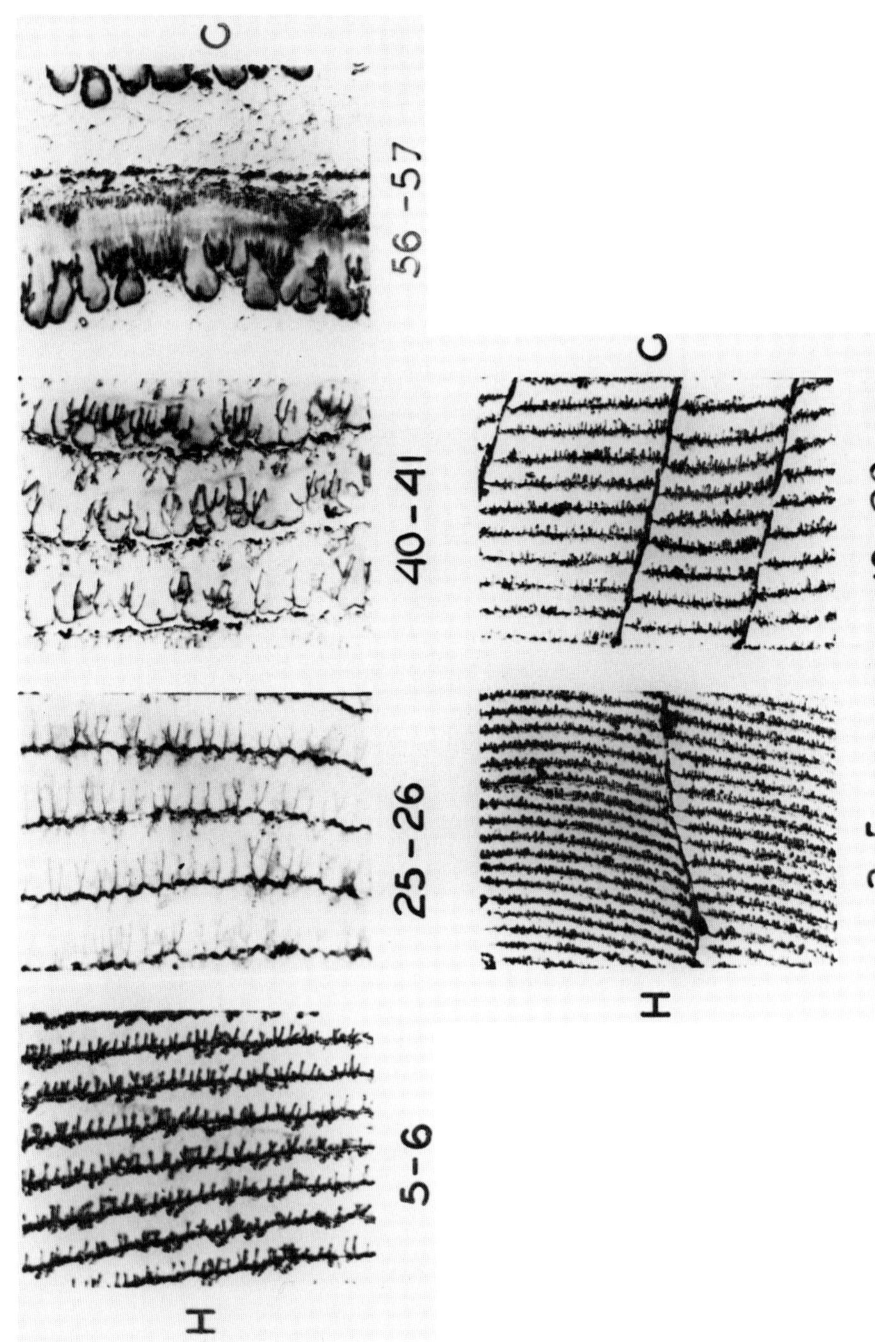
C
56-57
40-41
25-26
5-6
H
C
H

of the organ the electroplax are very thin; both innervated and noninnervated faces have very few folds. From the anterior to the posterior end the thickness of the cells gradually increases and becomes markedly greater; the folds at the noninnervated face become very deep, the surface is full of digitations, and thus the total surface area becomes very huge, but that of the conducting surface remains unchanged. Since voltage and enzyme concentration per cell remain the same, there is no proportionality between enzyme activity and surface, but there is a direct proportionality between enzyme and conducting surface. Thus, the relationship is not only structural but functional, suggesting an interdependence between the two events, i.e., chemical and electrical activity. It still would be possible to refer the relationship to localized processes instead of to voltage if, for instance, all or most of the enzyme would be present at the synaptic junctions. There are many thousands of synapses per cell at distances ranging from 5 to 30 μ. Assuming average distances of 15–20 μ, a cell with a conducting surface of 10 $mm.^2$ would have about 25000 to 50000 synapses. Assuming a surface area of 12 μ^2 per synapse, estimated on the basis of the electron-microscope pictures of Luft (1956), the synaptic junctions would form about 3–6 per cent of the total surface area of the conducting membrane. The enzyme concentration in synaptic regions is usually higher than in the conducting membrane outside the synapse by a factor of 3 to 5, as has been discussed in Chapter III. Since it is well known that the surface increases at synaptic junctions, the increase of enzyme concentration is probably just an expression of increased surface. The enzyme located at the synapses forms apparently a relatively small fraction of the total enzyme. The relationship between electrical and chemical activity must, therefore, be referred to the whole conducting membrane and not just to the junctions.

There are few examples in biology where direct proportionality between a physical manifestation and a specific chemical reaction has been demonstrated. The combination of the highly specialized function of electric organs in general with the unusual structural features of the organ of this particular species created a uniquely favorable material

---

FIG. 26. Changes of frequency of electroplax per centimeter in the electric organ of *Electrophorus electricus*.

Frequency decreases from anterior (*A*) to posterior end (*P*) and also depends on the size of the specimen: the longer the specimen, the lower the frequency. Upper sections taken from a specimen of 114-cm. length, lower section from one of 57-cm. length. Numbers below each section indicate distance in centimeters from the anterior end of the organ. Magnification of all histological preparations is the same (145 times).

for demonstrating the functional relationship, but, as always, the finding obtains its real significance only in combination with all other data.

The esterase activity in electric tissue of rays is low: about 10–20 mg. of acetylcholine is hydrolyzed per gram tissue per hour (Nachmansohn and Bettina Meyerhof, 1941). This is a concentration of the same order of magnitude as that in the bundle of Sachs of *Electrophorus*. Both are

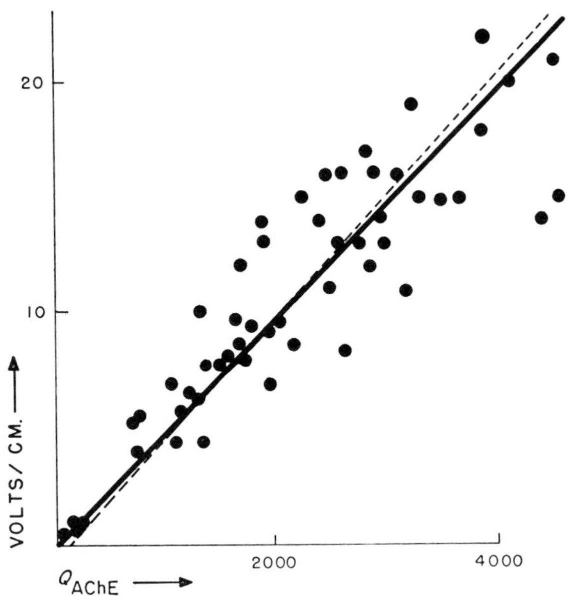

FIG. 27. Direct proportionality between voltage per centimeter and acetylcholinesterase concentration in electric organs of *Electrophorus electricus*.

Specimens varying greatly in size were used, and samples of many sections of each specimen were taken for determination. Dotted line calculated from the experimental data with method of least squares; solid line calculated on assumption that line goes through zero point.

relatively weak electric organs in terms of voltage per gram tissue. In one specimen of *Malopterurus* which we had the opportunity to study in 1944, we found in two samples 24 and 38 mg. of acetylcholine split per gram per hour. The total electric discharge of this specimen was 80 volts (D. Nachmansohn and C. W. Coates, unpublished experiments).

It has been stressed before that there are great differences of enzyme concentrations in various types of nerves, in nerve tissue of various species throughout the animal kingdom, in different parts of the brain of the same species, etc. We are unable to account at present for these quantitative variations. The acetylcholine system forms only one component of a complex membrane. The generation of bioelectricity must depend on a great variety of additional factors, even if the acetylcholine system is the trigger essential for the process and is inseparable from it. The complexity of the acetylcholine system will be discussed in connection with its four proteins; the complexity of additional chemical factors will be discussed in several other chapters. The direct proportionality found between electrical activity, active surface area, and enzyme concentration in *Electrophorus* is due, as explained above, to the special features of this organ which are exceptionally favorable for this correlation. One would not *a priori* expect this to apply to the relationship of voltage and esterase activity in different species. The electric cells of *Malopterurus* are known to differ greatly in structural and functional aspects from those of *Electrophorus*. Quantitative variations both of physical and chemical events, rather, would be anticipated from all our knowledge and experience in all fields of biology. The speed of conduction in different axons may vary by a factor of 100; nobody would exclude on this basis alone that electric currents are, in both types of axons, the propagating agents.

Recently, Augstinsson and Johnels (1958) measured cholinesterase activity in a few specimens of *Malopterurus*. The concentration was rather low, of the same order of magnitude as that which we had found in 1944. The voltage of the specimens varied from 100 to 400 volts. On the basis of these quantitative differences the authors arrive at the conclusion that the acetylcholine system cannot play the same significant role in the generation of electric potentials in *Malopterurus* that is claimed for it in other species. If this statement is meant to exclude an essential role of acetylcholine in the generation of bioelectric currents of *Malopterurus*, it would be a rather far-reaching conclusion unless much additional evidence of an entirely different kind would be offered. Since this monograph is concerned with the many problems involved in this particular question, a more detailed discussion of this statement appears to be unnecessary.

An organ extremely poor in protein, but at the same time extraordinarily rich in content of a specific enzyme, suggested itself as an unusual and remarkably good source for obtaining this enzyme in solution and for purifying it. It is quite rare that such a favorable biological material is offered to the enzymologist. Acetylcholinesterase was, for the first time, obtained in solution from the electric organ of *Torpedo* in 1938 (Nachmansohn and Lederer, 1939a, b). In the following years the enzyme was purified from the electric organ of *Electrophorus* by ammonium sulfate fractionation to a degree where the preparation showed only one component in the analytical ultracentrifuge run and in electrophoresis. The specific activity of these solutions was about 400 mmoles (70–75 gm.) of ester hydrolyzed per milligram protein per hour (Rothenberg and Nachmansohn, 1947). An improved procedure has been recently worked out by Lawler (1959). The availability of a highly purified and active preparation opened the way for an analysis of the properties of the enzyme, of the kinetics, of the molecular forces in the active site and the mechanism of action. These studies will be described in Chapters VII and VIII.

C. Phosphorylated Compounds As Source of Energy

The next step was the exploration of the source of chemical energy which would account for the total electrical energy released during the discharge and which should provide the energy for the resynthesis of acetylcholine if the assumption were correct that the action of the ester and its hydrolysis are associated with the primary event. The large amounts of electric energy released in electric tissue by the discharge makes this tissue particularly suitable for quantitative evaluations of energy relationships.

The most readily available source of energy for endergonic life processes is the energy of phosphorylated compounds rich in energy (Meyerhof, 1937, 1941). Today it is generally accepted that the hydrolysis of adenosine triphosphate (ATP) is the source of energy in the elementary process of contraction, as was proposed by Meyerhof and Lohmann in 1934. The adenosine diphosphate, ADP, formed is, in muscle, rephosphorylated by the breakdown of phosphocreatine, a transphosphorylation which occurs without loss of energy. Phosphocreatine thus acts as a "storehouse" for energy-rich phosphorylated compounds.

The presence of phosphocreatine in the electric organ of *Torpedo* was first demonstrated by Kisch (1930), who suspected that this compound might be connected with the activity of the organ. In the electric tissue of *Electrophorus electricus* the concentration of phosphocreatine is high, as high or even higher than in striated muscle in spite of the low protein and high water content. The concentration of ATP, although lower than in muscle, is also quite high (Nachmansohn *et al.*, 1943a, b).

In studies initiated in 1942 we found that the energy released by the hydrolysis of phosphocreatine is adequate to account not only for the electrical energy released externally during the discharge, but for the total electrical energy. Both chemical and electrical energy released in the electric tissue of *Electrophorus* per gram per impulse vary considerably, depending on the size of the specimen. The larger the size, the smaller the metabolic rate and, as described before, the smaller the potential difference developed per centimeter. The total electrical energy released per gram and impulse in large specimens of about 170–180 cm. length was found to range from 18 to 35 μcal., with an average of 24 μcal. The electrical energy released externally is under the most favorable conditions only one-sixth of the total. But, for the relationship between electrical and chemical energy, it is, of course, necessary to consider the total output. In medium-sized specimens of 90 to 120 cm. length, the total electrical energy per gram and impulse was found to be about 47 μcal., ranging from 29 to 71. These figures are based on some assumptions which have been fully discussed by Cox, Coates, and Brown (1946). The values given are based on the most *probable* assumption. If all *possible* assumptions are taken into account, the figures may have to be revised either downward by 50 per cent or upward to 100 per cent.

Tested on the same specimens under the same conditions, the energy released per gram and impulse by the hydrolysis of phosphocreatine has been found to be higher than the total electrical energy. The average figures are summarized in Table VIII. Previously, the figures were calculated on the assumption that —11,000 cal. are released per mole of phosphocreatine hydrolyzed. This figure was based on the values which were originally proposed by Meyerhof and his associates and which had been widely accepted (see for instance Meyerhof, 1937). During the last few years there were various attempts to reevaluate the free energy released by the hydrolysis of ATP. The enthalpy (ΔH) of ATP hydrolysis was found to be markedly smaller than had been assumed before, since in earlier determinations the neutralization heat was not adequately evaluated. The enthalpy was found to be about —4600 cal. per mole at 18° C. (Podolsky and Morales, 1956). However, the really important value for estimating the actually available energy is the free energy, ΔF, also referred to as the "Gibbs free energy" or ΔG. The standard free en-

TABLE VIII

TOTAL ELECTRICAL ENERGY RELEASED BY ELECTRIC DISCHARGE OF *Electrophorus electricus* AND CHEMICAL ENERGY RELEASED BY BREAKDOWN OF PHOSPHOCREATINE AND FORMATION OF LACTIC ACID[a]

Length of fish (cm.)	Energy released/gm./impulse (gm. calories × 10^{-6})			
	Electrical	Phosphocreatine	Lactic acid	Combined
103	*47* (9)	*62* (21)	*78* (21)	*140*
180	*24* (6)	*40* (15)	*26* (7)	*66*

[a] The average values per gram per impulse are based on a number of experiments indicated by the figures in parentheses. The total number of discharges in these experiments ranged from about 1500 to 1800 in a period of 2–5 min. There was no measurable difference between determinations carried out immediately at the end of activity or after a recovery period of about 3 min.

ergy, ΔF°, of ATP hydrolysis, has been estimated by Burton and Krebs (1953) to be about —9000 cal. per mole, or more precisely —8900 cal. (Burton, 1955), whereas Robbins and Boyer (1957) arrived at a value of about —8000 cal. The most reliable value is now considered to be —8400 cal. (Burton, 1958). However, in living cells we do not deal with standard states. Most satisfactory estimates as to the available free energy under the conditions of the living cell have been obtained with mitochondria preparations (see for instance Whittam, Bartley, and Weber, 1955). Under the conditions of these *in vitro* studies, ΔF, (or ΔG) for the hydrolysis of ATP is now assumed to be about —12,500 cal. This figure has been used for the calculation of the data presented in Table VIII.

In addition, there is energy released by the simultaneous formation of lactic acid. The energy derived from this reaction has been calculated on the assumption that —25,000 cal. per mole of acid formed are released (Nachmansohn, Coates, Rothenberg, and Brown, 1946b). The energy supplied by aerobic recovery has not been evaluated. The chemical determinations were made after about 1500 discharges within a relatively short period of time, usually 3–4 min. The tissue was frozen with liquid air immediately at the end of the stimulation period. The respiration in electric tissue and the increase of oxygen uptake following activity are probably small, as may be concluded on the basis of available data. It appears, therefore, likely that during the short period of the experiments the correction which has to be made due to increased respiration is small, probably only a few per cent of the two reactions measured. The amount found is therefore probably close to the total energy released anaerobically. The amount of energy supplied by the hydrolysis of phosphocreatine and the formation of lactic acid is much higher than would be required to account for the total electric energy, but it appears likely that the major part of this chemical energy is not used for the immediate process of recovery, but for the restoration of the ionic concentration gradient.

It was safe to assume that, as in muscle, the breakdown of ATP during nerve activity precedes that of phosphocreatine. But whereas it is today widely accepted that ATP reacts directly with the structural muscle protein in the elementary process of contraction, it appeared for many reasons unlikely that ATP is responsible for the change in permeability postulated in the elementary process of conduction, as has been proposed by some investigators. It would be, indeed, surprising that the cell should use the same compound for two so entirely different types of function which in regard to energy requirements and speed involved differ by several orders of magnitude. If, however, the action of acetylcholine were responsible for the alterations of the membrane required for the generation of the action potential, as the available evidence suggested, then these reactions should occur prior to the breakdown of ATP. The latter should then be the recovery process supplying the energy for the resynthesis of acetylcholine hydrolyzed during activity.

D. Discovery of Choline Acetylase

This assumption proved to be correct. In 1943 an enzyme was extracted from brain and electric tissue which in a cell-free solution acetylated choline on addition of ATP. The enzyme was referred to as choline acetylase (Nachmansohn and Machado, 1943). This was the first demonstration that the energy of ATP may be used for biosynthesis and the first observation of the function of ATP outside the glycolytic cycle. Since then a great number of endergonic reactions have been demonstrated to utilize the energy of ATP. It was the first time that an enzymatic acetylation was obtained in a soluble system. The observations opened, therefore, the way for a detailed analysis of the mechanism of acetylation in general. This mechanism and the role of acetate in intermediary metabolism became, in the decade following this discovery, one of the most actively studied fields in biochemistry.

At that time, in the early 1940's, the paramount importance of acetate in intermediary metabolism as a building stone of many cell constituents became increasingly apparent, mainly through the application of isotope techniques. The first indication of the physiological importance of acetate goes back to the investigations of Knoop (1905). He concluded that fatty acids were oxidized stepwise with the formation of a 2-carbon compound, presumably acetic acid, and postulated his well-known β-oxidation theory. This concept was supported by the findings of Dakin (1909) that the presumed intermediates in this oxidation gave similar yields of acetoacetic acid. For decades there was little further progress. This fact may be attributed to the lack of adequate methods for identifying and determining acetate in small amounts. The situation changed in the 1930's with the introduction of isotopes as a tool for intermediary metabolism. Schoenheimer and Rittenberg (1936) found in animals fed with D_2O, that about half of the fatty acid hydrogen was derived from the water. Schoenheimer (1942) suggested, therefore, that the fatty acids are formed by the coupling of small units. Shortly afterward, Bernhard (1940), fed deuterium-labeled acetate together with sulfanilamide to rabbits and rats and recovered in the urine acetylsulfanilamide labeled with deuterium. Thus he had shown that acetate may be used directly for acetylation, although only a small fraction was labeled; the greater part was apparently derived from endogenously formed acetate.

Experiments of Rittenberg and Bloch (1944, 1945; Bloch and Rittenberg, 1945) with doubly labeled acetate ($CD_3C^{13}OOH$) fed to rats further suggested the incorporation of acetate into fatty acids. Their results indicate that the acetic acid used in acetylation is indeed largely formed in intermediary metabolism, but they also confirmed Bernhard's finding that part of the acetic acid fed was used directly for acetylation. Experiments with labeled acetate have later shown that many important cell constituents use this 2-carbon unit as a building stone.

Acetylation with added acetate was demonstrated by Klein and Harris (1938) in tissue slices of guinea pig liver; they obtained acetylation of sulfanilamide in presence of oxygen. No acetylation was observed in anaerobic conditions. A formation of acetylcholine in brain slices in oxygen was observed by Mann, Tennenbaum, and Quastel (1938).

Both *in vivo* and *in vitro* observations had then established that acetate may be used directly for acetylation. The mechanism, however, was obscure. Its exploration became possible by the discovery that the energy of ATP hydrolysis is used in solution for the acetylation of choline.

Shortly after the discovery of choline acetylase, Nachmansohn and Machado observed, in 1943, that on dialysis the enzyme rapidly loses its activity. This observation was confirmed and extended by Nachmansohn, John, and Waelsch (1943). The findings suggested the presence of a coenzyme in the system, and in 1945 Nachmansohn and Berman (1946) obtained a purified preparation from *Kochsaft* of liver, heart, and brain of a coenzyme which reactivated completely a dialyzed choline acetylase solution prepared from acetone-dried powder of rat brain. In view of Nachmansohn's findings, Lipmann (1945) tested the effect of ATP on the acetylation of sulfanilamide by liver extracts and confirmed that ATP provides the energy of acetylation. Simultaneously with Nachmansohn and Berman, he prepared a coenzyme required for the sulfanilamide acetylating system. When the two findings were presented at the same meeting (at the Symposium on "The Physicochemical Mechanism of Nerve Activity," 1946) it became obvious that the formation of *N*- and *O*-acyl groups used a similar if not identical mechanism (Nachmansohn, 1946a; Lipmann and Kaplan, 1946). A coenzyme for choline acetylase was simultaneously and independently described by Lipton (1946) in Barron's laboratory. Soon it became clear that the coenzyme is used in acylations in general, and Lipmann referred to it as coenzyme A (CoA).

In the following years Lipmann and his associates greatly contributed to the elucidation of the chemical structure of CoA. In 1947 they found that CoA contains pantothenic acid linked to a nucleotide by phosphate groups (Lipmann, Kaplan, Novelli, Tuttle, and Guirard, 1947; Lipmann, 1950). Snell and co-workers (1950) reported the requirement of a compound, the *Lactobacillus bulgaricus* factor, for bacterial growth; in this substance β-mercaptoethanolamine is bound to the β-alanine of pantothenic acid giving *N*-pantothenyl-β-mercaptoethanolamine. This compound is referred to as pantetheine and may be obtained by the action of intestinal phosphatase on CoA (Brown, Craig, and Snell, 1950). The exact position of the phosphate groups was studied by Baddiley and Thain (1951); they found that, whereas two phosphate groups link the pantothenic acid to the ribose of the nucleotide, the third phosphate group must be attached to the second or to the third atom of the ribose. It was shown by Shuster and Kaplan (1953) with the aid of a specific enzyme that it is the third carbon to which the phosphate is linked. Novelli and his associates confirmed the structure by enzymatic degradation and resynthesis (Novelli, 1953; Novelli, Schmetz, and Kaplan, 1954). The active group of CoA was elucidated by Lynen, Reichert, and Rueff (1951). They found that acetate in the reaction with CoA forms an ester with an SH group. Thus it became apparent that the SH group of the β-mercaptoethanolamine is the functional group of CoA and that the acetyl CoA is the thioester. On the basis of all these investigations the structure (VIII) of CoA has been established.

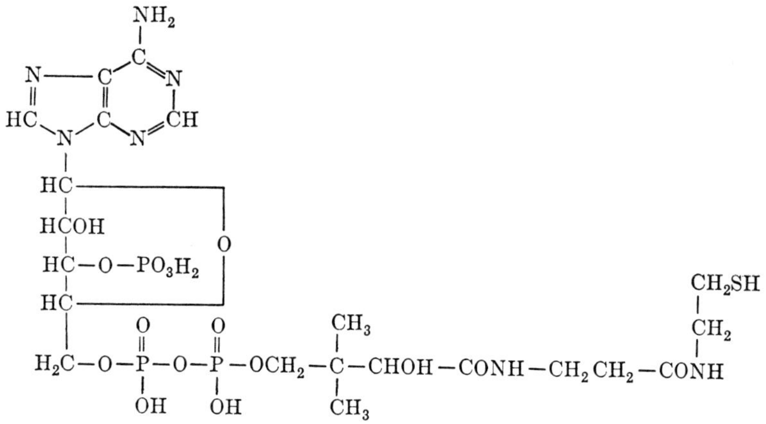

```
                  NH2
                   |
                   C
                 /   \\
     N ——— C          N
     ||        ||          |
    HC        C         CH
       \\    /   \\    //
         N        N
         |
        HC ————————┐
         |                  |
        HCOH             O
         |                  |
        HC—O—PO3H2        |                                            CH2SH
         |                  |                                               |
        HC ————————┘                                               CH2
         |        O       O           CH3                                   |
         |        ||       ||           |                                     |
      H2C—O—P—O—P—OCH2—C—CHOH—CONH—CH2CH2—CONH
                  |        |            |
                 OH      OH          CH3
```

Coenzyme A

(VIII)

The change of the standard free energy of the acetyl CoA hydrolysis is large, as was found with the aid of various equibrium reactions (Stern, Ochoa, and Lynen, 1952; Stadtman, 1952b). Thus, the thioester was shown to be a new type of a biologically important compound rich in energy. The ΔF° was estimated to be about —10,000 to —12,000 cal. per mole.

Acetylation then takes place in at least two enzymatic steps. The first is the formation of acetyl CoA. In animal tissue and in yeast this step is catalyzed by an "acetate-activating" enzyme referred to as acetyl kinase. Thioacetate may replace ATP acetate in the formation of acetyl CoA in the presence of the acetate-activating system prepared from pigeon liver or rabbit brain, although at a lower rate (Nachmansohn, Wilson, Korey, and Berman, 1952). Presumably, the reaction is:

$$CH_3\overset{\overset{\large O}{\|}}{C}-SH + HSCoA \rightleftharpoons H_2S + CH_3\overset{\overset{\large O}{\|}}{C}-SCoA$$

The formation of acetyl CoA in bacteria is catalyzed by an enzyme discovered and partially purified by Stadtman (1952a, b). This enzyme is referred to as phosphotransacetylase. Acetyl phosphate provides in this case the source of energy. The existence of this enzyme was suggested by observations of Novelli and Lipmann (1950) and Stern and Ochoa (1951). An exception is the photosynthetic bacterium *Rhodospirillum rubrum* (Eisenberg, 1953, 1955). These bacterial cells use ATP for acetate activation.

The mechanism of the enzymatic formation of acetyl CoA from acetate and ATP, the acetate activating reaction, is pictured as follows (Berg, 1956):

1. ATP + acetate ⇌ adenylacetate + P-P
2. Adenylacetate + CoA ⇌ acetyl-CoA + adenylic acid

Berg synthetically prepared adenylacetate and his data seem to support the assumption that this compound is an intermediate, although this view is not yet generally accepted (see, e.g., Dixon and Webb, 1958).

The mechanism of the two types of activation of the acetyl group in bacterial and animal cells has been frequently discussed in recent years. Although ATP (or more generally nucleotide triphosphate) is the major source of energy in both types, in the one group activation results from the transfer of the terminal phosphate of ATP; a phosphoryl derivative is formed by the displacement of ADP. In the second type the phosphorus (P) atom next to the ribose of the nucleoside reacts with the group. In both types a nucleophilic effect of the carboxyl group on a P atom takes place, in an S_N2 reaction. The two processes may be pictured, according to Lipmann (1957), as shown in reactions (1) and (2).

On the left (1) is shown the sequence of the reaction in which the terminal P is attacked by a carboxyl oxygen; the carboxyl group and

ADP are connected to that phosphorus in the transition state, resulting in acyl phosphate and ADP. In the reactions shown on the right hand (2) the carboxyl group attacks the phosphorus linked to adenosine. In the transition state the carboxyl and pyrophosphate groups are linked to the P of adenosine monophosphate (AMP), yielding eventually AMP acylate by the displacement of pyrophosphate.

```
O                 / |  |              O                 / |  |
 ↖                                     ↖
  CO:  ⟶  POPOPOAd                      CO:  ⟶  POPOPO
 /          /\ |  |                    /          /\ |  |
R                                     R             O
                                                     \
                                                      Ad

O         |   |  |                    O         |    |  |
 ↖                                     ↖
  CO---P---OPOPOAd                      CO---P---OPOPO
 /    /\    |  |                       /    /\    |  |
R                                     R         O
                                                 \
                                                  Ad

O     \         |  |                  O     |            |  |
 ↖                                     ↖
  CO·P  ⟶  :OPOPOAd                     CO·POAd  ⟶  :OPOPO
 /   /\        |  |                    /    |            |  |
R                                     R
         (1)                                    (2)
```

Once the acyl group is activated, a group transfer reaction to an acceptor takes place. In analogy to substitution reactions known from organic chemistry, Koshland (1954) has proposed to refer to enzymatically catalyzed group transfer reactions, such as transacylation, transphosphorylation, etc., as "enzymatic substitution reactions." In the reaction (3) between CoA and acyl adenylate the break takes place between carbon and oxygen.

```
                     O    |
                      ↖
CoA·S:    ⟶            COPOAd
                      /   |
                     R

          O
          ↑    |
CoA·S---C---OPOAd
          |    |
          R

          O          |
         ↗
CoA·SC       ⟶   :OPOAd
         \           |
          R

            (3)
```

Both activation and transfer reactions seem frequently to be catalyzed by the same enzyme.

In the decade following the discovery of the role of ATP and CoA in acetylation, a great number of key reactions in intermediary metabolism of the cell, of carbohydrates, lipids, and amino acids, have been established to be mediated by CoA. A great variety of compounds may be donors of the acetyl, or more generally of an acyl group, of acetyl or acyl CoA, respectively. ATP and acetyl phosphate are in many cases the compounds providing the energy, but in other reactions they are not required for the acyl CoA formation. The transfer of the acyl group from acyl CoA to the acceptor is the second enzymatic step. In the second step the enzyme is more or less specific for the acceptor. These developments had a great impact on our knowledge of intermediary metabolism in general. They form an integral part of biochemistry, and more detailed discussion may be found in standard books. Among the investigators who took a leading part in this development may be mentioned Lipmann, Ochoa, Lynen, Green, Stadtman, Gunzalus, Lehninger, and their associates, although significant contributions also came from other laboratories.

When it became apparent that there are at least two enzymatic steps in the process of acetylation, choline acetylase was redefined as the enzyme which transfers the acetyl group from acetyl CoA to choline (Korey, Braganza, and Nachmansohn, 1951). The specific reaction (4) catalyzed by choline acetylase is then as follows:

$$CH_3-\overset{+}{\underset{CH_3}{\overset{CH_3}{\overset{|}{\underset{|}{N}}}}}CH_2CH_2OH + CH_3\overset{O}{\overset{\|}{C}}SCoA \rightleftharpoons CH_3-\overset{+}{\underset{CH_3}{\overset{CH_3}{\overset{|}{\underset{|}{N}}}}}CH_2CH_2O\overset{O}{\overset{\|}{C}}CH_3 + HSCoA \quad (4)$$

It is obvious from the discussion above that acetyl CoA may be derived in intact cells from various sources, since it is an intermediate form of many reactions. Therefore, coupling of choline acetylase may take place with various acetyl donor systems in presence of catalytic amounts of CoA. Such coupling was demonstrated with a partially purified choline acetylase preparation, obtained from squid ganglia, by Korkes, del Campillo, Korey, Stern, Nachmansohn, and Ochoa (1952a). The acetyl donors used were acetylphosphate, pyruvate, and citrate in the presence of phosphotransacetylase (Stadtman, Novelli, and Lipmann, 1951), the pyruvate oxydation system (Korkes *et al.*, 1951, 1952b) and the condensing system (Stern and Ochoa, 1949, 1951), respectively. Although the activation of acetate, with ATP as energy source, may be the main pathway of acetyl CoA in the formation of acetylcholine, it must be kept in mind that it need not be the only source; especially under unusual conditions other alternatives must be taken into consideration.

E. Occurrence and Concentration of Choline Acetylase

Having described the discovery of the role of ATP and CoA in the process of acetylation and briefly referred to its impact on some general aspects of biochemistry, we return to the special problem of this chapter. The observations indicate that the action and hydrolysis of acetylcholine precede the hydrolysis of ATP in the sequence of energy transformations; they have made it possible to integrate the ester in the intermediary pathways. Some special properties of the choline acetylase molecule will be discussed later. For the interpretation of its physiological role the occurrence and concentration of the enzyme are pertinent. The ability of conducting tissue to form acetylcholine is an obvious corollary to the postulate of its essential role in the elementary process just as is the presence of cholinesterase and the rapid removal of the ester.

The ability of conducting tissue to synthesize acetylcholine has not been investigated as extensively as the hydrolysis by esterase, but whatever conducting tissues, nerve and muscle, were tested, the enzyme was found to be present. The enzyme system is much more complex, much more labile, and requires more material than the tests for acetylcholinesterase. It is present in much lower concentration. In some cases these limitations, especially the amount of material required, are not easy to overcome. Nevertheless, in the years from 1943 to 1946 we have demonstrated the presence of choline acetylase in a great variety of conducting tissues and in different types of fibers which, in view of the underlying physiological problem, offered a special interest. Since the special reaction of choline acetylase is the formation of acetylcholine from choline and acetyl CoA, the maximum rate of activity may not be achieved if the supply of acetyl CoA is not adequate. For testing the actual concentration of choline acetylase and its maximal rate in the test tube, it is necessary to have a potent acetyl CoA forming a system in the reaction mixture. In the early observations the experimental conditions were not always optimal. No purified CoA was available. ATP was used exclusively as energy source. It was later found that in some cases the optimal rate of choline acetylase was higher when acetyl phosphate and partially purified phosphotransacetylase prepared from *Escherichia coli* were used. When the reaction mechanisms had been analyzed and were better understood, we retested some earlier values under optimal conditions (Cohen, 1956). In most cases there was either a small difference or none. The figures given in Table IX give a few examples of choline

TABLE IX
CONCENTRATIONS OF CHOLINE ACETYLASE IN A FEW TYPES OF CONDUCTING TISSUES, COMPARED TO THOSE OF ACETYLCHOLINESTERASE[a]

Tissue	Species	Temperature (° C.)	ACh formed (mg./gm./hour)	ACh split (mg./gm./hour)	Ratio
Brain	Rat, guinea pig, rabbit	37	2.0–2.5	80–100	200
Sciatic nerve	Rabbit	37	1.0–1.3	12–15	50–80
Ventral roots	Ox	37	7.0–12.0	12–15	6–7
Dorsal roots	Ox	37	0.03–0.05	9	800–1000
Dorsal roots	Dog	37	0.015–0.02*	11–15	600–1000
Optic nerve	Rabbit	37	0.015–0.02*	18–25	800–1000
Head ganglia	Squid	25	10.0–15.0	3–4000	1000
Abdominal chain	Lobster	25	0.05–0.09*	20–30	300–400
Electric tissue	Electric eel	25	0.3–0.5*	3–4000	1000
Striated muscle	Pigeon	37	0.03–0.05	—	—
Heart muscle[b]	Rabbit	37	0.02–0.04	5–7	1000
Striated muscle	Goldfish	25	3.0–4.0	900	1000

[a] Choline acetylase was tested in extracts of acetone-dried powder; the values given indicate the amount formed per gram powder except those marked by an asterisk, where fresh tissue was used and the data were calculated per gram fresh tissue. The activity of acetylcholinesterase was always determined in homogenized tissue suspensions and the figures of the hydrolytic power refer to grams fresh tissue. In calculating the ratio it was assumed that 1 gm. powder corresponds approximately to 5 gm. fresh tissue.

[b] Apex only, free of nerve endings.

acetylase activity in some characteristic conducting tissues; most of them were tested under optimal conditions; only a few older data were not checked but have been nevertheless included in the table. For comparison the concentrations of acetylcholinesterase are given; the approximate ratios of esterase : acetylase are found in the last column. Among other tissues in which the presence of choline acetylase was shown, may be mentioned those of worms (*Neanthes virens, Euplanaria maculata*) and coelenterates (*Tubularia crocca*), by Persky and Gold (1948). Bueding (1952) found choline acetylase in *Schistosoma mansoni.*

The role of acetylcholine in sensory fibers has been under discussion for a long time. Loewi and Hellauer (1938) reported that sensory fibers in contrast to motor fibers do not contain acetylcholine. This report initiated some speculation about the question whether acetylcholine has any function in sensory fibers. In the following year, however, Chinese investigators succeeded in demonstrating the presence of acetylcholine in sensory fibers, dorsal roots, and optic nerve of dog (Chang *et al.*, 1939). Their finding has been repeatedly confirmed (e.g., Lissak and Pasztor, 1941). Brecht and Corsten (1941) were even able to show that acetylcholine is released from cut sensory fibers on stimulation (see also Chapter XIV).

Following the discovery of choline acetylase, the presence of this enzyme was demonstrated in rabbit optic nerve (see Table IX) (Nachmansohn and Berman, 1946). Shortly afterward experiments were reported in which the presence of choline acetylase appeared to be irregular in sensory nerves and occasionally absent (Feldberg and Mann, 1946). Feldberg based rather sweeping conclusions on this failure. This failure was, however, due to inadequate techniques. De Roetth (1951) not only confirmed the positive results but found with the methods improved in the meantime even higher values. Feldberg's claims seemed to be supported by Wolfgram (1954). This author found cholinesterase in dorsal as well as in ventral roots of ox, but he did not find any choline acetylase in the dorsal roots. He therefore suggested that the theory of an essential role of acetylcholine in conduction should be withdrawn.

When we determined, on the suggestion and with the collaboration of Dr. J. Frumin, the concentration of choline acetylase in dorsal and ventral roots of the ox and dog in a proper reaction mixture and with adequate methods, the following results were obtained: in homogenized suspensions of dog dorsal roots about 15–20 μg. of acetylcholine were formed per gram (fresh weight) per hour; with ox dorsal roots the values obtained in three experiments with fresh tissue suspensions were 19.0, 7.4, and 13.0 μg. per gram per hour; with extracts of acetone-dried powder of ox dorsal roots the values of three experiments were 48.5,

33.0, and 49.0 μg. per gram powder per hour (see also Table IX) (Cohen, 1956). The methods of Wolfgram were inadequate on three counts: (1) Although homogenized suspensions of fresh tissue were used, no fluoride was added, without which ATP is rapidly hydrolyzed by ATPase in extracts of fresh tissue; in absence of fluoride only insignificant yields are obtained, if any, as was emphasized in the first publication (Nachmansohn and Machado, 1943). (2) The experiments were carried out over a period of 8 hours, whereas in extracts of fresh tissue the enzyme activity falls rapidly after 30 minutes and the ester formed is slowly hydrolyzed. (3) There was a small increase in color observed after 8 hours' incubation when ventral roots were used; no attempt was made to identify the change of color with the formation of acetylcholine.* The values which Wolfgram reports for ventral roots correspond to less than 1 per cent of the actual activity. The negative results have been considered, and still are presented in some text books, as a serious difficulty for the assumption of an essential role of acetylcholine in conduction. It seemed, therefore, useful to discuss the basis for the claim of the absence of choline acetylase in sensory fibers. The high concentration of acetylcholinesterase in dorsal roots should have been a warning not to reject so categorically the possibility of the substrate being formed there. The presence of the esterase in sensory fibers is all the more pertinent since its activity cannot be separated from conduction, thus indicating that the substrate too must be formed and be present in these cells.

A sharp distinction must be made between the significance of the concentration of acetylcholinesterase and that of choline acetylase. There is a fundamental difference as to the function of these two enzymes. The action of the ester on the receptor, postulated to increase sodium conductance, must be rapidly reversed, as discussed before. The hydrolysis of the ester must take place within a millisecond or a fraction of it. The enzyme must act fast, but only for this very short period of time. Evidence for an adequate concentration of the hydrolytic enzyme and a high velocity of the reaction were therefore of paramount importance if the theory were to be accepted. The synthesis, on the other hand, takes place in the recovery and may be a relatively slow process which may go on continuously over longer periods of time, over seconds or even minutes. In axons, the esterase is about five times in excess. The excess of acetylase is not known, but let us assume that it is about the same. A concentration of the esterase which, in some tissues, is several hundred to a thousand times as high as that of the acetylating enzyme is therefore not at all surprising and does not present a difficulty.

* For these determinations, the colorimetric method of Hestrin (1949a) was used.

A few other quantitative aspects may be briefly discussed. At the Neurochemistry Symposium of the IVth International Congress of Biochemistry in Vienna, Keynes (1958) rejected categorically the requirement of metabolic activity associated with nerve conduction and specifically a role of acetylcholine in electrical activity. As the main argument he presented the observation that squid giant axons may still conduct up to one hundred thousand impulses after depletion of ATP and arginine phosphate by the exposure of the fiber to cyanide or dinitrophenol, thus blocking respiration or oxidative phosphorylation, respectively. Since there should be, during activity, a rapid breakdown of acetylcholine and since ATP is required for acetylcholine synthesis, the depletion of ATP and the inability to form it should have made conduction impossible. We have seen that the amount of acetylcholine required per impulse per square centimeter is of the order of $5–10 \times 10^{-15}$ mole. Assuming a surface of about 80 $cm.^2$ per gram nerve, one hundred thousand impulses would require 7–14 μg. of acetylcholine per gram. The amount of acetylcholine per gram squid axon has, to my knowledge, not been determined. An amount of 5 μg. per gram may well be present in resting state, as is the case in many other nerves, so that the reserve available in the beginning would account for a large fraction, if not for most, of the acetylcholine required for one hundred thousand impulses. But even if additional synthesis of a few micrograms of acetylcholine per gram should be necessary, it would mean very little indeed in terms of ATP and be hardly detectable. Some ATP still may be derived from anaerobic glycolysis under the conditions reported. However, let us assume that there was really no ATP formed or present, the argument of Keynes still would have no validity. As pointed out before, acetylcholine may be provided by some of the coupling reactions discussed in the preceding section, and especially in conditions of ATP shortage, the system may well use some of the other sources available. In a sample of a giant axon of 100 mg. weight, 1 μg. of such acetyl groups capable of forming 4–5 μg. of acetylcholine, would be an ample supply, considering the low order of magnitude of acetylcholine required for function.

The low requirements are also an important factor in evaluating the relatively low acetylase concentration in dorsal roots and other sensory fibers, especially if they are myelinated, because the amounts required may be markedly smaller than in the unmyelinated owing to the saltatory type of conduction. There is virtually no information about the actual requirements in these fibers. The concentration of acetylcholine in the ox dorsal roots has been reported to be 0.2 μg. per gram. These fibers are able to synthesize 15–20 μg. per gram per hour. The whole reserve may then be restored in less than 1 min. It appears not unreasonable to assume that the fiber has stored an essential compound in amounts adequate to satisfy its requirements for such a short period of time. We know too little about these requirements which are necessary for an evaluation of the significance of the differences in enzyme activity, be it for the same type of fiber or for different types of fibers. All available data show that the actual requirements of acetylcholine, be it at junction or in the axons, are of an extremely low order of magnitude. The demonstration that both esterase and acetylase are present in purely sensory fibers and that rapid block of conduction may be achieved by specific inhibitors of the esterase, just as in motor nerve fibers, is sufficient evidence that the acetylcholine system is essential for conduction in this type, as well as in other types, of fibers.

The exceptionally high concentration of choline acetylase in ventral roots appears surprising for two reasons: (1) The concentration of the esterase is only a few times higher and therefore the *ratio* of the activities is extremely low compared to that in all other tissues tested. (2) The concentration is much higher than in the brain, even of small mammals (by a factor of more than 5). What then is the explanation for such an extraordinarily high acetylase activity in ventral roots? As we just said, the significance of variations of enzyme concentration is still poorly understood in general. We certainly would not plead, merely on the basis of a difference in concentration, that acetylcholine plays a role in axonal conduction of the ventral roots but not in synaptic transmission of neighboring tissues, as has been repeatedly done in cases where the reverse condition prevails.

Another problem of physiological interest may be briefly discussed, namely, the rate of disappearance of choline acetylase activity in a degenerating nerve fiber and the relationship to the ability of conducting impulses. Feldberg (1943) claimed that, in degenerating fibers, the ability of synthesis of acetylcholine ceases before that of conducting impulses. He considers this as an important argument against an essential role of acetylcholine in conduction, and this argument has since then been frequently used in the literature. Unfortunately, these findings too were based on inadequate techniques. Feldberg used finely cut nerve fibers of sheep. Respiration in nerve fibers, particularly those of warm-blooded animals, stops when they are treated in this way. Therefore there was no source of energy present under the conditions of his experiment. Feldberg found in normal nerves after a 2-hour period of incubation, 1–2 μg. of acetylcholine formed per gram, but even this amount must be attributed to the fact that he had eserine in the test experiment but none in the control. When the acetylase activity in rabbit sciatic nerve was tested under proper conditions, with ATP present as energy source, it was found that acetylcholine was formed at a rate of about 100 μg. per gram fresh tissue per hour (Nachmansohn, John, and Berman, 1946c). Two days after section of the fiber, about two-thirds of the initial concentration of choline acetylase was still present in the degenerating sections. Conduction usually ceases on the third day. Even at that time the enzyme activity is still about one-third of the initial. When the acetylase activity was retested with a still stronger acetate-activating system, it was found that more than 1 gm. of acetylcholine was formed per gram acetone-dried powder prepared from rabbit sciatic nerve (Cohen, 1956). Assuming 1 gm. powder to be equivalent to 5 gm. fresh tissue, the value found corresponds to more than 200 μg. per gram fresh tissue per hour.

F. Depolarizing (Electrogenic) Action of Acetylcholine

In regard to the sequence of energy transformations, the question remained whether acetylcholine itself is responsible for the permeability change associated with the generation of bioelectric potentials, or whether it is preceded by another chemical reaction. If its action is the primary event it should have a depolarizing, i.e., an electrogenic, action.

The possibility of a depolarizing action of acetylcholine has long been suspected. Dubuisson and Monnier (1934) and Cowan (1936) were the first to consider specifically such a mode of action. At that time the end plate potential was not yet explored and techniques using intracellular microelectrodes were not yet developed. The prerequisite for a compound to have a brief and quickly reversible depolarizing, i.e. electrogenic, action, is a correspondingly rapid removal of the active agent. When we demonstrated, in 1937, the extraordinarily high concentration of cholinesterase in the electric organ of *Torpedo*, Auger and Fessard (1939), under the impact of this finding, tested the effect of eserine on the action potential of the electric organ of *Torpedo marmorata.* They found it was markedly depressed and the duration of the descending phase was considerably prolonged.

In view of their corresponding bioelectrical and biochemical findings Fessard and the writer decided to test whether acetylcholine injected into the electric organ may produce a depolarization and have an electrogenic action, an effect to be expected if acetylcholine is a compound responsible for the change of permeability of the membrane during activity. In experiments carried out in Arcachon, in 1939, on *Torpedo marmorata,* in which Fessard and the writer were joined, on their invitation, by Feldberg, they were able to demonstrate that injections of acetylcholine into the electrical organ perfused with a solution containing low concentrations of eserine, does indeed generate electric potentials (Fig. 28) (Feldberg, Fessard, and Nachmansohn, 1940).* Without the addition of eserine, much higher amounts (up to 200 μg.) of acetylcholine injected had only a very weak effect. The depolarizing action makes it difficult to assume that the action of the ester is required for recovery and supports the assumption that the action of the ester is responsible for the change in permeability associated with the nerve impulse. Other compounds will, of course, produce similar effects. But acetylcholine is the only compound found in conducting tissue which fulfills the requirements for the assumption of the physiological role proposed, and all the known facts about the system are consistent with this conclusion.

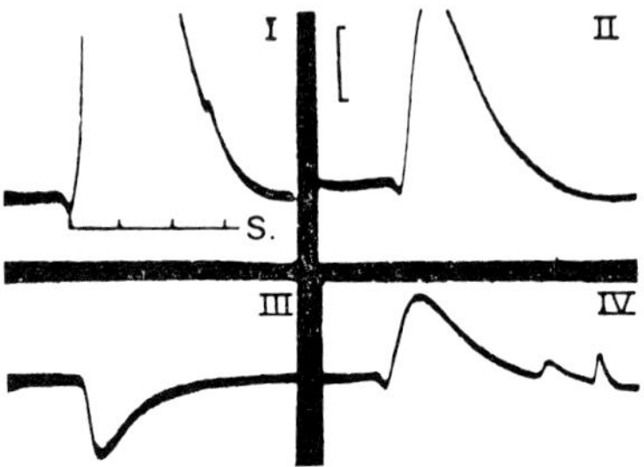

FIG. 28. Depolarizing action of acetylcholine.

The potential changes are produced by intra-arterial injection of acetylcholine into the electric organ of *Torpedo marmorata*, in the presence of eserine. *I*, *II*, and *IV* present the effects of 10, 5, and 2.5 μg. of acetylcholine; at *III* only perfusion fluid was injected. 0.5 Millivolt indicated at *II*. Time in seconds.

The physiological discharge and the electrogenic effect of injected acetylcholine differ considerably in respect to the voltage and duration of the spike. This is to be expected. An injection experiment is at best a crude method for reproducing anything which, under physiological conditions, takes place within milliseconds and within a membrane of less than 100 A. thickness. If acetylcholine generates bioelectric potentials it must be postulated that it is released within the active membrane, acts on some receptor protein at a few Angstroms' distance, and having produced its effect is hydrolyzed in microseconds. Under such conditions, the action may be tremendously more powerful and more

* The expedition to Arcachon was made possible by a special grant which the writer received from Professor Henri Laugier. The full paper describing the experiments, which was jointly written by the three authors in August, 1939, was later published by Feldberg with the modification that the writer's name was omitted (Feldberg and Fessard, 1942, *J. Physiol.*, **101**, 200). This omission was made without knowledge and authorization of Fessard, who at that time was cut off by the German occupation of France (see also Fessard, 1946).

rapid than in experiments in which the active compound is merely injected into blood vessels at a rather great distance. Moreover, during the physiological discharge the electroplax which are arranged in series, are simultaneously activated, whereas by injection into a vessel probably very few plates will be reached at the same moment. There are other factors which make the reproduction of the physiological process *a priori* unlikely as far as the quantitative aspects are concerned.

The significant result of our experiments was the demonstration of the depolarizing ability of acetylcholine and its electrogenic effect, even if duration and strengths of action were not comparable to the physiological event. Since then the discovery of the end plate potential, the development of methods applying microelectrodes for intracellular recording, and refined methods for application of acetylcholine have made it possible to confirm in many ways the depolarizing action of acetylcholine, at least at the synapse. This aspect will be further discussed in Chapter XVI.

G. Role of Acetylcholine in Sensory Endings

To a biochemist accustomed to the notion of biochemical unity of life it will *a priori* appear likely that acetylcholine plays a role basically similar in all types of neurons in the generation of bioelectric potentials. As just discussed, the presence of the ester and the two enzymes in sensory fibers and the block of conduction by specific inhibitors of esterase are well established. For a more complete picture it appeared of interest whether or not sensory endings react to acetylcholine and analogous structures. The primary effect in different sensory receptors would be expected to be a specific reaction in response to specific stimuli such as light, touch, pressure, etc. But this reaction should, directly or indirectly, activate the acetylcholine system.

Several types of sensory nerve endings and sensory receptors were found to respond to acetylcholine and to related structures, known to react with the acetylcholine system, such as curare, decamethonium, eserine, etc., in such a way as would be expected by analogy with their action on motor end plates (Bing and Skouby, 1950; Skouby, 1951; Zotterman, 1953; Dodt, Skouby, and Zotterman, 1953; Granit *et al.*, 1953; Buchthal, 1954). These and other observations indicate that in sensory endings a cell consitutent is present which responds to acetylcholine and has receptorlike properties. The effect is all the more interesting because in sensory endings there are no synaptic junctions in the usual meaning of the word. Many of these endings are quite simple structures, within which the impulse is initiated; in this case the system must then be intracellular.

A pertinent recent development concerning this question has been the studies of W. R. Loewenstein on the concentration and distribution of cholinesterase in Pacinian corpuscles. The value and significance of this type of information have been previously discussed. The corpuscles, obtained from the mesentery and the pancreas of cat, were found to contain rather large amounts of cholinesterase (Loewenstein and Molins, 1958). However, the most significant finding is the uneven distribution of the enzyme inside the structures of the corpuscle. The various parts of the corpuscle were separated by dissection. Thus a direct determination of enzyme activity in the different structures of the corpuscle became possible. By far the largest fraction of the enzyme was found to be localized in the axon and ending and in the thin hull

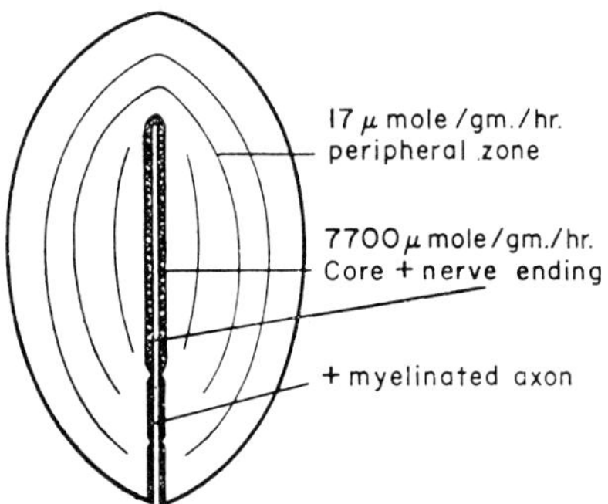

FIG. 29. Distribution of cholinesterase in a Pacinian corpuscle.

The figures to the right indicate the amount of acetylcholine which is split in 1 hour by 1 gm. of the corresponding tissue of the sense organ. Note the very high cholinesterase activity around the nerve ending; this activity is of the same order of magnitude as that of the motor end plate of frog sartorius muscle. (From Loewenstein and Molins, 1958.)

of core structure which surrounds the ending, whereas the remaining tissue of the corpuscle contains only little of the enzyme (Fig. 29). The activity in the "core fraction"—containing the nerve ending plus the myelinated stump plus the core—amounts to 1390 μg. of acetylcholine split per gram fresh weight per hour. In contrast, the capsular fraction, which amounts to 99 per cent or more of the mass of the corpuscle, splits as little as 3 μg. of acetylcholine per gram per hour. On the basis of their data, the authors calculated that the structures around the nerve ending, precisely there where the generation of electrical activity of the receptor has been shown to take place, are able to hydrolyze 0.7×10^9 molecules per millisecond. This is about the same figure as obtained in a single motor end plate of frog's sartorius muscle, which is 1.6×10^9 molecules per millisecond.

The esterase found in the corpuscles is a cholinesterase. It does not split any detectable amounts of monobutyrin. On the other hand, the investigators found some deviations from the usual characteristics of acetylcholinesterase. The significance of the enzymatic features remains still to be established. But the remarkable achievement reported adds a new and pertinent fact to the other findings supporting the assumption of an essential role of the enzyme and its substrate in the activity of the sensory nerve endings comparable to that at the motor end plate.

CHAPTER VI

Tentative Picture of the Role of the Acetylcholine System in the Permeability Change

The picture of the role of the acetylcholine system in the elementary process which has emerged from the data accumulated may be described as follows (Fig. 30) (Nachmansohn, 1955a): In the resting condition acetylcholine (O┬) is present in an inactive form probably bound to a

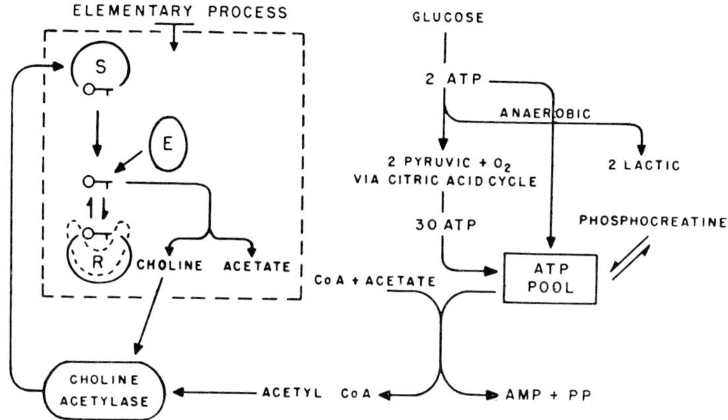

FIG. 30. Sequence of energy transformations associated with conduction and integration of the acetylcholine system into the metabolic pathways.

The role of acetylcholine in the elementary process may be pictured as follows: (1) In resting condition acetylcholine (O┬) is bound to a storage protein (S). The membrane is polarized. (2) Acetylcholine is released by current flow (possibly hydrogen ion movements) or any other excitatory agent. The free ester combines with the receptor protein (*R*). (3) The receptor changes its configuration (symbolized by dotted line). This process increases sodium conductance; it is the trigger action by which the potential primary source of EMF, the ionic concentration gradient, becomes effective and by which the action current is generated. (4) The ester-receptor complex is in dynamic equilibrium with free ester and receptor; the free ester is open to attack by acetylcholinesterase (*E*). (5) The hydrolysis of the ester permits the receptor to return to its original condition. The sodium conductance decreases; the membrane returns to its original condition.

protein (or conjugated protein). The complex may be tentatively called the storage form (S). Excitation of the membrane by current or any other disturbance leads to a dissociation of the complex, and acetylcholine is released. The free ester acts upon a receptor (*R*), a specific protein, and this action upon the receptor is essential for the change of ionic permeability, the increased sodium conductance, and thus the generation of the bioelectric potential. Some facts to be discussed later suggest the possibility that the effect of acetylcholine may be a change in configuration of the protein.

The complex between acetylcholine and the receptor is in a dynamic equilibrium with the free ester and the receptor. The free ester will be susceptible to attack by acetylcholinesterase. The enzymatic hydrolysis of acetylcholine will permit the receptor to return to its resting condition. The barrier to the rapid ion movements is thereby reestablished. The action of the enzyme leads then to immediate recovery and ends the cycle of the elementary process. It is the rapidity of this inactivation process which makes rapid restoration of the membrane possible, by the decrease of sodium conductance to its resting value, and permits the nerve to respond to the next stimulus within milliseconds. All these events controlling the ion movements during activity must take place within the structural or functional barrier for these ions in rest. The further recovery leads to the resynthesis of acetylcholine in its bound form catalyzed by choline acetylase and the other components of the system. From there on the cyclic processes known from other cells enter the picture. They provide the energy for the restoration of the ionic concentration gradients.

Many years ago Otto Meyerhof raised the interesting question: Living cells are able to utilize the energy of various substances such as carbohydrates, proteins, and lipids for a great variety of different and highly specific functions with about the same efficiency. How does the cell accomplish the differentiation of the common energy source? The phenomenon is explained by Meyerhof in the following way: There is a "specific operative substance" for each specific function. This substance exists only in small quantities, but its concentration does not decrease substantially during the performance of its specific function because it is at once restored. The recovery and the maintenance of the concentration of this specific operative substance is performed by a sequence of cyclic reactions which are partly consecutive and partly in parallel, but which are all chemically and energetically coupled. The further the cycles are removed from the specific process, the larger may be the energy supplied, and in later stages of recovery several energy-yielding processes are available and may be used either simultaneously or in turn. The specific operative substance for the elementary process of contraction is ATP, as has been proposed by Meyerhof and Lohmann and their associates and borne out by later developments. The reaction of ATP with the structural proteins, actin and myosin, appears to be the primary chemical event in muscular contraction, as has been ascertained by the work of many investigators, although here too the detailed mechanisms of this process are still far from being elucidated (for references see, e.g., Weber and Portzehl, 1954; Dubuisson, 1954; Weber, 1958).

The evidence summarized in the preceding chapters has established a solid basis for the assumption that acetylcholine is the specific operative substance in the conduction of nerve impulses, its action being essential for the generation of bioelectric currents. Studies on the detailed mechanism, as will be discussed more fully in following chapters in connection with the analysis of the molecular forces acting in the acetylcholine system, are still in their initial phase. Many additional factors about which virtually nothing is known must be important. Suggestions as to the detailed mechanism of action of acetylcholine and its relationship to other chemical and structural factors must clearly be only tentative. But a comprehensive theory of conduction cannot ignore the huge amount of data and the great variety of physical and chemical facts which have been accumulated to support the essential role of acetylcholine in the electrical activity associated with conduction: the many physiologically pertinent features of acetylcholinesterase, the presence of cholinesterase and of choline acetylase in all conducting tissues throughout the animal kingdom, the impossibility of separating chemical and electrical activity, the evidence that the hydrolysis and the formation of acetylcholine are the first reactions in the chain of events associated with activity, that in the sequence of energy transformations they precede the other reactions as indicated by the depolarizing (electrogenic) action of the ester. In view of the complexity of biological processes, one type of evidence would be inadequate, whereas by the combination of all the facts the case has become very strong. The data appear to be as conclusive as those in favor of any biological concept or, in fact, of any scientific concept: all are apt to be modified.

CHAPTER VII

Mechanism of Reactions Catalyzed by Acetylcholinesterase

The stage of development outlined in the preceding chapters was reached about 1948, as far as the fundamental features are concerned. It then became apparent that it was necessary, as the next step for a better understanding of the precise mechanism of conduction, to study in greater detail the proteins of the acetylcholine system. The situation bears some analogy to that prevailing in the late 1930's in the investigations of muscular contraction, when by the observations of Engelhardt and Ljubimova (1939; Engelhardt, 1942) the interaction between myosin and ATP moved into the center of interest.

As we have seen, there are four proteins (or conjugated proteins) which are directly tied to the function of acetylcholine. Two of them, the enzymes acetylcholinesterase and choline acetylase, were isolated in 1938 and 1943, respectively, and are available for analysis in highly active and purified form in solution. The existence of a receptor protein has been experimentally established on living cells (Altamirano, Schleyer, Coates, and Nachmansohn, 1955). Recently, Ehrenpreis (1959a, b) achieved its isolation and identification in solution (see Chapter XI). Virtually nothing is known about the storage form.

A molecule such as acetylcholine has only a limited number of possibilities of reacting with a protein; the molecular forces acting between small and macromolecules of the system must therefore be similar. Relatively small modifications in the surface of the protein may lead to important changes of function. Information obtained by the analysis of the molecular forces of one of these proteins may, therefore, offer useful indications for a better understanding of the reactions of the ester with the other proteins. The most favorable protein for these studies is acetylcholinesterase. It is available in a stable and highly active form. Its activity can be easily measured by precise, simple, and rapid methods. It reacts with several substrates and a very great number of inhibitors. Its specificity is of a type which favors the analysis of its active site: it is not too specific, which would greatly restrict the number of variations of reacting molecules, nor is it too unspecific, which would decrease the value of information derived from such analysis. The combination of the various features of this enzyme greatly facilitated the analysis of the molecular forces in the active surface.

A. Molecular Forces Acting between Substrates and Enzyme

Considerable information has been obtained during the last decade about the molecular forces in the active surface of acetylcholinesterase and the mechanism of hydrolysis in general. Summaries of this development may be found in several review articles (Nachmansohn and Wilson, 1951, 1955, 1956; Wilson, 1954b; Nachmansohn, 1955a, b, 1959a; Wilson and Nachmansohn, 1954).

The process of hydrolytic enzyme action takes place in two stages and may be expressed in the usual way:

$$E + S \rightleftharpoons ES \rightarrow E + \text{products}$$

In the first phase the enzyme (E) combines with the substrate (S) to form the enzyme-substrate complex (ES), the quite generally postulated Michaelis-Menten complex. In the second phase the hydrolytic process takes place. The two phases will be considered separately.

The formation of an enzyme-substrate complex is today generally postulated. It is difficult to conceive how an enzyme could function without interaction with a reactant, although this interaction need not result in the formation of a stable intermediate. The assumption of a stable enzyme substrate complex is valid at least for those cases where the apparent dissociation constant is small. The distinction between a critical or transition complex and an enzyme substrate complex is one of stability; the former lies at a thermodynamic potential energy maximum and the latter at a minimum. The apparent dissociation constant for the acetylcholine-acetylcholinesterase complex is about 1×10^{-4} and, therefore, the concept of a stable reversible complex is applicable to this system.

The types of interaction which may take part in holding the complex may be the usual ionic and covalent chemical bonds and the less specific van der Waals' or London dispersion forces. All these types of interaction have been experimentally shown to occur in the formation of the acetylcholine-enzyme complex. The positive electric charge of the substrate immediately suggests that the enzyme might contain a suitably located and properly shaped negatively charged region which increases the enzyme activity by contributing to the attraction, orientation, and fixation of the substrate molecule on the enzyme surface. The existence of such a negative site has been demonstrated in various ways with the aid of competitive inhibitors and appropriate substrates (Wilson and Bergmann, 1950a; Wilson, 1952a).

As an illustration of the result with competitive inhibitors may be mentioned the difference of interaction between the enzyme and Prostigmine and eserine, respectively. We have seen that these two reversible competitive inhibitors are of comparable effectiveness and have some similarity in structure. Prostigmine, however, is a quaternary ammonium ion and is, therefore, positively charged at all pH values whereas eserine, a tertiary amine whose conjugate acid has a dissociation constant of 8×10^{-9}, changes from almost exclusively cation at pH 6 to equally predominantly neutral molecules at pH 10. The pH dependence of enzyme inhibition by these two substances is shown in Fig. 31. Since Prostigmine inhibition does not change with pH, we may conclude that the changes of the inhibitory effect observed with eserine are due to changes in the eserine molecule, namely, the association of a proton and the simultaneous acquisition of a positive charge in acid media

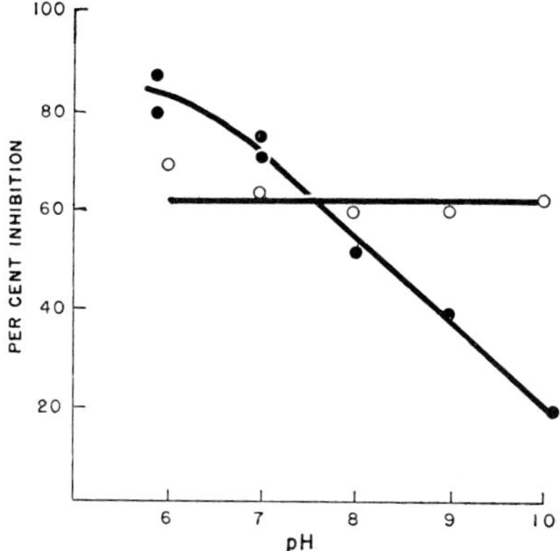

FIG. 31. Inhibition of acetylcholinesterase by Prostigmine (O) and eserine (●) as a function of pH.

(Wilson and Bergmann, 1950a). The much greater inhibitory effectiveness of the positively charge molecule thus indicates the existence of some negatively charged site in the enzyme which aids in the formation of the complex.

Another comparison of the inhibitory effectiveness of charged and uncharged molecules is the difference found between dimethylethanol ammonium ion and the structurally similar but uncharged isoamyl alcohol: the charged molecule is a thirtyfold stronger competitive inhibitor. Similarly, nicotinamide, which exists at neutral pH as the uncharged base, is only one-eighth as effective an inhibitor as the positively charged *N*-methyl nicotinamide.

The effect of electric charge has also been demonstrated with substrates by comparing the rates of hydrolysis of dimethylaminoethyl acetate at different pH values. The dissociation constant of the conjugate acid is about 5×10^{-8}, so that the substrate exists below pH 8.3

predominantly in the cationic form and at a higher pH in uncharged form. The hydrolytic rate measured as a function of pH is high in the acid range; it decreases rapidly between pH 8 and 9, reflecting the change in electric charge (Fig. 32). The enzyme efficiency changes at different pH values (Hestrin, 1949b; Wilson and Bergmann, 1950b). For proper evaluation of the effect of charge on the rates of hydrolysis the data must be corrected by taking into account the pH dependency of the enzyme toward acetylcholine. In Fig. 32 only the relative rates are given after the correction has been made.

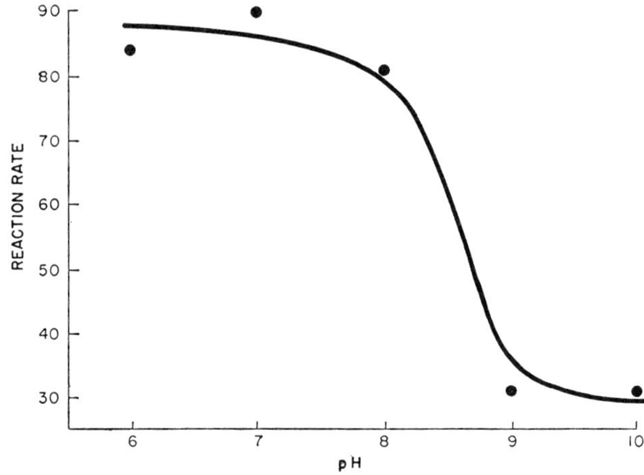

FIG. 32. Rate of hydrolysis of dimethylaminoethyl acetate in per cent of that of acetylcholine as a function of pH.

The effect of electric charge has been evaluated quantitatively from electrostatic theory (Wilson and Bergmann, 1950a). The data for eserine indicate about a twentyfold greater affinity of the cation. The Bronsted equation relates the dissociation constants of ionic complexes to the electric charges and their distance of separation. It also includes corrections for the salt effect derived from the Debye-Hückel theory:

$$\log K = \log K' - \frac{Z_A Z_B \varepsilon^2}{2.3\,DkTr} + \frac{Z_A Z_B \varepsilon^2}{2.3\,DkT} \times \frac{\kappa}{1 + r\kappa}$$

where K is the affinity constant, K' is the affinity constant for $D = \infty$, $\kappa = 0$, Z_A and Z_B are the number of electronic charges for ions A and B respectively, D is the dielectric constant, ε is the electronic charge, k is the Boltzmann constant, T is the absolute temperature, μ is the ionic strength, r is the distance of closest approach, and $\kappa = \sqrt{(8\pi\varepsilon^2/DkT)\mu}$.

Let Z_B apply to the enzyme and $Z_A = +1$ apply to dimethylaminoethyl acetate cation. For the free base set $Z_A = 0$. Then:

$$\log \frac{K\,(\text{base})}{K\,(\text{cation})} = \frac{Z_B \varepsilon^2}{2.3\,DkT} \times \frac{1}{1 + r\kappa}$$

For $\mu = 0.2$, water at 25°, this equation reduces to:

$$r = \frac{2.1 \times 10^{-10}}{\log \dfrac{K\,(\text{base})}{K\,(\text{cation})}} \times \frac{Z_B}{D}$$

where the value of D is to be estimated from the Schwarzenbach (1936) approximation.

Assuming the charge of the anionic site as one electronic charge, Wilson calculated the distance of closest approach of the opposite charges to be 6.3 A. The radius of the tetramethylammonium group is about 3.5 A. The negative charge of the anionic site is associated with some unknown grouping and its size is, therefore, unknown. However, negative central atoms of large groupings do not occur and it is, therefore, some small group, perhaps an oxygen atom. Using the unbonded oxygen radius of 1.5 A., the minimum distance of approach would be 5 A., which is in good agreement with the value estimated by Wilson on the basis of the experimental data. Adams and Whittaker (1950) investigated the possibility of an anionic site by comparing the dissociation constant for complexes of acetylcholinesterase with acetylcholine and with an uncharged geometrical analog, 3,3-dimethylbutyl acetate. They could account for the observed greater binding of acetylcholine of a uni-univalent ionic bond with the distance of separation of charge of 5.4 A.

On the basis of all the data presented, the effect of electric charge upon molecular species capable of combining with the enzyme can be explained as due to ionic bond formation.

The hydrolysis of dimethylaminoethyl acetate at pH 10 cannot be ascribed alone to the small fraction of molecules in the cationic form. The concentration of such ions is about $10^{-5}\,M$; at this concentration the rate of hydrolysis would be negligibly small. Therefore, we may conclude that the neutral molecules can be hydrolyzed by the enzyme at a rate which, though considerably lower than that of the corresponding cation, is still faster than that of the simple esters such as ethyl acetate.

In addition to the contributions by Coulombic forces, the methyl groups on the cationic portion of the molecule contribute to the binding by unspecific van der Waals' forces. This was shown by Wilson (1952a), using methylated competitive inhibitors of the ammonium and hydroxyethyl ammonium series at pH 7, where all these inhibitors are cationic. The results are shown in Table X. Except for the methyl group, which becomes the fourth alkyl group, each methyl group has binding properties amounting to about 1.2 kcal. per mole and on the average increases the potency of an inhibitor about sevenfold. It is reasonable to assume that neither the changes in hydration characteristics attending binding, nor the entropies of binding, differ markedly for any member of the series. Therefore, the additional binding associated with each methyl group may be attributed to van der Waals' attraction between the methyl group and a hydrocarbon group of the protein. The latent energy of evaporation of methane is about 2 kcal. per mole, so energies of this force are of a suitable magnitude for explaining the observed binding property of a methyl group. In accordance with expectation large alkyl or aryl groups improve the binding properties of ammonium ions.

TABLE X

INHIBITORY POTENCY OF METHYLATED AMMONIUM IONS

—N—		—N—C_2H_4OH	
Number of methyl groups	M^a	Number of methyl groups	M^a
4	0.018	3	0.005
3	0.015	2	0.005
2	0.12	1	0.07
1	0.7	0	0.28

[a] M = molar concentration to produce 50 per cent inhibition when the acetylcholine concentration is 4 x 10^{-4} M.

In addition to the activating anionic site, there must be a region in the active surface in close vicinity which acts upon the ester linkage, an ester-breaking or esteratic site.* The carbonyl group has a marked polar character: the positive carbon and the negative oxygen contribute about 50 per cent of the bond strength (Hammett, 1940). This electrophilic carbon is a site of attack for basic reagents. It is a point of attack by OH^- in the base-catalyzed hydrolysis of esters. On the basis of modern organic theory one may assume that a basic group in the enzyme forms a covalent bond with the acyl carbon of the ester and that this binding in combination with Coulombic and van der Waals' forces at the anionic

* The term "esteratic" rather than "esterolytic" site has been selected. The site as will be seen, may catalyze the formation as well as the splitting of the ester link, depending upon the experimental conditions. A term indicating the reactivity with ester linkages but not specifying the effect appeared therefore preferable.

site constitute the main forces which stabilize the enzyme-substrate complex. The importance of an electrophilic carbon is illustrated by the effects of a series of nicotinic acid derivatives (Bergmann, Wilson, and Nachmansohn, 1950a). The order of increasing inhibition must be compared with the order of increasing electrophilic character of the carbonyl atom. The more electrophilic members would be expected to form more stable complexes with the basic group of the enzyme as a result of stronger binding. The order of increasing electrophilic character

$$-\overset{\overset{\displaystyle O}{\|}}{C}-O^- < -\overset{\overset{\displaystyle O}{\|}}{C}-NH_2 < \overset{\overset{\displaystyle O}{\|}}{C}-N(C_2H_5)_2 =$$

$$-\overset{\overset{\displaystyle O}{\|}}{C}-CH_3 < -\overset{\overset{\displaystyle O}{\|}}{C}-O-C_2H_5$$

parallels the observed order of inhibition (Fig. 33) and suggests the formation of a covalent bond between the carbon and some basic group

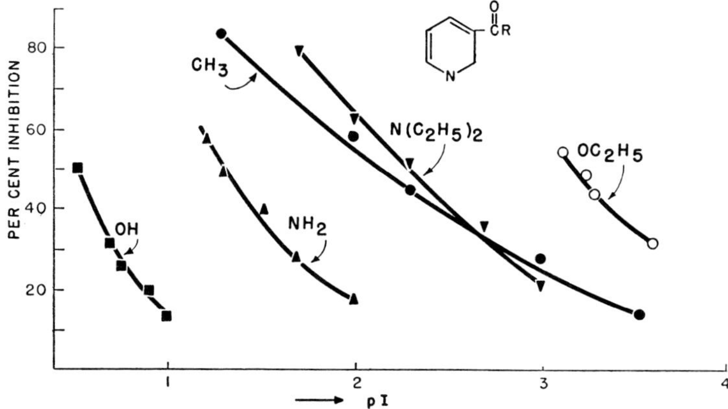

Fig. 33. Effect of substitution of the carboxyl group of nicotinic acid upon the inhibitory strength in the reaction with acetylcholinesterase.

in the enzyme. Similar effects were observed by Wilson (1952a) with substrates. Hydrogen bonding does not seem to be of importance in the enzyme-substrate complex, since choline and trimethylaminopropane inhibit equally (Wilson, 1952a).

In neutral media nicotinic and picolinic acids are highly ionized. The resulting anions have very little electrophilic character, but the undissociated molecules should be as effective as the esters. The concentration of undissociated acid molecules is such that the observed inhibition may be entirely, although not necessarily, attributed to this form (Bergmann, Wilson, and Nachmansohn, 1950a). For instance, nicotinic acid with an ionization constant of 1.4×10^{-5} inhibits 50 per cent at a concentration of 0.3 *M* at pH 7.2. The concentration of undissociated molecules under these circumstances (including 0.1 *M* salt concentration) is less than 1.3×10^{-3} *M*, or about the concentration of ethyl nicotinate required to cause 50 per cent inhibition.

A carbonyl group is not essential; an electrophilic central atom, however, is. Amino nitriles, for example, which contain an electrophilic carbon, also inhibit (Bergmann, Wilson, and Nachmansohn, 1950a).

A further clue as to the forces in the active enzyme surface was obtained by Wilson in the pH dependence of enzyme-catalyzed hydrolysis. The structure of acetylcholine is not altered by pH changes, so that changes in activity must be attributed to changes in protein structure. As shown in Fig. 34, the activity is at a maximum between pH 8 and 9,

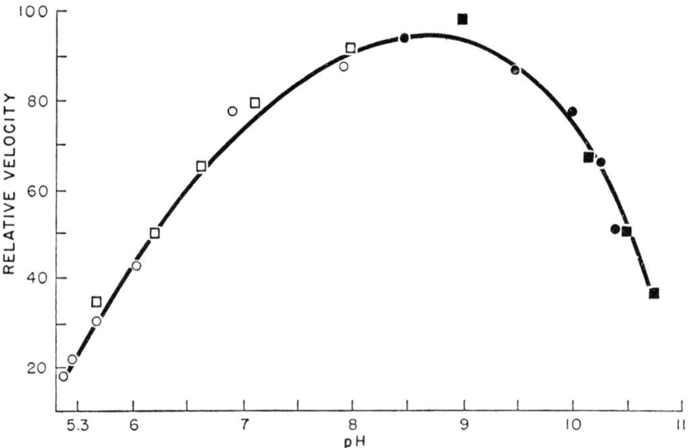

FIG. 34. Velocity of acetylcholine hydrolysis by acetylcholinesterase from electric tissue as a function of pH.

Each symbol indicates a different experiment.

and declines in more alkaline and more acid media. These changes may be interpreted in terms of the dissociation of acidic and basic groups and represented schematically as follows:

$$\underset{\text{inactive}}{EH_2^+} \overset{+H^+}{\rightleftharpoons} \underset{\text{active}}{EH} \overset{+OH^-}{\rightleftharpoons} \underset{\text{inactive}}{E^- + H_2O}$$

where EH, the active enzyme, is arbitrarily assigned a relative charge of 0. The forms EH_2^+ and E^- either cannot form complexes at all, or such complexes are inactive. On the basis of this concept Wilson suggested the following equilibria and dissociation constants:

$$
\begin{array}{rll}
EH_2^+ \rightleftharpoons EH + H^+ & K_{EH_2^+} \\
EH \rightleftharpoons E^- + H^+ & K_{EH} \\
EHS \rightleftharpoons ES^- + H^+ & K_{EHS} \\
EHS + S \rightleftharpoons EHS_2 & K_2
\end{array}
$$

and the rate equation

$$
EH + S \underset{k_2}{\overset{k_1}{\rightleftharpoons}} EHS \xrightarrow{k_3} EH + \text{products}
$$

where S is the substrate and EHS_2 an inactive supercomplex, which may account for substrate inhibition observed at concentrations higher than $4 \times 10^{-3}\ M$. Mathematical analysis of these relations led Wilson to the equation for the velocity, v:

$$
\frac{v^0}{v} = 1 + \frac{K_1}{K_{EH_2^+}\left(K_1 + (S) + \dfrac{(S)^2}{K_2}\right)}(H^+) + \frac{K_1K_{EH} + (S)K_{EHS}}{\left(K_1 + (S) + \dfrac{(S)^2}{K_2}\right)}\frac{1}{(H^+)}
$$

where v^0 is the reaction velocity at optimum pH and K_1 is the apparent dissociation constant $(k_2 + k_3)/k_1$.

The equation shows that v^0/v varies linearly with H^+ in the acid region and with OH^- in the alkaline region. This prediction is borne out experimentally (Figs. 35 and 36).

These experimental observations are thus in agreement with the concept that certain basic and acidic groups are essential for enzyme activity, and also with the previous conclusion that the interaction of a basic group in the enzyme with the carbonyl carbon atom of substrates and inhibitors contributes to the binding of these compounds. The decline of enzyme activity in acid media can be attributed, in part, to poorer binding caused by the conversion of the basic group of the esteratic site to the conjugated acid, and, in part, by a similar conversion of the negatively charged groups of the anionic site. The binding of inhibitors containing a carbonyl group as well as a methylated ammonium structure declines much more rapidly than that of those containing the ammonium structure alone (Bergmann and Shimoni, 1952). The decline in alkaline media is not, however, caused by poorer binding. This is indicated by the fact that noncompetitive inhibition by Prostigmine, although it declines in acid media at precisely the same rate as acetylcholine hydrolysis, remains constant in alkaline media even up to pH 11, where this hydrolysis has fallen to 30 per cent (Wilson, 1951a). The fact that the binding is just as good in alkaline media indicates that the acid group is not involved in binding; it suggests that the decline in enzyme activity is due to the requirement of the acid group in the hydrolytic process.

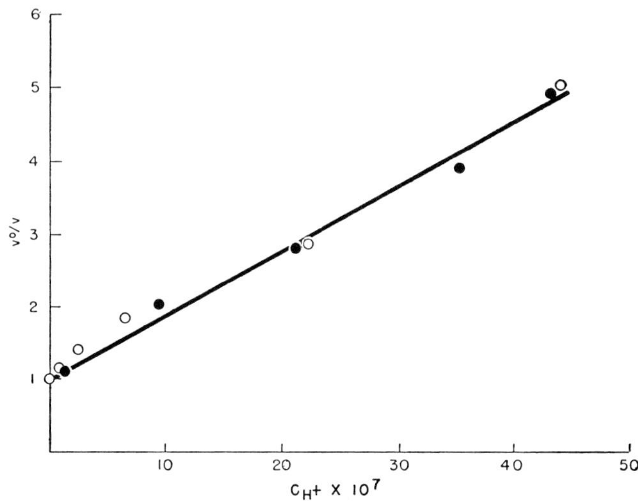

FIG. 35. $\frac{v^0}{v}$ of acetylcholine hydrolysis as a function of hydrogen ions (v^0 = reaction velocity of optimum pH).

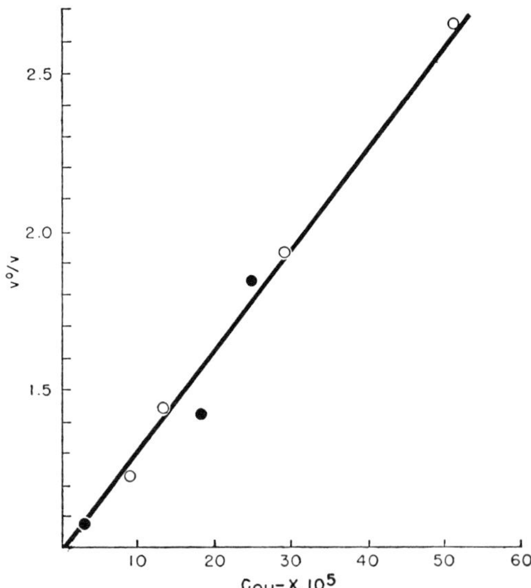

FIG. 36. $\frac{v^0}{v}$ of acetylcholine hydrolysis as a function of hydroxyl ions.

On the basis of these studies Wilson was able to calculate the pK's of the acidic and basic groups. The pK of the acidic group is 9.2, that of the basic group about 6.5, the latter suggesting the possibility of an imidazole ring being an active group in this process (Wilson and Bergmann, 1950b). Further details may be found in the original papers quoted and in the summaries (Nachmansohn and Wilson, 1951; Wilson, 1954b).

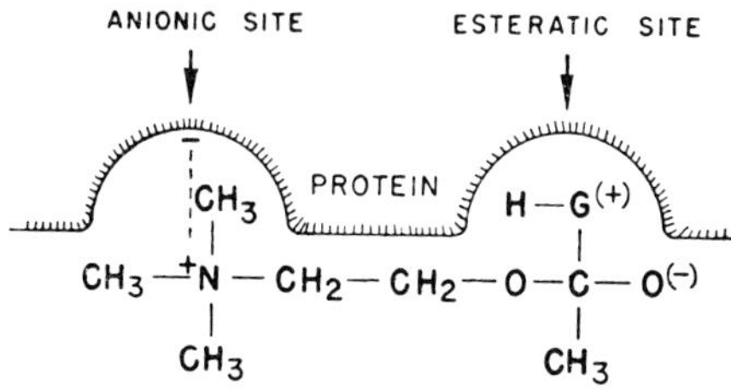

FIG. 37. Schematic presentation of the interaction of the active groups in the surface of acetylcholinesterase and the substrate; the Michaelis-Menten complex.

The enzyme-substrate complex, schematically presented in Fig. 37, is then stabilized by Coulombic and van der Waals' forces at the anionic site and by covalent bond formation between the carbonyl carbon and the basic group of the esteratic site; the latter is symbolized by G (for basic group), H representing a dissociable hydrogen atom, not involved in binding.

B. The Hydrolytic Process

Let us now turn to the mechanism of the hydrolytic process. Wilson proposed (Wilson, Bergmann, and Nachmansohn, 1950; Nachmansohn and Wilson, 1951; Wilson, 1954b) the mechanism illustrated (5).

$$\mathrm{H{-}\ddot{G} + R'O{-}\underset{\displaystyle R}{\underset{|}{C}}{=}O \rightleftharpoons R'\ddot{O}{-}\overset{\displaystyle H{-}G^{(+)}}{\overset{|}{\underset{\displaystyle R}{\underset{|}{C}}}}{-}O^{(-)} \rightleftharpoons \overset{\displaystyle G^{(+)}}{\overset{\|}{\underset{\displaystyle R}{\underset{|}{C}}}}{-}O^{(-)} + R'OH}$$

$$(A) \qquad\qquad (B)$$

$$\mathrm{H{-}\underset{..}{\overset{\displaystyle H}{\overset{|}{O}}}{:} + \overset{\displaystyle G^{(+)}}{\overset{\|}{\underset{\displaystyle R}{\underset{|}{C}}}}{-}O^{(-)} \rightleftharpoons HO{-}\overset{\displaystyle H{-}G^{(+)}}{\overset{|}{\underset{\displaystyle R}{\underset{|}{C}}}}{-}O^{(-)} \rightleftharpoons H{-}\ddot{G} + HO{-}\underset{\displaystyle R}{\underset{|}{C}}{=}O}$$

$$(B) \qquad\qquad (C)$$

$$(5)$$

H symbolizes again the acidic and G the nucleophilic group in the esteratic site. The pair of electrons symbolizes the electron-transmitting properties of the group. Acetylcholine forms the Michaelis-Menten complex (*A*). The proposed mechanism assumes a process taking place in two consecutive steps. The first step is the acetylation of the enzyme with simultaneous elimination of choline. (*B*) shows the acylated enzyme depicted as enolate ion, which is one of the resonance forms. (*C*) is an acid enzyme complex similar to the ester enzyme complex and leads to regenerated enzyme and acetic acid. The mechanism follows from the structure of the enzyme-substrate complex and assigns a positive role to the enzyme in effecting a combined acid-base attack. The acetyl enzyme reacts with water or other nucleophilic agents, such as hydroxylamine, or an alcohol, e.g., choline, to yield an acid or an ester. It is possible to start with acids or esters, but only the undissociated acid molecules have the electrophilic carbon atom necessary for the enzyme-substrate complex. Since at pH 7 the fraction of undissociated acid molecules is small, the intermediate is far more rapidly formed from esters than from carboxylic acids. Any reaction which can be carried out with the acid will occur much more rapidly with the corresponding esters. When the enzyme-catalyzed formation of hydroxamic acid and choline esters from simple esters and the corresponding acids were compared, the reaction with the esters was indeed found to be about a hundred times faster (Wilson, Bergmann, and Nachmansohn, 1950).

TABLE XI
HYDROXAMIC ACID FORMATION FROM ACIDS AND ESTERS CATALYZED BY ACETYLCHOLINESTERASE[a]

Substrate	Concentration *M*	Incubation (min.)	Hydroxamic acid formed (μmoles/ml.)	
			pH 6.5	pH 7.5
Sodium acetate	0.5	60	2.47	—
		120	4.10	—
Sodium propionate	0.5	60	0.77	—
		120	1.39	—
Sodium butyrate	0.5	60	0.05	—
		120	0.05	—
Ethyl acetate	0.17	15	10.4	16.0
		30	14.5	21.4
		60	17.5	25.3
Ethyl acetate	—	3	1.4	3.0
		6	2.8	6.2
Ethyl propionate	0.05	60	2.02	2.63
		120	2.91	3.66

[a] The incubation mixture contained, per milliliter, 0.25 ml. enzyme, prepared from electric tissue of *Torpedo*. The enzyme solution contained about 1.8 mg. protein per milliliter and hydrolyzed about 4.6 gm. acetylcholine per milliliter per hour. In the second experiment with ethyl acetate, the enzyme was more dilute than in the others.

Table XI shows some observations of the rate of hydroxamic acid formation from acids and esters catalyzed by the enzyme.

The mechanism proposed has been confirmed in several ways (Wilson, 1951b, c). Using thioacetic acid as substrate, Wilson found that H_2S is evolved and acetic acid is formed as predicted by theory (Fig. 38). The reaction is completely inhibited by Prostigmine. In his observations on the reaction of thioacetate with the enzyme, Wilson was also able to demonstrate that anionic and esteratic sites are not only functionally, but also spatially, separated. When a small-sized cationic

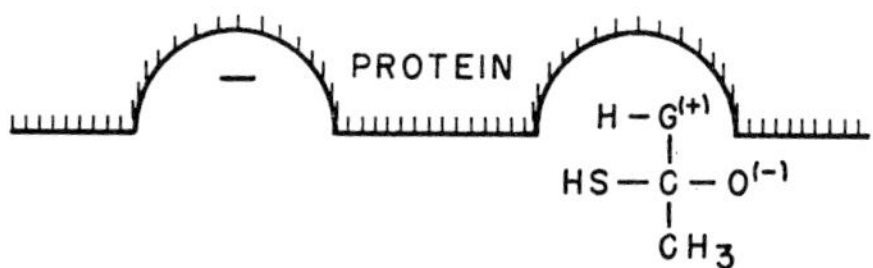

FIG. 38. Reaction of thioacetic acid with the esteratic site of acetylcholinesterase.

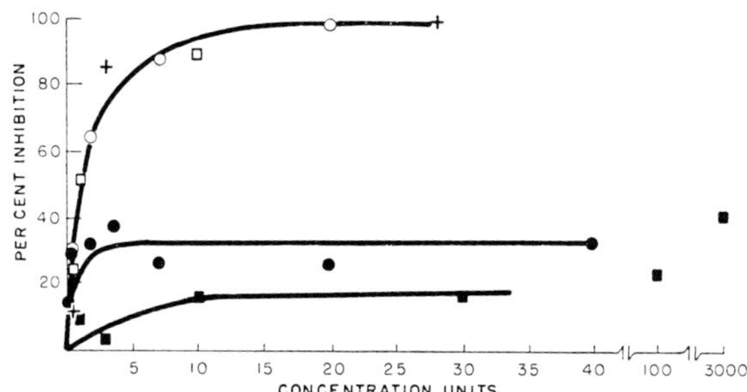

FIG. 39. Effect of cationic inhibition upon the hydrolysis of thioacetic acid by acetylcholinesterase.

The unit of the concentration is that required for 50 per cent inhibition with acetylcholine as substrate. This value is 1.5×10^{-2} for trimethylammonium chloride, 2.5×10^{-5} for decamethonium bromide, and 7×10^{-7} for Prostigmine bromide. Prostigmine is included to demonstrate the effect of an inhibitor which acts on both the anionic and the esteratic sites. Thioacetic acid: (+) Prostigmine, (●) trimethylamine, (■) decamethonium. Ethyl acetate: (O) trimethylamine, (□) decamethonium. Trimethylammonium, which at neutral pH is a cation and reacts with the anionic site only, is unable to inhibit the hydrolysis of the small thioacetate molecule more than 30 per cent, even at high concentration. If, however, the substrate molecule is larger, such as, e.g., ethyl acetate, the inhibition is complete only at higher inhibitor concentrations than those required for blocking the hydrolysis of acetylcholine. Even if the inhibitor is a rather large molecule, such as, for instance, decamethonium, it does not strongly inhibit the splitting of thioacetate, only that of ethyl acetate. In contrast, Prostigmine blocks even in small concentrations the splitting of thioacetate. These experiments show the spatial separation of the anionic and the esteratic site in the surface of the protein.

inhibitor, such as trimethyl ammonium ion, is used, the hydrolysis of ethyl acetate is completely inhibited, that of thioacetate, even in high inhibitor concentrations, not more than 30 per cent (Fig. 39). The binding of the small molecule to the anionic site apparently does not interfere markedly with the reaction at the esteratic site, provided the reacting molecule is as small as thioacetate.

Further support is the evidence of oxygen exchange between acids and water, as postulated by theory, shown with the use of isotopic oxygen (O^{18}) (Sprinson and Rittenberg, 1951; Bentley and Rittenberg, 1954). The theory has also been confirmed by observations of Stein and Koshland (1953). Wilson's idea of an acylated enzyme being the intermediary form in the hydrolytic process has since been widely accepted by many enzyme chemists in the analysis of various enzymatic mechanisms.

Knowledge of the molecular forces acting between acetylcholine and the enzyme protein has greatly contributed to a better understanding of several problems of nerve function in general, as will be seen in the chapters that follow.

CHAPTER VIII

Nerve Gases, Insecticides, and Antidotes

A. Mechanism of Inhibition by Organophosphorus Compounds

The irreversible inhibition of acetylcholinesterase by DFP and other organophosphorus compounds and their fatal effects were briefly discussed before. The findings presented a twofold challenge to the biochemist. First, what is the mechanism of these reactions between enzyme and irreversible inhibitor? The type of reaction appeared unusual and raised interesting problems for the enzymologist. Second, in case the mechanism can be explained, is it possible to reverse the reaction and to develop eventually an antidote capable of an efficient protection?

An important clue was the finding that in a highly purified enzyme solution DFP competes with the reversible inhibitors Prostigmine and eserine for the active surface of acetylcholinesterase: on incubation of the solution with Prostigmine prior to the addition of DFP in equimolar concentrations it was found, by Nachmansohn and Rothenberg, that on dilution no inhibition had taken place. In the control experiment without Prostigmine, the same molar concentration of DFP under the same experimental conditions had inactivated 50 per cent of the enzyme. The details of the experiment have been described in a paper by Augustinsson and Nachmansohn (1949b). The complete protection obtained with Prostigmine on mole-to-mole basis against the action of DFP clearly indicated that the two types of inhibitors were competing for the same site in the enzyme surface. In view of the competitive nature of the inhibition of acetylcholine hydrolysis by Prostigmine, it became apparent, for the first time, that the site must be identical with the active surface of the enzyme. Eserine also did protect against the action of DFP in equimolar concentration although not quite as efficiently as Prostigmine. In further observations with a whole series of competitive inhibitors it was found that the weaker the binding forces of a competitive inhibitor to the enzyme, the higher is the concentration required for complete protection against DFP (unpublished observations of the writer's laboratory).

An explanation of the precise mechanism of the reaction between organophosphorus compounds and enzyme became possible after the elucidation of the mechanism of enzymatic hydrolysis discussed in the preceding chapter. The enzyme attacks the electrophilic P atom in an S_N2 reaction eliminating an acidic group (Wilson and Bergmann, 1950a).

In this bimolecular nucleophilic substitution reaction the P atom forms a covalent bond with the nucleophilic group in the enzyme. Instead of an acetylated, a phosphorylated enzyme is formed (Fig. 40). But,

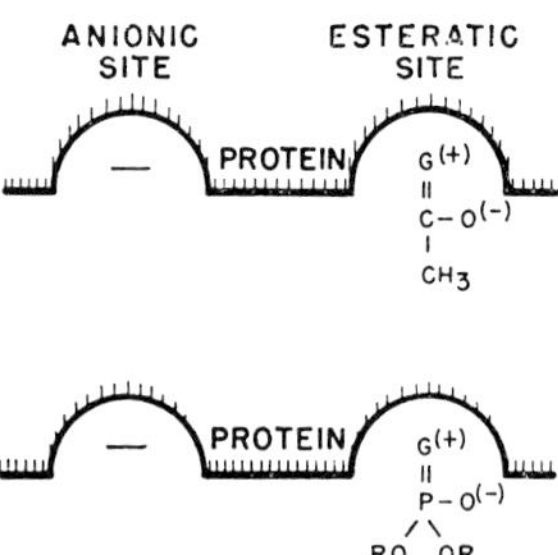

FIG. 40. Schematic presentation of the acetylated enzyme (upper row) compared to the phosphorylated enzyme (lower row).

whereas in the physiological process the acetylated enzyme reacts rapidly with water (in a few microseconds), the phosphorylated enzyme reacts extremely slowly with water or not at all, depending on the type of phosphoryl group of which there exists a very large number of varieties. In principle however, the structure is always similar. The organophosphorus compounds, which are inhibitors of cholinesterase, have the general formula (IX).

$$\begin{matrix} RO & & O \\ & \diagdown \; \diagup\!\!\!\diagup & \\ & P & \\ & \diagup \; \diagdown & \\ RO & & X \end{matrix} \quad \text{or} \quad \begin{matrix} RO & & O \\ & \diagdown \; \diagup\!\!\!\diagup & \\ & P & \\ & \diagup \; \diagdown & \\ R & & X \end{matrix}$$

(IX)

X is an acidic group. It may be F^- or Cl^- or CN^- or nitrophenol or another alkyl phosphate as, for instance, in tetraethylpyrophosphate (TEPP). When an enzyme molecule attacks its physiological substrate it is capable of hydrolyzing a huge number of molecules, millions per minute. But when attacking an organophosphorus compound it hydrolyzes one molecule, because then it becomes phosphorylated and thus is inactivated.

The idea that inhibitors of this type are actually substrates and are hydrolyzed by the enzyme is supported by the observation that the inhibitory power measured as a function of pH is greatest at pH 8, i.e., it coincides with the optimum of enzyme activity (Fig. 41) (Wilson and Bergmann, 1950a). Further support was the observation that in the reaction of chymotrypsin with DFP, labeled with radioactive phosphorus, the P atom remains attached to the protein, whereas the fluoride is split off (Jansen, Nutting, and Balls, 1949; Jansen, Nutting, Jang, and Balls,

1950). The significance of this finding for the mechanism of the reaction was, however, not recognized by these investigators. A fixation of the phosphorus to acetylcholinesterase prepared from electric tissue, was subsequently reported by Michel and Krop (1951).

In the following years the mechanism proposed has found increasing recognition. Additional evidence was presented by Aldridge and Davison (1953). Today this reaction mechanism is generally accepted.

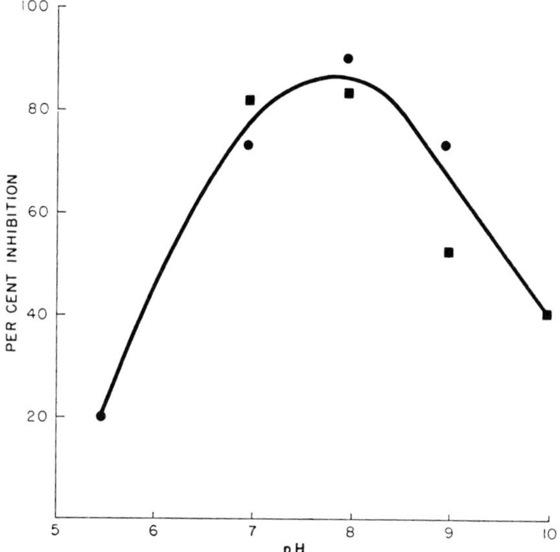

FIG. 41. Inhibition of acetylcholinesterase by alkyl phosphate as a function of pH.

Concentrated enzyme was exposed to tetraethylpyrophosphate (TEPP) at various pH's, then rapidly diluted and the activity tested at pH 7. Symbols represent two different sets of data.

B. EFFICIENCY OF AN ANTIDOTE ON THE BASIS OF MOLECULAR COMPLEMENTARITY

If the mechanism of enzyme inhibition is based upon the attachment of the P atom to the nucleophilic group in the active enzyme surface, nucleophilic compounds might be expected to remove the phosphoryl group in the enzyme surface by a displacement reaction; the nucleophilic group in the enzyme surface would become free and the enzyme reactivated. This assumption was supported by the finding that an enzyme solution inactivated by TEPP regains within 3–4 weeks about 30–50 per cent of its activity when kept in the refrigerator, although water is a weak nucleophilic agent.

Hestrin (1949b, 1950) has observed that purified acetylcholinesterase prepared from electric tissue catalyzes the formation of acethydroxamic acid from acetate and hydroxylamine. He found a lower reaction rate with propionic acid; with butyric acid the hydroxamic acid formation was negligible. The reaction was inhibited by Prostigmine and TEPP. He assumed that the enzyme was activating the carboxylic acid and that the active acetyl group reacted with hydroxylamine. It thus appeared possible that hydroxylamine would also react with the phosphoryl group, although at a much lower rate than with acetyl enzyme, and that hydroxylamine would, therefore, greatly accelerate the reactivation of the enzyme inhibited by TEPP by displacing the phosphoryl group from the enzyme.

This expectation was borne out by experiment (Wilson, 1951a). With 0.7 *M* solution of hydroxylamine more than 50 per cent of the enzyme activity was restored within 5 hours. Choline was also a reactivator, but only about half as efficient. Wilson (1952b) then tested a great number of other nucleophilic agents, compounds containing amino, amidino, guanidino, pyridyl, hydroxyl, or mercaptyl groups, such as creatine, ethyl glycinate, nicotinamide, glycine amide, pyridine, 3-acetylpyridine, 2 OH- and 2 NH_2 pyridine, and others. All these compounds were able to reactivate TEPP-inhibited enzyme, a diethylphosphoryl enzyme, although not as effectively as hydroxylamine. When the enzyme was inhibited by DFP, resulting in a diisopropylphosphoryl enzyme, only a negligible reactivation was found or none.

The question arose whether it was possible to accelerate the reactivation by nucleophilic agents and eventually develop an antidote against nerve gas or insecticide poisoning. For antidotal action the effectiveness of the compounds mentioned above was not adequate because the rate of reactivation was too slow and the concentrations required much too high.

In the phosphorylated enzyme the anionic site is more or less free. The fact that choline is a fairly good reactivator although a very weak nucleophilic agent, suggested that in this case the reaction was promoted by the attraction of the cationic group to the anionic site, an assumption supported by the inhibition of the reactivation by tetra- or trimethyl ammonium ions. Wilson reasoned, therefore, that if an attacking agent such as hydroxylamine should be linked to a cationic nitrogen group at an appropriate distance, this group might greatly promote the reaction, just as the methylated nitrogen in acetylcholine makes this ester a many thousand times better substrate of the enzyme than ethyl acetate. The first compound synthesized, mainly because it was easy to prepare, was nicotinohydroxamic acid methiodide (Wilson and Meislich,

1953; Wilson, 1955a). Hydroxamic acids such as acethydroxamic acid or glycylhydroxamic acid are poorer reactivators than is hydroxylamine, but in combination with a quaternary nitrogen group located approximately at the right distance to react with the anionic site and to promote the attack of the hydroxamic acid, the reactivating power was several times greater than that of hydroxylamine. It thus proved in principle the soundness of Wilson's argument. Moreover, whereas hydroxylamine reactivated only diethylphosphoryl enzyme resulting from the inhibitory action of TEPP, nicotinohydroxamic acid methiodide reactivated for the first time enzyme inhibited by DFP, the diisopropylphosphoryl enzyme (Wilson and Meislich, 1953; Wilson, Ginsburg, and Meislich, 1955). In this case the promotion is, however, not as evident and clear as with diethylphosphoryl enzyme, because the anionic site, being in the close vicinity of the esteratic site, is partly shielded owing to the largeness of the phosphoryl group. The chemical properties of the functional group, the hydroxamic acid function, are altered by the presence of the cationic center in the molecule, and the question arose whether the promotion is actually due to the attraction of the cationic group to the anionic site or to the intrinsic change of the properties of the molecule. A strong support for the effectiveness of promotion is, however, obtained if the ratio of the rate constant for the reactivation of diethylphosphoryl enzyme is compared with that of diisopropylphosphoryl enzyme by reactivators with or without a cationic group. The latter are in general two to nine times more effective in reactivating the diethyl- than the diisopropylphosphoryl enzyme; but compounds with a cationic group are very much more effective in reactivating diethylphosphoryl enzyme than in reactivating diisopropylphosphoryl enzyme (Wilson, 1955a).

A still stronger reactivation was obtained when the hydroxamic acid was in a 2- instead of a 3-position on the ring: picolinohydroxamic acid methiodide proved to be about ten times as active as nicotinohydroxamic acid methiodide (Wilson and Ginsburg, 1955a).

When tested on animals, some antidotal properties of these compounds were observed, but they were not impressive. Moreover, it is not sure, and has not been tested, whether in these cases the protection is due essentially to the reactivation of the enzyme by the hydroxamic acids, which was the aim of these studies, or due to a direct reaction between the alkyl phosphates and the hydroxamic acids.

A much more powerful reactivator was synthesized early in 1955: 2-pyridine aldoxime methiodide (2-PAM) (Wilson and Ginsburg, 1955b). In suggesting this compound which was synthesized by Sara Ginsburg, Wilson originally intended to increase the structural rigidity of the molecule. If the quaternary group is bound to the anionic site, a molecule with one conformation, or at most a few, would be expected to be favorable for orienting the active nucleophilic atom toward the phosphorus. The 2-position on the ring was suggested by the superiority of picolino- over nicotinohydroxamic acid methiodide. 2-PAM proved to be about a million times better than hydroxylamine. It reactivates in 1 minute a large fraction of acetylcholinesterase inhibited by TEPP or DFP, in concentrations of 10^{-5} *M*. The tertiary analog of 2-PAM, although active, is incomparably weaker. The active form has the (anti) configuration (X).

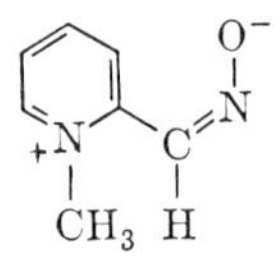

(X)

Because of resonance, the molecules tend to be planar.

In the following years Wilson and his associates studied the geometry of the enzyme in order to find out the basis for the exceptionally great reactivating power of 2-PAM. They tested the inhibition of acetylcholinesterase by a number of phenyltrimethyl ammonium derivatives, their dimethyl carbamates and other related compounds as to the degree of binding in terms of the respective positions of various atoms of the inhibitor and of the enzyme. The results revealed an extraordinarily high degree of molecular complementarity between 2-PAM and the inhibited enzyme (Wilson and Quan, 1958; Wilson, Ginsburg, and Quan, 1958; Wilson, 1959). Only a few data will be briefly presented to illustrate the procedure applied.

Phenyltrimethyl ammonium iodide was used as the reference compound. The remarkably strong binding of the drug Tensilon, (3-hydroxy)-phenylethyldimethyl ammonium chloride, indicates that a phenolic hydroxyl group may greatly contribute to binding. The binding strength of (3-hydroxy)phenyltrimethyl ammonium ion was found to be 120 times as high as that of the reference compound. This is equivalent to a decrease of 2.9 kcal. per mole in the free energy of binding. Such a large effect strongly suggests that a hydrogen bond is involved. This assumption is supported by the fact that the binding of the 3-methoxy derivative is 26 times poorer than of the 3-hydroxy compound.

Resonance between the phenolic oxygen and the ring introduces a strong tendency for the phenolic hydrogen to lie in the same plane as the ring. This leads to two possible conformations (XI, *a* and *b*) of the

H atom, and the question arose in which direction does the hydrogen bonding atom of the enzyme, Z, lie with respect to the anionic site.

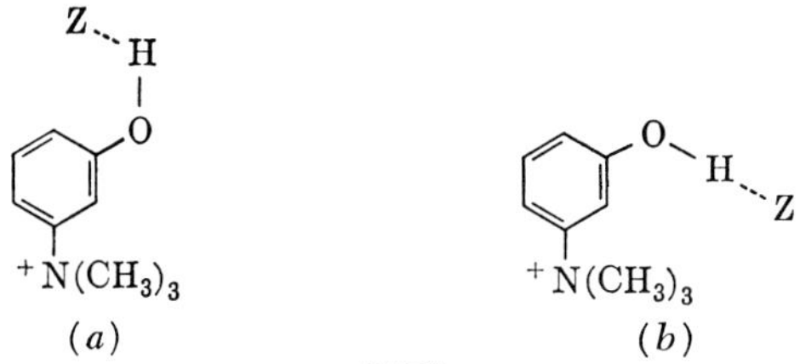

$^{+}N(CH_3)_3$ (*a*) $^{+}N(CH_3)_3$ (*b*)

(XI)

To test which of the two dispositions is the correct one, Wilson studied the effect of a compound containing a 4-CH_3 in addition to the 3-OH group. In this compound most molecules may be expected to have disposition (*b*). Therefore, considering just this factor, this compound should be superior to the 3-hydroxy derivative without the 4-CH_3 by a factor of 2. However, the alignment of the O—H dipoles in the field of the cation favors disposition (*a*). Wilson calculated that owing to this factor the compound should be better bound by a factor of 2.5. In addition, the 4-CH_3 group was found to increase the binding by a factor of 1.25. Therefore, the total improvement should amount to a factor of $2.5 \times 1.25 = 3.1$. Exactly this increase in binding strength was found experimentally.

The relative binding strength of the dimethyl carbamate of (3-hydroxy) phenyltrimethyl ammonium ion (Prostigmine) is about ten times higher than that of (4-methyl-3-hydroxyphenyl)trimethyl ammonium bromide. The disposition of this compound may be assumed to be that shown in formula (XII). It appears probable that the carboxyl carbon

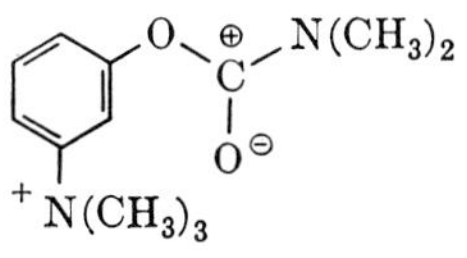

$^{+}N(CH_3)_3$ $N(CH_3)_2$

(XII)

atom forms a covalent bond with the same basic group in the esteratic site which is acetylated in the normal catalytic process and is phosphorylated by the organophosphorus inhibitors. Therefore, the position of this carbon must be a probable position for the P atom in the phosphorylated enzyme. Since the quaternary ammonium group must be bound to the anionic site, this molecule gives the essential spatial relationships of the phosphoryl enzyme. Taking the nitrogen as the origin of the coordinates and the axis of the ring as the *y* axis, Wilson calculated the coordinates of the P atom to be 3.4 and 3.3 A.

Comparing 2-, 3-, and 4-PAM, Wilson found that 4-PAM is still a good reactivator, only about forty times poorer than 2-PAM, whereas 3-PAM is very poor. This is easily understood from the dispositions of these three compounds in the light of the information obtained. In 2-PAM the nucleophilic oxygen is directed against the electrophilic P atom attached to the enzyme; in 4-PAM one of the dispositions is directed this way; in 3-PAM neither disposition is suitable (XIII).

2-PAM 4-PAM 3-PAM

(XIII)

There is no great difference in intrinsic nucleophilicity among the compounds as was shown by their reaction with sarin and with *p*-nitrophenyl acetate. The difference of reactivating power thus must be attributed to the spatial arrangements and the various degrees of molecular complementarity. Figure 42 gives a schematic presentation of the spatial

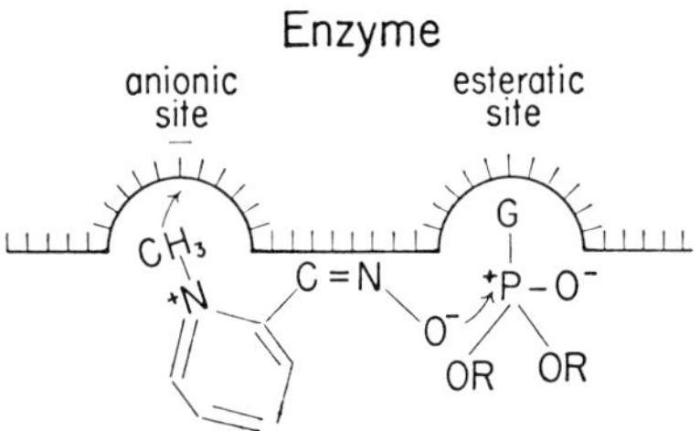

FIG. 42. Schematic presentation of the complementary fit of PAM to acetylcholinesterase, according to I. B. Wilson.

arrangement according to Wilson. When the reactivating power of these compounds was studied with other enzymes, where no molecular complementarity is to be expected, their effectiveness was found to be low. Inhibited chymotrypsin, for instance, is very poorly reactivated; 3-PAM is the best in the group. Although the reactivating power is much greater with human serum esterase than with chymotrypsin, it is still very low when compared with acetylcholinesterase.

These few examples may suffice to illustrate the approach and procedures used by Wilson and his associates for demonstrating that the high degree of molecular complementarity between 2-PAM and the phosphorylated enzyme is the basis of the extraordinary reactivating power. The analysis not only has been of great importance for the understanding of this particular problem, but is, in addition, a significant and valuable contribution to the problem of the analysis of geometry of enzyme surfaces and of active groups of proteins in general.

It may be mentioned that shortly after the description of PAM by Wilson and Ginsburg (1955b), Childs, Davies, Green, and Rutland (1955) described a series of oximes which they had synthesized independently, in view of Wilson's earlier studies with hydroxamic acids. One of them was PAM. The authors recognized the improvement of the oximes over the hydroxamic acids, but they did not recognize the extraordinary power of this particular oxime compared to other active oximes.

Once the extraordinary power of 2-PAM to reactivate acetylcholinesterase had been established, Kewitz and Wilson (1956) applied the compound to mice in order to test its efficiency as an antidote against alkyl phosphate poisoning. The results were striking. In the first experiment a sure lethal dose (LD_{100}) of Paraoxon was injected into twenty mice. Ten of them received subsequently an injection of PAM. All ten mice treated with PAM survived, all the others were dead within 30–40 min.; no statistics were required for evaluation.

According to theory, PAM reactivates the alkyl phosphate-inhibited enzyme. However, the immediate cause of death is in most cases respiratory failure as a result of bronchial spasms primarily due to peripheral action, although in many instances central effects may be an additional factor. This action must be attributed to an accumulation of acetylcholine, which follows logically from the specific biochemical lesion postulated. In the sequence of events presented in Chapter VI, the dissociation of acetylcholine from the storage protein and its combination with the receptor are responsible for the permeability change in the activated membrane. The action of acetylcholinesterase is the primary event in recovery, making it possible for the receptor protein to return to its resting condition. Consequently, if the inhibition of the enzyme is the specific chemical lesion, persistence of the acetylcholine released should result in hyperactivity, and if this hyperactivity interferes with a vital function, such as respiration, as in the case with bronchial spasms, the condition becomes incompatible with life, and death ensues. Thus theory requires that the specific biochemical lesion postulated should lead to hyperactivity by unremoved acetylcholine, although in some cases, as will be discussed later, the organophosphorus compounds may directly affect the receptor. As the logical consequence of this concept, it should be expected that in the treatment of intoxication by organophosphorus compounds protection of the receptor would be an important and desirable and possibly a necessary corollary of the repair of the chemical lesion. Without this protection the organism may not survive until the damage to the enzyme is repaired. These considerations suggested that PAM should be applied in combination with atropine, which is a receptor inhibitor (see Chapter X) and therefore competitively protects this protein against the action of acetylcholine. Atropine alone does not protect against a single lethal dose of alkyl phosphates such as Paraoxon or DFP. When atropine was used combined with PAM, a complete survival of all animals injected with ten- to twentyfold lethal doses of Paraoxon and DFP was obtained (Table XII) (Kewitz, Wilson, and Nachmansohn, 1956).

TABLE XII

PROTECTION AGAINST PARAOXON AND DFP BY PAM IN COMBINATION WITH ATROPINE[a]

Expt. no.	Atropine (mg./kg.)	PAM (mg./kg.)	Paraoxon (mg./kg.)	Multiple of LD_{50}	No. of Animals	Survivors
1	0	30	1.0	1.5	20	20
2	10	0	3.5	5	5	0
3	0	90	3.5	5	5	3
4	0	90	7.0	10	5	2
5	10	90	7.0	10	10	10
6	10	45	7.0	10	10	9
			+ 3.5	5	9	8
7	10	50	10.0	15	20	20
			DFP (mg./kg.)			
1	0	30	5	1.2	20	20
2	5	0	20	5	10	0
3	0	90	20	5	9	2
4	5	90	20	5	10	10
5	10	90	40	10	24	13
6	10	50	50	12.5	10	9
7	10	50	75	19	10	4

[a] Atropine was always injected 30 min., PAM usually 1–2 min., before the alkyl phosphate. Paraoxon: In experiment 7, PAM was injected 5 min. before Paraoxon; in experiment 6, the survivors got a second dose 1 hour after the first injection. DFP: In experiments 6 and 7, PAM was injected 5 min. before DFP. LD_{50} of Paraoxon = 0.7 mg./kg.; 0.9 mg./kg. is a sure lethal dose. LD_{50} of DFP = 4 mg./kg.; 5 mg./kg. is a sure lethal dose.

These developments have laid the foundation for a rational and efficient treatment of insecticide and nerve gas poisoning. They are illustrations of how studies on the fundamental aspects of nerve activity, on the relationships between metabolic activity and function, on the enzymes and proteins involved, and the molecular forces acting in the proteins, may contribute to the solution of extremely difficult practical problems of toxicology and pharmacodynamics. When it was proposed more than twelve years ago that the toxic effects of DFP must be attributed to a specific biochemical lesion, namely the inhibition of acetylcholinesterase (Nachmansohn and Feld, 1947; Nachmansohn, 1947) this view was vigorously criticized. The complexity of the symptoms seemed to contradict this notion. The inhibition of a single enzyme as an explanation for the great variety of effects seemed an oversimplification to many critics, who, at the International Congress of Physiology in Oxford in 1947 and on many other occasions, rejected this idea (see e.g., Gerard, 1950). But it was this notion and the underlying concepts which eventually resulted in PAM. Since its powerful antidotal effect must be attributed exclusively to the reactivation of acetylcholinesterase, it is now certain that the inhibition of the enzyme is responsible for the toxic effects. The repair of the biochemical lesion by PAM has been, as we will see below, experimentally confirmed. Thus the correctness of the notion has been borne out by the developments.

C. Biochemical and Pharmacological Aspects of PAM Action

For the understanding of the antidotal action of PAM, from the biochemical point of view, the question is pertinent whether a reactivation of phosphorylated acetylcholinesterase can be demonstrated *in vivo* and, if so, to what extent and in what parts of the organism the reaction takes place. Obviously, not all foci will be equally affected and not all are equally essential for life. Information as to where PAM acts effectively and where its action is limited may provide clues as to what organs or centers were exposed but remained unprotected. A better picture of the site of action both of the poison and of the antidote may thus be valuable in devising additional means of counteraction.

Impairment of respiration is one of the earliest symptoms of poisoning by organophosphorus compounds. It precedes the effects on circulation. Paralysis of respiration appears to be the decisive cause of death in all cases of alkyl phosphate poisoning tested (Frawley, Hagan, and Fitzhugh, 1952). There is some difference of opinion whether this respiratory failure is due mostly to a neuromuscular paralysis of the diaphragm or whether central effects on respiratory centers are additional factors. It may be that this factor is a variable. But there is no question that alkyl phosphates have a strong peripheral action on respiration, produce severe bronchial spasms, lung edema, and a paralysis of the diaphragm. The motor end plates of the diaphragm are blocked at a very early stage in many species (Lundholm, 1949). In some species the paralysis of the diaphragm is supposed to be the main factor in the respiratory failure (De Candole *et al.*, 1953). It thus appeared logical to use the diaphragm for testing the effect of PAM on cholinesterase in animals exposed to severe alkyl phosphate poisoning, since this material should be the most sensitive indicator of reactivation *in vivo* and therefore suitable for demonstrating the repair of the biochemical lesion. Such experiments were carried out by Kewitz (1957a) on the diaphragm of mice.

The evaluation of the degree to which the inhibition of acetylcholinesterase actually takes place during the lifetime of the animal after exposure to alkyl phosphates offers considerable difficulties. It has been previously discussed (Chapter IV) that even in isolated nerve fibers which had been exposed to alkyl phosphates, an exact evaluation of the enzyme activity at a given period is quite an intricate undertaking. The complexity increases manifold in dealing with animals exposed to poisoning. There are many pitfalls and it is not surprising that in the early investigations many confusing and contradictory reports appeared in the literature. A procedure widely used was the removal of an organ and the preparation of homogenized suspensions of the tissue. This procedure will not indicate the actual activity in the intact cell at the time of death because portions of the injected alkyl phosphate are localized outside the cell in interstitial tissue. This fraction does not, of course, inhibit the intracellular fraction of cholinesterase during lifetime. But during the preparation of a homogenized suspension, the cell walls are broken down, the inside and outside media are mixed, and an additional inhibition occurs in the test tube. For that reason the observed activities may be considerably lower than those actually present in the living cell and a misleading picture may be obtained unless special precautionary measures are taken. Moreover, some of the alkyl phosphates, such as Paraoxon and DFP, are much more soluble in lipids than in water. Therefore, they accumulate in the lipids outside the cell interior and are not readily removable.

It appears reasonable to assume that the blood stream removes with good efficiency the alkyl phosphates stored extracellularly. This may account for the apparent ready recovery observed in an early phase. But 10–15 hours after the injection this factor will probably no longer play an essential role. Removal of extracellular DFP by washing thin slices with Ringer's solution was found to improve the yield of enzyme activity in tissues examined at an early stage; but even with this procedure the removal was still incomplete (Nachmansohn and Feld, 1947).

It seemed possible that extraction of lipids with chloroform would remove extracellular alkyl phosphates stored in lipids in a more efficient way than the procedure used previously. Applying this method, Kewitz (1957a) indeed found markedly increased enzyme activities during the first 10–15 hours but not after 24 hours. His data support the assumption that the earlier phase of recovery must be attributed to removal of extracellular alkyl phosphate and does not represent a true recovery of the inhibited intracellular enzyme.

For evaluating the fraction of cholinesterase reactivated by PAM it was first necessary to establish the rate of spontaneous recovery of the enzyme inhibited after the injection of alkyl phosphates. Figure 43 shows the results obtained by Kewitz (1957a) when he tested spon-

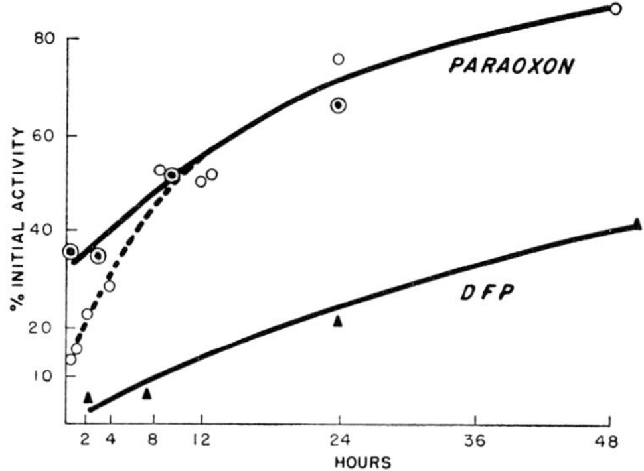

FIG. 43. Spontaneous recovery of esterase activity in the diaphragm of mice after injection of LD_{50} of either Paraoxon or DFP.

The dotted line indicates figures obtained after Paraoxon without using chloroform extraction, whereas this procedure was used for all other determinations within the first 24-hour period. The figures represent the average value. The average value 3 days after DFP injection was 56 per cent of the initial activity; after 5 days, 92 per cent.

taneous recovery in the diaphragm of mice after injection of LD_{50} of Paraoxon; 90 per cent of the original enzyme activity was restored after about 2 days. The inhibition after injections of DFP is stronger, and it takes 5 days to restore 90 per cent of the initial activity.

For the problem of the reactivation of inhibited enzyme by PAM it is necessary to consider a direct reaction between aldoxime and alkyl phosphates in the blood stream or in the tissue. This reaction, however, even *in vitro* is very slow and could, therefore, probably not contribute much to an apparent recovery. However, a direct reaction must be excluded for the demonstration of the reactivation of inhibited enzyme by PAM. In order to avoid interference by a direct reaction between PAM and alkyl phosphate, Kewitz injected PAM into surviving mice only 1 hour after the application of LD_{50} of paraoxon or 1½ hours after the LD_{50} of DFP. Within these periods of time the maximum inhibition of enzyme activity had taken place and the possible reaction of PAM with free alkyl phosphate present would not decrease the enzyme inhibition. The results of Kewitz' experiments are summarized in Table XIII. The

TABLE XIII
EFFECT OF PAM ON CHOLINESTERASE IN DIAPHRAGM OF MICE POISONED BY ALKYL PHOSPHATES. REPAIR OF THE CHEMICAL LESION[a]

Compound injected	Hours after injection	Per cent enzyme activity (average)		Per cent of possible recovery
		Control	+PAM	
Paraoxon (8)	24	70	97	90
DFP (8)	7	7	28	25
DFP (4)	48	40	62	41
DFP (4)	48	40	70	50

[a] The dose used in all experiments was the LD_{50} (0.7 mg./kg. of Paraoxon and 4 mg./kg. of DFP). The figures given are average values based on the number of experiments indicated in brackets. To the animals surviving the injection of Paraoxon, one dose of PAM (90 mg./kg.) was injected 1 hour after the alkyl phosphate. In the first series of experiments with DFP the animals were treated with one dose of PAM (90 mg./kg.) and sacrificed after 7 hours. In the two other series PAM was repeatedly injected, either 75, 50, and 75 mg./kg., 6, 8, and 16 hours after DFP, or 75 mg./kg., 1½, 2, 4, and 14 hours (last row).

significance of this work is the demonstration of the ability of PAM to repair *in vivo* the biochemical lesion produced by the alkyl phosphates.

Exposure to alkyl phosphates may produce, as mentioned before, strong central effects; the activity of acetylcholinesterase in brain has been found to be very low. For this and other reasons it appeared of interest to learn whether PAM has the ability to reactivate the enzyme in the brain *in vivo*. Added to homogenized suspension of brain tissue the activity of alkyl phosphate-inhibited enzyme is, of course, readily restored. PAM is, however, a quaternary ammonium ion. This raises the question of whether it penetrates the blood-brain barrier and, if so, to what extent. Studies on this problem, in which mice again were used as experimental animal, have provided interesting information, although they have not yet given an unequivocal answer (Kewitz and Nachmansohn, 1957; Kewitz, 1957b).

It was again necessary first to establish the rate of spontaneous recovery of the enzyme before testing the effect of PAM. Chloroform extraction was applied as with diaphragm. Figure 44 shows the data obtained with the whole brain of mice after injection of LD_{50} of either Paraoxon or DFP. The values after DFP injection are, as in the diaphragm, much lower than those after injection of Paraoxon. PAM injected in amounts adequate to protect the animals against lethal doses of alkyl phosphate had no significant effect on the level of enzyme activity. However, after injection of LD_{50} or LD_{100} of Paraoxon, a significant increase of enzyme activity (about 25 per cent above that to be expected without PAM) has been obtained when PAM was repeatedly injected in high concentrations: 50mg./kg. five times at intervals of 1 hour. Since no definite effect was observed with smaller concentrations of PAM, the assumption appears justified that PAM may penetrate through the blood-brain barrier, but at a slow rate. Therefore, concentrations adequate to raise the activity of the esterase in the periphery do not reactivate the enzyme in the brain. No reactivation by PAM was

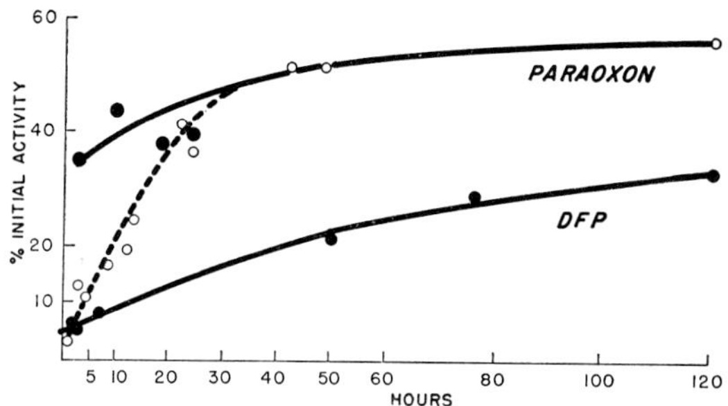

FIG. 44. Spontaneous recovery of esterase activity in brain of mice after injection of LD_{50} of either Paraoxon or DFP.

The dotted line indicates the values obtained after Paraoxon injection without using chloroform extraction, whereas this procedure was used for all other determinations within the first 24-hour period. The figures represent average values. The activities 7 and 15 days following DFP injection were found to be 40 ± 4 per cent and 50 ± 5 per cent, respectively.

observed in similar experiments after injection of DFP. This is in line with the observations *in vitro,* viz., that diisopropylphosphoryl enzyme is much more difficult to reactivate than diethylphosphoryl enzyme.

The efficiency of PAM as an antidote against alkyl phosphate poisoning in the experiments reported thus appeared to be based mostly on its peripheral action. However, in those experiments the whole mouse brain was used for determining the enzyme activity. This is a rather crude procedure. Even in this way reactivation of the enzyme activity of the whole brain did take place with higher amounts and frequent injections of PAM, at least after Paraoxon poisoning. It is known that many drugs selectively affect certain parts of the brain. This may apply to alkyl phosphates and to PAM. Moreover, we know from experience with peripheral nerve fibers that the enzyme concentration compatible with function is rather critical. Therefore, a localized small increase at some vital points, such as respiratory centers, may decide between life and death. A mouse brain is too small for detailed studies of this kind. Furthermore, in experiments which are still unpublished, Bovet and Longo found an action of PAM on the electroencephalogram of rabbits exposed to sarin. Investigations were, therefore, carried out by P. Rosenberg in which the effects of PAM on the esterase activity in different parts of rabbit brains after exposure to Paraoxon were studied. With an improved chloroform extraction procedure he found considerable variations as to the degree to which the esterase activity in various parts of the brain are affected by the antidote (unpublished experiments). PAM caused a significant increase in the cholinesterase level in the cerebral cortex and in the cerebellum and a marked one in medulla and pons including the region of area postrema. Rosenberg's data show that sufficient amounts of PAM are able to penetrate the brain thereby reactivating significant amounts of diethylphosphoryl enzyme. It is apparent that these central effects must play an important role in the antidotal properties of PAM.

The availability of PAM offers a possibility to test the amount of cholinesterase in brain still compatible with life, since multiple lethal doses may be applied without fatal consequences. The previous measurements of the esterase activity in brain with LD_{50}'s of alkyl phosphates were much too low since the procedures applied were inadequate on several accounts. The question was studied in a few preliminary experiments on the brain of mice using the improved techniques (Kewitz and Nachmansohn, 1957). Repeated injections of LD_{50} of DFP did not suppress the enzyme activity to low values. But with a single injection of a tenfold sure lethal dose (50 mg./kg. of DFP), low values of esterase activity were found in the whole brain, about 2 per cent of the initial activity. The significance of this low value is, however, not clear. Within a few hours the figures were several times as high. It appears, therefore, doubtful that the increase in such a short period of time represents a real recovery since DFP inactivation is irreversible and spontaneous recovery, slow. It is more likely that the removal of the huge amount of DFP from brain tissue, which is rich in lipids, was incomplete with the procedure used. This assumption is supported by the results of recent systematic studies on the effect of various solvents on the extraction and enzyme activity by P. Rosenberg (unpublished data).

Because of the wide variety in chemical structure of the toxic alkyl phosphates a great number of different phosphorylated enzymes can be produced. Two types have been previously considered, the diethylphosphoryl enzyme which results from the reaction with Paraoxon, TEPP, diethylfluorophosphonate and others, and the diisopropylphosphoryl enzyme resulting from the reaction with DFP. PAM does not reactivate the various phosphorylated enzymes at the same rate; some of them are not reactivated at all. The phosphoryl enzyme formed by the potent nerve gas sarin, isopropoxymethyl fluorophosphonate, is readily reactivated by PAM, whereas that formed by tabun, dimethylaminoethoxyphosphoryl cyanide, is not at all reactivated. Another compound of some interest is the insecticide octamethylpyrophosphoramide (OMPA). This compound does not inhibit esterases *in vitro,* but it is converted in animal tissue to an effective inhibitor (Du Bois, Doull, and Coon, 1950; Casida, Allen, and Stahmann, 1954). OMPA produces either tetramethyldiaminophosphoryl enzyme, or its amine oxide derivative. The inhibited enzyme is not reactivated by PAM.

If the antidotal action of PAM is based on its ability to reactivate phosphoryl enzyme, it should not be effective against compounds which form a type of phosphoryl enzyme which cannot be reactivated *in vitro.* This proved to be the case. PAM in combination with atropine is a powerful antidote against sarin. No protecting effect is obtained against OMPA. In the case of tabun a weak but definite protection is observed; however, this slight antidotal effect may be attributed to a direct reaction between PAM and tabun in the blood stream, since this reaction takes place at a much faster rate than with the other alkyl phosphates tested (Kewitz, Wilson, and Nachmansohn, 1957; Wilson and Sondheimer, 1957; Kewitz, 1957b).

The early experiments with PAM were all carried out with mice. The sensitivity of different species toward alkyl phosphates shows great variations. Mice are, for instance, particularly sensitive to sarin. In the last few years PAM has been tested by many investigators as an antidote against alkyl phosphate poisoning on a great number of species in combination with atropine and other types of treatment. With many species much more spectacular effects were obtained than with mice. In experiments with guinea pigs, Bethe, Erdmann, Lendle, and Schmidt (1957) found a strong protection, especially when the use of PAM and atropine was combined with artificial respiration. The 160 times lethal dose of paraoxon and the 20 times lethal dose of DFP were tolerated. The improvement of the effect by artificial respiration is attributed by the authors to the beneficial action on the central respiratory centers, since PAM and atropine do not sufficiently counteract the central effects. Cats treated with PAM and atropine are protected against twentyfold lethal doses of sarin (Wills, Kunkel, Brown, and Groblewski, 1957).

A further increase of the antidotal power of PAM against sarin has been obtained by the use of bis quaternary derivatives (Poziomek, Hackley, and Steinberg, 1958). A systematic study on the reactivating power of diquaternary derivatives of 2- and 4-PAM *in vitro* was carried out by Wilson and Ginsburg (1958). Reactivation of acetylcholinesterase inhibited by either TEPP or DFP was determined. Although 2-PAM is much more active than 4-PAM, the bis quaternary derivatives of the latter were considerably more active in regard to DFP inhibition and slightly more active in regard to TEPP inhibition. The bis compound of 4-PAM containing five methylene groups between the two rings was, for instance, a thousand times better than 4-PAM for reactivating DFP inhibited enzyme. The data suggest that the extra binding by the second quaternary group has altered and obviously improved the binding orientation of the 4-PAM. In the case of 2-PAM the derivatives are poorer except those with five and six methylene groups.

Sarin produces quite marked central effects. Since PAM penetrates the brain apparently quite slowly, it appeared interesting to test whether a lipid-soluble analog of PAM would be able to reach with greater ease the central nervous system and thus become a still more effective antidote. If the methyl group on the nitrogen of the pyridine ring is replaced by a dodecyl group, the compound becomes lipid soluble: pyridine aldoxime dodeciodide (PAD). By a combination of PAD, PAM, and atropine it is possible to improve significantly the protection of mice against sarin (Wilson, 1958). Whether the compound in the small amounts used actually does penetrate the brain has not been ascertained. The improvement may just be due to an increased range of peripheral action due to the better lipid solubility of PAD. However, lipid-soluble quaternary ammonium ions are quite toxic, in contrast to the remarkably small toxicity of PAM (D. Bovet, personal communication). It is, therefore, an open question whether these compounds are of practical value.

PAM was first tried by Namba and Hiraki (1958) on humans suffering from acute alkyl phosphate poisoning in Japan. In that country alkyl phosphates, particularly Parathion, have been widely used to protect the rice crop which is of vital importance there. The number of victims estimated for a period of five years seems to be over 6000. Victims with severe symptoms of poisoning were treated by intravenous injections of 1 gm. PAM. The clinical effects were striking. The patients recovered promptly. Determinations of red cell and serum esterase revealed that the activity of the former was immediately restored; the effect on the serum esterase was small and transient. A dramatic case of severe acute poisoning and successful treatment with PAM and atropine was recently reported by Erdmann, Sakai, and Scheler (1958).

In view of the great practical and theoretical interest of organophosphorus compounds there exists a huge literature on their chemistry, biochemistry, pharmacology, and toxicology. Although PAM was developed only about four years ago, the volume of work done and published on this antidote is already considerable. Only a few selected aspects have been discussed. The reader interested in this field is referred to the books and special reviews dealing with this problem (see, for instance, Schrader, 1952; Koelle and Gilman, 1949; Gunther and Blinn, 1955; Erdmann and Lendle, 1958).

CHAPTER IX

Properties of Choline Acetylase

A. Test System

In contrast to acetylcholinesterase, which does not depend for the hydrolytic process on either a coenzyme or on specific ions, choline acetylase requires for full activity, as we have seen, a complex system. Before discussing some characteristic properties of this enzyme, it appears desirable to describe the conditions of the test system used in the experiments.

In the cell-free extracts, in which synthesis of acetylcholine was first accomplished (Nachmansohn and Machado, 1943), the enzymatic activity was very unstable. Moreover, in view of the presence of adenosinetriphosphatase, a satisfactory and reproducible rate of acetylcholine formation became possible only by the addition of sodium fluoride, which inhibits the activity of ATPase. But even under these conditions, the rate of formation decreased rapidly and leveled off after 30 min. In the following year, however, extracts were prepared from acetone-dried powder of brains of rats and guinea pigs in which the enzyme system had a fair degree of stability (Nachmansohn and John, 1944, 1945a). Nevertheless, the amounts formed were still rather small, even when, since 1945, coenzyme was added to the reaction mixture: about 0.05–0.1 μmole per milligram of protein per hour (2–3 mg. of acetylcholine per gram powder per hour) (Nachmansohn and Berman, 1946). The activity was tested by bioassay, which for many reasons is not quite satisfactory for enzyme studies.

The situation was improved by several developments. In 1940 it was found that head ganglia of squid have a very high concentration of acetylcholinesterase: 1 gm. of tissue (fresh weight) is capable of hydrolyzing 3 gm. of acetylcholine per hour (Nachmansohn and Bettina Meyerhof, 1941). Assuming that this tissue has a high rate of acetylcholine metabolism and might therefore be a good source of choline acetylase, an acetone-dried powder was prepared from this material. A very high activity in extracts was found: about 5–10 μmoles of acetylcholine were formed per milligram protein (Nachmansohn and Weiss, 1948). The specific activity was 100–200 times as high as in the extracts prepared from rat, rabbit, or guinea pig brains. The tissue appeared then to be a good starting material for purification, a necessary prerequisite for the study of enzyme properties. Electric tissue may also be a good source, and the presence of the enzyme there was found in the very first observations in 1943. However, attempts to obtain active extracts from acetone-dried powder were not successful.

Another advance was the elaboration of a simple and sensitive colorimetric method for determination of acetylcholine, which was worked out by Hestrin (1949a) when he was working in the writer's laboratory. The procedure is based upon the reaction of hydroxylamine with *O*-acyl derivatives to form hydroxamic acids, a reaction known for more than half a century. All hydroxamic acids give a red or red-purple color with ferric chloride in acid solution. With the ferric ion a soluble inner complex salt is formed. These reactions have been discussed by Feigl, Auger, and Frehden (1934, Feigl, 1946) and used for spot tests under proper conditions with carboxylic acids, esters, and anhydrides. On the basis of this reaction, Lipmann and Tuttle (1945) have worked out a quantitative method for acyl phosphates and Hestrin for *O*-acyl esters. Hereby, a convenient analytical procedure became available for testing acetylcholine formation. It may also, of course, be applied for determination of acetylcholinesterase activity, and we have indeed extensively used the method under conditions in which manometric methods are inconvenient or inadequate.

A prerequisite of efficient purification was the separation of the two enzymatic steps of acetylation, discussed in Chapter V: the formation of acetyl CoA and the subsequent transfer of the acetyl group to the acceptor. Both the purified phosphotransacetylase obtained from *Escherichia coli* with acetyl phosphate as energy source and ATP plus the acetate-activating enzyme purified from extracts of acetone-dried powder of pigeon liver proved to be extremely active in forming acetylcholine when added to CoA and choline acetylase (Korkes, del Campillo, Korey, Stern, Nachmansohn, and Ochoa, 1952a).

By all these advances it became possible to obtain from squid ganglia an enzyme preparation with an adequate degree of purity (Berman, Wilson, and Nachmansohn, 1953). The enzyme preparation had a specific activity of about 40–80 μmoles per milligram protein per hour. The usual purification procedures were applied: fractionation with ammonium sulfate, treatment with protamine sulfate, adsorption by calcium gel, and subsequent elution. Although the degree of purity is not very high, it proved sufficient for the information desired, especially after the removal of deacylase which interfered with some of the tests (Berman-Reisberg, 1957).

Whereas the phosphotransacetylase system was used for the purification, for the study of specificity it was, of course, necessary to have choline acetylase as the only active enzyme in the reaction mixture. This became possible by the availability of reasonably pure CoA and the elaboration of methods to prepare acetyl CoA nonenzymatically (Wilson, 1952c; Simon and Shemin, 1953). Whereas the rate of formation with the phosphotransacetylase system is linear, that with acetyl CoA is not, except for relatively short periods of time (Fig. 45) (Berman-Reisberg, 1957). The reasons are not quite clear, but the system is nevertheless satisfactory if used for short periods of time only. The stability of the enzyme is quite satisfactory; it remains stable for 6–8 weeks when kept in the deep-freeze. The dissociation constant with

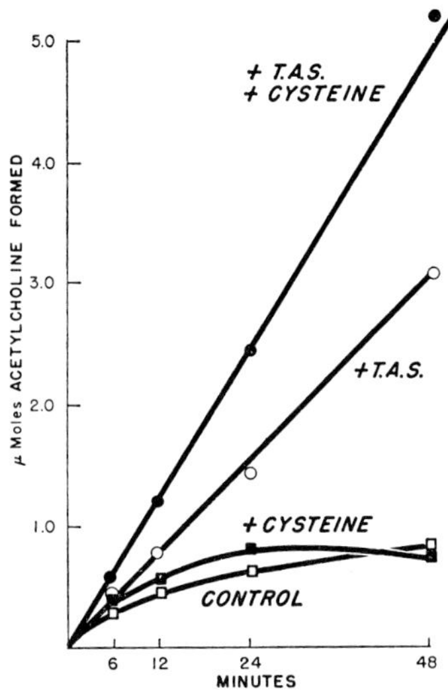

FIG. 45. Rates of acetylcholine synthesis by choline acetylase in presence and in absence of phosphotransacetylase.

The control reaction mixture contains in addition to choline acetylase and choline only acetyl CoA. The transacetylase system (T.A.S.) contains (instead of acetyl CoA) acetyl phosphate, CoA, and magnesium. The rates of the two systems are shown with and without addition of cysteine.

choline was found to have a value of 5×10^{-4} M, with acetyl CoA of 1.6×10^{-3} M. The dissociation constant between choline and choline acetylase is of the same order of magnitude as between acetylcholine and acetylcholinesterase.

The enzyme activity is readily blocked by compounds reacting with SH groups, as shown in Table XIV (Berman-Reisberg, 1954). The sensitivity of the enzyme system toward compounds reacting with SH groups was found in the very first observations (Nachmansohn and Machado, 1943); however, after the functional group of CoA was identified as an SH group (Lynen, Reichert, and Rueff, 1951) a reexamination became necessary to ascertain whether the enzyme itself has an active SH group. According to the new data, the enzyme must be classified among the enzymes which have functional SH groups. Addition of ethylenediamine tetraacetate and cysteine is necessary for optimal rates and enzyme stability.

TABLE XIV

EFFECT OF SH INHIBITION ON THE ACTIVITY OF CHOLINE ACETYLASE[a]

Compound	Concentration (μmoles/ml.)	Per cent inhibition
Iodoacetate	0.5	36
	1.5	63
	4.5	78
p-Chloromercuribenzoate	0.002	40
	0.01	69
	0.04	96
Iodosobenzoate	0.01	50
	0.1	69
$CuSO_4$	0.0025	38
	0.005	60
	0.01	84

[a] The enzyme used was a purified preparation obtained from acetone-dried powder of squid head ganglia. The test mixture contained, in micromoles per milliliter, the following components: choline chloride 20; acetyl CoA 3.5; potassium phosphate buffer 100, pH 7; TEPP 0.13. Inhibitor as indicated. Enzyme (0.01 ml.) containing 150 μg. protein was added to the test mixture. The final volume was 0.5 ml. When iodoacetate was the inhibitor, the test mixture contained in addition 4 μmoles/ml. Versene (ethylenediamine tetraacetate); in the case of *p*-chloromercuribenzoate and iodosobenzoate, 2 μmoles/ml.; no Versene in the case of $CuSO_4$. The acetylcholine formed was determined by bioassay. The incubation time was 10 min. The activity of the control was about 80 μg. of acetylcholine formed during that period.

B. SPECIFICITY

The specificity of choline acetylase has been tested with the purified enzyme preparation in reaction mixtures which contained, in addition to the enzyme, acyl derivatives of CoA and ethanolamine, with one, two, or three methyl groups on the nitrogen, as substrates. The interesting and pertinent question of these studies is that of whether the protein has some similarities with the characteristics of acetylcholinesterase, and to what extent.

Although the analysis of the reaction between the protein and small molecules has not been carried as far as that of the esterase, some pertinent similarities have been found. Virtually no difference exists as to the rate of synthesis between acetyl and propionyl CoA substrates. Butyryl CoA, on the other hand, is a very poor substrate; the rate of synthesis is only about 10 per cent of that of acetyl CoA as substrate. The parallelism with acetylcholinesterase is apparent. It may be mentioned that butyrylcholine has been found in brain (Holtz and Schueman, 1954). Although the synthesis of this ester is markedly slower than that with acyl groups containing two and three carbons, it must be kept in mind, in considering possible functional relationships, that the rate of hydrolysis is quite low, in fact less than 1 per cent of that of acetylcholine.

Of particular interest, however, is the behavior of the protein toward the nitrogen group when the number of methyl groups is decreased, as will be discussed in the next chapter. The dimethyl group is acetylated at a rate which is only 8 per cent of that of choline. Still lower is the rate of acetylation of the monomethyl derivative; in this case the rate of formation is only 2 per cent compared with that of choline. The difference between the mono- and the dimethyl derivatives may possibly be attributed to better binding by van der Waals' forces, but the strong effect of the third methyl group requires another explanation.

Experiments with a few quaternary ammonium ions, testing the competitive action with choline, revealed a rather weak inhibitory effect of the compounds tested. To obtain a 50 per cent inhibition of enzyme activity by tetramethyl ammonium ions, a concentration of 30–40 μmoles per milliliter is required. The tertiary analog is even a weaker inhibitor. The inhibitory power of acetylcholine itself was found to be even smaller than that of the tetramethyl ammonium ion. Prostigmine is a very poor inhibitor: 80 μmoles per milliliter inhibit the activity by only 50 per cent. Its tertiary analog, on the other hand, becomes a relatively potent inhibitor on preincubation: 5 μmoles per milliliter inhibit 60–70 per cent after 15–30 min. incubation. This is in contrast to the inhibitory action of these two compounds on cholinesterase. On that enzyme, the quaternary analog acts one hundred times more strongly than the tertiary compound. Apparently, there is an interaction of the tertiary base with the acyl site, since on preincubation acetyl CoA protects against the inhibitory effect.

No increase of inhibitory power was observed with diquaternary compounds such as curare, decamethonium, or succinylcholine, compared to monoquaternary nitrogen derivatives. Apparently, there is no second negative charge on the protein located at the proper distance. Inhibition by *d*-tubocurarine increases on preincubation.

Benzoyl CoA in a concentration of 0.4 μmole per milliliter was found to compete, after incubation, with the acyl group. The inhibitory effect of sodium hippurate was small (Berman-Reisberg, 1957).

CHAPTER X

Action of Acetylcholine on the Receptor in Intact Cells

A. Difference between Tertiary and Quaternary Nitrogen Derivatives in Their Reactions with Esterase and Acetylase

An illustration of how the analysis of molecular forces of the proteins in solution has been helpful in interpreting relationships between chemical forces and electrical activity is the information obtained in studies on the receptor.

From the analysis of the role of van der Waals' forces involved in the formation of the enzyme substrate complex the remarkable fact had emerged that if one substitutes the protons of an ammonium ion by methyl groups, each of the first three methyl groups increases the *binding* by a factor of about 7, while the fourth group is without effect. Similar results are obtained with the hydroxyethyl ammonium ion: the fourth alkyl (third methyl group) is without effect (Wilson, 1952a).

In striking contrast the difference in *enzymatic activity* is extremely marked between tertiary and quaternary nitrogen derivatives; the rate of formation of acetyl enzyme by acetylcholinesterase was found by Wilson and Cabib (1956) to be about ten times higher with acetylcholine as the substrate than with its tertiary analog, dimethylaminoethyl acetate. We have just seen that a similar difference is observed in the activity of choline acetylase toward choline and its tertiary analog, dimethylethanolamine (Berman, Wilson, and Nachmansohn, 1953; Berman-Reisberg, 1957).

How can one explain these striking differences in enzyme activity due to the presence of one extra methyl group? The binding forces, as we have seen, are not increased. The quaternary nitrogen is a saturated group and is less reactive than a tertiary nitrogen because at neutral pH the latter is in equilibrium with a small amount of conjugated base which has a free pair of electrons. Chemical reactivity then cannot be the answer and another explanation must be found. A clue may be the tetrahedral structure of the quaternary nitrogen group. Such a structure is more or less spherical. If such a molecule is attracted to a protein surface, direct contact of the fourth alkyl will not be possible since it is located in the direction of the solution away from the protein. One way in which the protein could have simultaneous contact with all the methyl groups would be by enveloping the interacting molecule. This implies a change of configuration of the protein during its active state in the enzymatic process.

This possibility has recently found some experimental support. Wilson and Cabib (1956) studied the enthalpies and entropies of activation, $\Delta H^{\ddagger}$ and $\Delta S^{\ddagger}$, of the ester of ethanolamine and its methylated derivatives, using purified acetylcholinesterase from electric tissue. Substitution of the first two protons by methyl groups produced little change in the activation energies. But the extra methyl group has a very pronounced effect. The enthalpy of activation, $\Delta H^{\ddagger}$, of the hydrolysis of acetylcholine is about 14,000 cal. as compared with about 8000 cal. for that of the tertiary analog. On the basis of this finding one could expect that the hydrolysis of the quaternary ester would be less favored than that of the tertiary. But the entropy of the activation, $\Delta S^{\ddagger}$, is extremely favorable for the quaternary compared to that of the tertiary compound. The value for the tertiary is —7 to —9 entropy units, whereas $\Delta S^{\ddagger}$ is strongly positive with the quaternary as substrate, about +15 to +25 entropy units. This extraordinary difference of the entropy of activation produced by the presence of the extra methyl group can be explained most simply in terms of a rearrangement of the protein molecule, i.e., a change in configuration.

B. Evidence for the Existence of a Receptor

The information obtained with the enzymes in solution proved useful for interpreting the interaction of acetylcholine with the receptor. A change of configuration of proteins with a rearrangement of acidic and basic groups was proposed by Kurt H. Meyer (1937) as the factor which is responsible for the change of permeability to ions during conduction. Does acetylcholine produce such an effect when reacting with the receptor?

The existence of a receptor has been postulated on the basis of physiological and pharmacological observations since the time of Langley (1907). An experimental distinction between compounds acting predominantly on the receptor and those acting on the enzyme has been achieved by a combination of two methods. The result permitted the study of properties of the receptor in the intact cell. Pieces of electric tissue containing one to three rows of electroplax were isolated (Fig. 46). Microelectrodes were inserted into the interior of a cell; the other electrode was inserted into the tissue on the innervated outside of the same electroplax (Fig. 47). This arrangement permits direct measurements of the potentials across resting and active membranes and in this

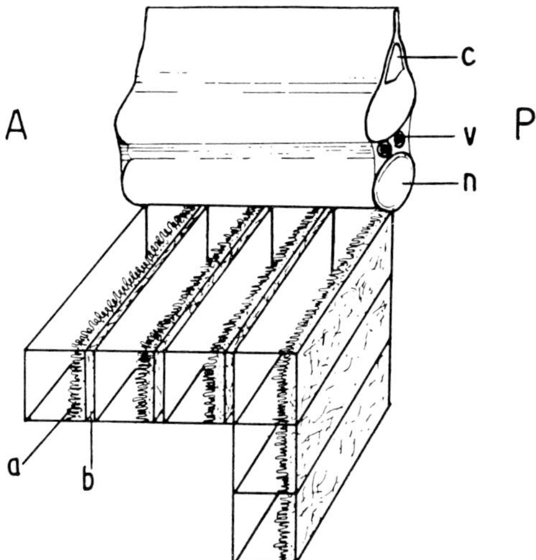

FIG. 46. Diagram of a fragment of electric organ (bundle of Sachs) of *Electrophorus electricus.*

The drawing shows the relationship between electroplax and compartment and the position of the conducting (caudal face: *b*) and nonconducting (rostral face: *a*) membranes. Although the drawing is not in scale, it shows that the ratio extracellular over intracellular space is high. *A,* anterior (rostral) end; *P,* posterior (caudal) end; *n,* swim bladder; *v,* dorsal blood vessel; *c,* spinal cord.

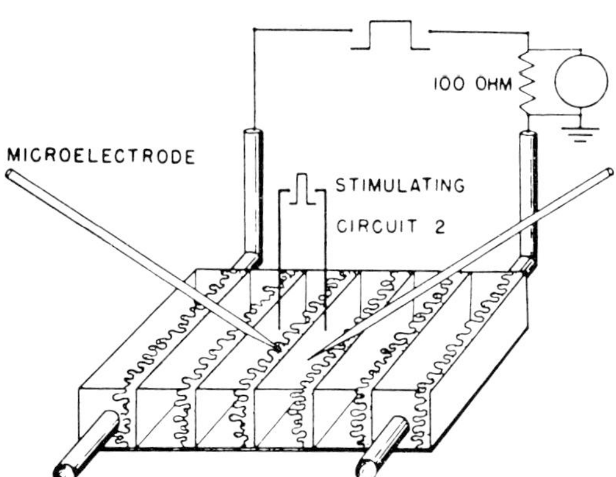

FIG. 47. Diagram of a monolayer of electroplax with stimulating and recording arrangement.

The electrical activity was recorded by means of KCl-filled micropipettes. One microelectrode is inserted into the interior of an electroplax though the non-innervated membrane, the tip of the other microcelectrode is placed just outside the innervated membrane in front of the tip of the internal electrode.

way one can study how compounds acting on the acetylcholine system affect these potentials. At the same time the esterase activity in the intact electroplax has been determined, using ethyl chloroaectate as substrate (Schleyer, 1955). Neither acetylcholine nor dimethyl amino-ethyl acetate were found to be adequate substrates, since their penetration into the tissue is too slow. The discrepancy between "true" and "apparent" Michaelis constant, determined by comparison of the enzyme activity of intact and ground tissue is too great to permit an evaluation of the esterase activity in the intact cell. Ethyl chloroacetate, on the other hand, is a fairly good substrate for acetylcholinesterase and permits estimates of the enzyme activity in the intact cell. There is still a discrepancy of about 30 per cent between true and apparent Michaelis constant. However, use of this substrate made possible the demonstration that some compounds may block conduction while having almost no effect on the esterase (Altamirano, Schleyer, Coates, and Nachmansohn, 1955; Nachmansohn, 1954, 1955a, b, c). Other compounds block only if the esterase activity has fallen to very low values; but in the latter case an estimate of the actual level of remaining activity is not possible owing to the inability of the substrate to reveal the total enzyme activity of the intact cell. Table XV gives a few data. Compounds such as carbamyl-

TABLE XV

RELATIONSHIP BETWEEN BLOCK OF ELECTRICAL AND CHOLINESTERASE ACTIVITY

Compound	Blocking concentration (μmoles/ml.)	Concentration tested (umoles/ml.)	Enzyme activity in % of initial
Group A[a]			
Carbamylcholine	0.05	0.05	99
		10.0	83
Decamethonium	0.05	0.05	86
		10.0	60
Procaine	1.0	2.0	91
		40.0	51
Group B[b]			
Eserine	2.0	2.0	28
Tertiary analog of Prostigmine	2.0	2.0	9
		4.0	4
DFP	2.0	1.5	4

[a] Group A: Compounds which block response to direct stimulation at concentrations which have either a small effect or none on the esterase activity.

[b] Group B: Compounds which block response to direct stimulation with a parallel inhibition of esterase activity.

choline, decamethonium, and procaine block conduction without markedly decreasing the enzyme activity. These compounds even at a 200-fold concentration above that producing block still leave the enzyme activity at a high level. Consequently, the blocking action cannot be attributed to the effect on the esterase, but to that upon a similar cell constituent, the long-postulated receptor. On the other hand, with compounds such as DFP, and the tertiary analog of Prostigmine, the enzyme activity is at a low level when electrical activity ceases, indicating that in these cases there is a strong action upon the enzyme and that the level of enzyme activity has probably fallen to a degree incompatible with electrical activity. Prostigmine blocks electrical activity at a level of enzyme activity (51 per cent of the initial) which must definitely be attributed to its action on the receptor rather than to that on the enzyme. This applies also to eserine: when conduction ceased the enzyme activity was found to be 28 per cent of the initial. Although this figure may appear low, we have to keep in mind that the measured activity does not represent the total cellular activity in view of the discrepancy between the rate of hydrolysis in intact and ground tissue; the actual concentration of the enzyme still left inside the cell must have been higher than the figure indicates. It appears, therefore, unlikely that the low enzyme activity was the factor responsible for block of electrical activity; the data suggest that eserine acts on the receptor as well as on the enzyme.

If the acetylcholine system is essential for the generation of the electric currents, all the protein members must be localized in close vicinity in a molecular layer of about 100 A. thickness. Such structural organization would make possible a rapid and efficient action requiring extremely small amounts of the ester. But, for the same reason, all compounds having structural features resembling those of acetylcholine and penetrating into this molecular layer should react with other members of the system. However, the binding constants, i.e., the affinities, may vary greatly. It is known, for instance, that the K_M, the dissociation constant, of acetylcholine as well as of the carbamylcholine-enzyme complex is about 10^{-4}, whereas their dissociation constants with the receptor are about 10^{-7}. With other substances, the constants vary in the opposite direction. One might, therefore, have predicted—as has now been experimentally demonstrated—that some substances will act preferentially on the receptor, others on the enzyme. It appears likely that in many cases both proteins are affected, although to different extents. The observations, of course, did not demonstrate the existence of two different proteins; they were evidence only for two different sites of action. Physiologically, this distinction was of great importance. Since then the receptor protein has been isolated and identified by Ehrenpreis (1959 a, b) (see next chapter).

C. Receptor Activators and Inhibitors

The combination of the two methods described above permits one to test the action on the electric response of the electroplax of those compounds which act predominantly on the receptor and to determine whether the extra methyl group in quaternary compounds acting on the receptor also causes in this case a special effect compared with that of their tertiary analogs; such tests should give a clue whether or not the receptor shows some analogy with the behavior of the enzyme proteins in solution. This proved to be the case. The compounds acting upon electrical activity may be divided into two distinctly different types: those which block conduction without depolarization, and those which block and simultaneously depolarize the membrane. In one case the barrier for ion movements remains closed when activity stops, in the other it remains open. Quaternary compounds such as acetylcholine, carbamylcholine, decamethonium, etc., belong to the latter. Figure 48 shows the effect of carbamylcholine and is typical for compounds blocking the response with simultaneous depolarization. On the other hand, tertiary acetylcholine analogs such as procaine and eserine block the activity but do not depolarize. An interesting example of how the effect on electrical activity may depend on the presence of the extra methyl group is the difference between the effect of Prostigmine (Fig. 49) and that of its tertiary analog (Fig. 50), although in this case there is an additional charge effect. There is no depolarization in the absence of the extra methyl group (Altamirano, Schleyer, Coates, and Nachmansohn, 1955; Nachmansohn, 1955 a, b, c).

If one associates the transient change of permeability in conducting tissues with the reaction between acetylcholine and the receptor, one must assume that acetylcholine not only combines with the receptor, but produces a simultaneous change. This was postulated and lucidly explained by Clark (1937) some twenty years ago. We have, therefore, introduced the distinction between receptor activators which effect this change, and receptor inhibitors which combine with the receptor but do not produce a change. The latter apparently block the access of acetylcholine to the active surface. These two different types of interaction are analogous to those in enzyme chemistry where we distinguish between enzyme substrates and inhibitors.

If one accepts the idea that the molecular forces acting between acetylcholine and the proteins associated with its function are more or less similar, one may visualize that the receptor in the active state undergoes a change in configuration comparable to that which has been suggested to occur in the enzyme during activity. Such an action may well be associated with the change of permeability taking place in the active membrane. One possible picture would be that positively charged amino

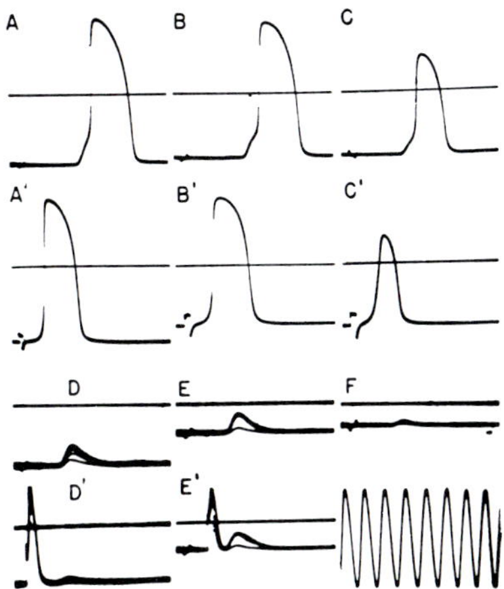

FIG. 48. Effect of carbamylcholine on the resting potential and the action current of an electroplax.

In this and in Figs. 49 and 50 the arrangement was as follows: The electrical activity was recorded with a cathode ray oscilloscope by means of two microelectrodes as shown in the diagram of Fig. 47. The straight upper horizontal line in the oscilloscope corresponds to zero potential difference between the recording electrodes when both are just outside the electroplax. When one electrode is inside the cell, the second line appears in the oscilloscope. The value of the resting potential is measured by the distance between two lines; it is usually about 80–82 millivolts. During the discharge there is an "overshoot," the potential difference between the inside and the outside does not merely disappear, but is reversed. *A–F*, neural; *A′–E′*, direct stimulation of the electroplax. In the former there is a longer latency period and the response shows the synaptic potential. *A,A′*: Control. *B,B′*: 6 min. after the addition of 10 μg./ml. of carbamylcholine to the Ringer's solution. The resting potential and the spike are already slightly decreased. *C,C′*: after 18 min. *D,D′*: after 23 min. Only postsynaptic potentials are obtained. *E.E′*: after 35 min. *F*: after 53 min. The records show that carbamylcholine not only blocks the propagated spike, but decreases the potential difference in resting condition, i.e., has a strong depolarizing action. Calibration for all recordings: 1000 cycles and 100 millivolts.

groups would prevent the passage of Na^+. A small, even very limited but strategically located, change in a long protein chain may alter the position of the positive charges by a few Angstroms and thereby open the passage. Folding or unfolding of a small part of a protein chain, helical or nonhelical, may be enough for the trigger action. The action of one molecule of acetylcholine may permit, as we have seen, the passage of about 500–1000 Na^+. This is a relationship of an order of magnitude consistent with that which one would expect for a trigger process in which the controlling energy required must be small.

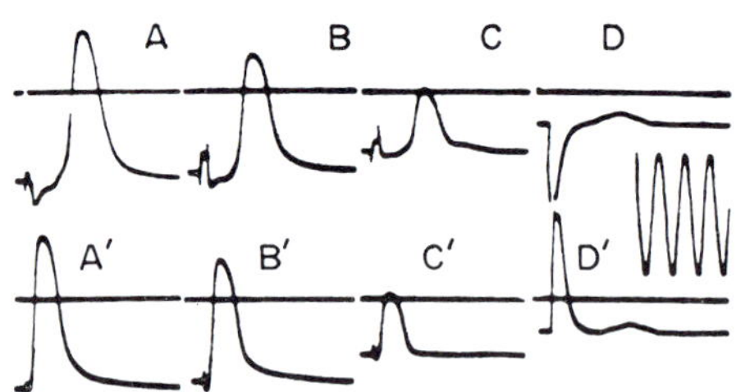

FIG. 49. Effect of Prostigmine on the resting potential and the action current of an electroplax.

A–D: neural, *A′–D′*: direct stimulation. *A,A′*: Control. *B,B′*: 1 min. after addition of 8 μmoles/ml. of Prostigmine to the Ringer's solution. *C,C′*: after 2 min.; the small spikes were the last obtained. *D,D′*: after 9 min.; resting potential strongly reduced. See Fig. 48 legend for details of arrangement.

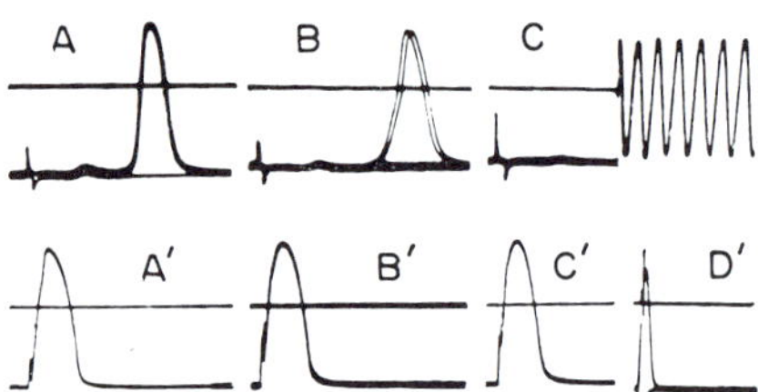

FIG. 50. Effect of the tertiary analog of Prostigmine on the resting potential and the action current of an electroplax.

A–C: neural, *A′–D′*: direct stimulation. *A,A′*: Control. *B,B′*: 2 min. after addition of 3 μmoles/ml. of the compound to the Ringer's solution. *C,C′*: 7½ min. afterward. The propagated spike was blocked after 15 min. *D′*: after 176 min. Resting potential same as in the control, in spite of the addition of 10 μg./ml. of carbamylcholine 124 min. before the recording of *D′*. See Fig. 48 legend for details of arrangement.

Since both the receptor inhibitor and activator must react with the same receptor site, it follows that prior application of an inhibitor should prevent the activation of the receptor by the subsequent addition of an activator. This postulate has been verified by the finding that depolarization by carbamylcholine, a receptor activator, is antagonized by either procaine, eserine, *d*-tubocurarine or the tertiary analog of Prostigmine. Since the receptor inhibitors and activators are competitive, the effect of one should be overcome by increased concentrations of the other, and this indeed proved to be the case (Schoffeniels and Nachmansohn, 1957).

The features required for a molecule to be a receptor activator are still poorly understood. If we refer again to the knowledge acquired by enzyme chemistry, we know that enzyme inhibitors require much less specificity than substrates. We have seen that if we compare, for instance, the rates of hydrolysis of acetyl- and propionylcholine by acetylcholinesterase, there is either little difference or none. Butyrylcholine, on the other hand, is very poorly hydrolyzed although it is bound twice as strongly to the enzyme as is acetylcholine. No information is available to explain why the addition of one methyl group should so effectively depress the enzyme activity, in spite of the better binding; steric hindrance may be a factor.

The methylated quaternary group seems to be an important factor in promoting receptor activation; the carbonyl group is quite evidently another one. Moreover, as in the case of the two enzymes of the system where the tertiary analogs are much poorer substrates by a factor of 10 and 15, respectively, dimethylaminoethyl acetate is also a receptor activator, but only at markedly higher concentrations; in the case of the electroplax the effective concentration is about twenty times higher, whereas in some muscles it is much higher (by a factor of 100 or more) than that of acetylcholine, as has long been known. Of course, the nitrogen of this tertiary analog is cationic since at the usual pH it accepts a proton, its pK being high. It is also well known that nicotine, although it has no quaternary nitrogen, has a strong depolarizing action, i.e., is a powerful activator. On the other hand *d*-tubocurarine, although containing two quaternary nitrogens, is a receptor inhibitor. In this case the reason is not difficult to understand. The molecule is rather large; it contains six rings and the two nitrogens are members of heterocyclic rings. The compounds with curarelike action will be discussed in more detail in Chapter XIV.

The similarities between receptor and enzyme offer important physiological and pharmacodynamic problems. Of special interest are certain observations with organophosphorus compounds. In the earlier phase of the development two findings in addition to those discussed in Chapter IV were considered as serious difficulties for the theory proposed: block of conduction produced by DFP is, for quite some time, readily reversible. Since the inhibition of cholinesterase is irreversible, this phenomenon appeared puzzling and was considered to be evidence for a general toxic effect rather than a specific effect on the acetylcholine system. However, only the reaction with the enzyme leads to the irreversible phosphorylated protein formed by the elimination of the acidic group attached to the phosphorus atom. If we assume, as appears possible, that DFP also affects the receptor as an inhibitor, one should expect a reversible block of conduction as long as the enzyme has not reached a critically low level. In the case of DFP this takes quite some time since the inactivation of the enzyme is a relatively slow process. The experimental findings described in Chapter IV are in agreement with expectation. The second finding which appeared difficult to explain was the absence of a depolarizing effect for a certain period of time after conduction had been blocked. This was reported by Toman, Woodbury, and Woodbury (1947) in experiments with frog sciatic nerve, and has been considered to be another difficulty for the theory. This finding has been confirmed with the electroplax (Altamirano, Schleyer, Coates, and Nachmansohn, 1955). Since DFP has no nitrogen group, its combination with the receptor would not be expected to effect an activation (Nachmansohn, 1955b, c). The subsequent depolarization, observed both in nerve fiber and in electroplax, may be the result of acetylcholine accumulation due to block of its hydrolysis. Although this explanation seems to be a good possibility, another explanation cannot be excluded at present: DFP is highly lipid soluble; it may react in a readily reversible process with some lipid components of the conducting membrane, thereby blocking conduction without depolarization. Only the subsequent irreversible inhibition of the enzyme leads to the persistence of acetylcholine and thus to depolarization.

CHAPTER XI

Isolated Single Electroplax Preparation

The results discussed in the preceding chapter were obtained with isolated pieces of electric tissue consisting of one to three rows of electroplax. Studies with these multicellular preparations frequently offered considerable difficulties for the interpretation of the results due to the complexity of the preparation. A multicellular preparation containing several types of membranes makes it impossible, in studies involving permeability problems, to distinguish between the different types of membranes affected. The large extracellular spaces and extracellular structural barriers preclude, for instance, a study of ion flux and an evaluation of rates across the conducting membrane. Several results appeared contradictory; some observations indicated that the experimental conditions were unsatisfactory. So, for instance, all the effects obtained with acetylcholine itself or analogous chemical structures were irreversible, in contrast to their action on other excitable structures.

For these and other reasons it appeared desirable to devise a method in which a single cellular unit could be used for the study of the properties of the cellular membrane. Preparations of single cellular units have frequently been of paramount importance for the analysis of cellular function.

A. Method of Separating Two Pools of Fluid by a Single Electroplax

In the last three years Schoffeniels (1957a, b, 1958a, b, 1959; Schoffeniels and Nachmansohn, 1957) succeeded in developing a method in which he used a single isolated electroplax. The new preparation has greatly contributed to overcome previous difficulties; it has provided valuable information and has proved to be a promising tool for further studies.

The electroplax is particularly favorable for a unit cell preparation for several reasons. The cell is a syncytium of rather large dimensions, suitable for a monocellular preparation. At the level of the organ of Sachs the compartments are more or less rectangular in shape. The electroplax there, in medium-sized specimens (80–110-cm. lengths), is about 5–10 mm. long, about 1–1.5 mm. high, and about 0.3–0.5 mm. thick. The extracellular space forms about 95 per cent of the compartment, and this facilitates the dissection of a single unit.

The following procedure has been developed by Schoffeniels (1957b, 1959; Schoffeniels and Nachmansohn, 1957). A single electroplax is dissected and mounted between two chambers so that the cell separates two pools of fluid. The cell is kept between a nylon sheet, with a window adjusted to the dimensions of the cell, and a grid consisting of nylon threads. Figure 51 shows the respective positions of the sheet of nylon, one of the two cells, and the grid; Fig. 52 is a cross section of one of the two chambers; and Fig. 53, a photograph of the chambers. A more de-

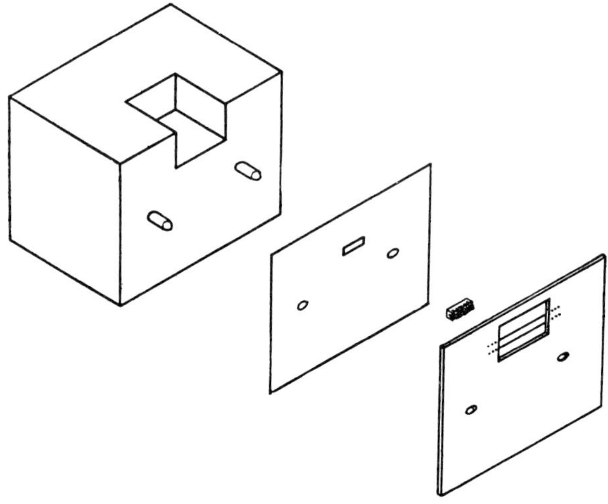

FIG. 51. Diagram presenting the arrangement in which a single isolated electroplax separates two pools of fluid.

The drawing shows the respective position of one chamber for the pool of fluid, the sheet of nylon containing a window, the single electroplax, and the grid used for pressing the cell against the window. The other chamber is not shown in the drawing.

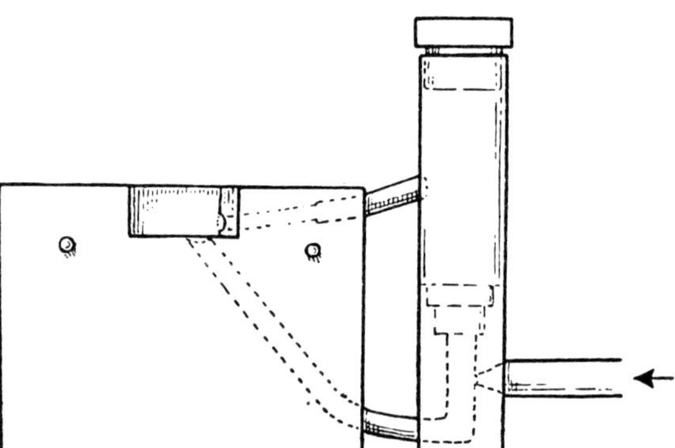

FIG. 52. Cross section of one of the chambers showing the relationships between the pool and the air lift.

The arrow indicates the O_2 inlet.

tailed description may be found in Schoffeniels' original papers. It may be recalled that only one face of the electroplax is innervated and has a conducting membrane, while the other face is not innervated and non-conducting. Therefore, the fluid of one chamber is bathing one type of membrane; that of the second chamber, the other type. The innervated membrane of the electroplax has a rectangular shape and is, therefore, uniquely suitable for the study of ion movements across such a membrane. The preparation is patterned after that developed by Ussing and his associates with the frog skin for studies of ion transport (see, for instance, Ussing, 1954). With the use of radioactive material and appropriate arrangements, the preparation permits one to follow the rates of

FIG. 53. Photographs of the chamber used.
The window between the threads of nylon is seen in the left picture. The position of the window with respect to the pools is seen in the right picture.

ion flux across the two types of membranes separately, to study physical and chemical factors affecting it, and to measure simultaneously electrical manifestations.

B. Effects of Acetylcholine and Analog Compounds

With the new preparation, markedly lower concentrations of acetylcholine and analogous structures were required for blocking electrical activity than in previous experiments. Moreover, for the first time the effects produced with many of the compounds tested were reversible. The action of the following compounds could be reversed: acetylcholine, carbamylcholine, eserine, Prostigmine, *d*-tubocurarine, and procaine. No reversibility was obtained with decamethonium and stilbamidine. One of the factors responsible for the previous failure to reverse the action might have been the high concentrations required to produce the effects and the low rate of diffusion taking place across the connective tissue surrounding the electroplax. This is supported by the recent observation that, with a still further improved preparation (Schoffeniels, 1957b) in which a membrane adjacent to the conducting membrane is removed, the effective concentration of the compounds tested has been found to be still lower, in the case of *d*-tubocurarine, for instance, by a factor of 10, 0.0075 μmoles/ml. or about 5 μg./ml. (unpublished experiments of Maxime Guinnebault).

The data obtained with the new preparation confirmed the idea of two categories of compounds affecting electrical activity: receptor inhibitors and receptor activators. In Table XVI a few observations on the effects of depolarizing compounds are summarized.

TABLE XVI

EFFECTS OF SOME TYPICAL DEPOLARIZING AGENTS ON THE ELECTRICAL ACTIVITY AND ON THE RESTING POTENTIAL OF THE SINGLE ISOLATED ELECTROPLAX[a]

Compound	Conc. (μmoles/ml.)	Min. of exposure	RP (mV.)
Acetylcholine	0.02	7	65
		27	35
Carbamylcholine	0.027	7	64
Prostigmine	0.2	13	68
		30	35
	0.0666	17	50
Decamethonium	0.022	5	63

[a] All compounds were applied to the solution bathing the innervated membrane. The underlined figure (under minutes of exposure) indicates the time when the electrical activity was completely blocked, i.e., no response could be obtained to either neural or direct stimulation. The resting potential (RP) of the controls varied between 80 and 84 mV. In the experiment with acetylcholine the solution contained eserine at a final concentration of 0.07 μmole/ml.

There is another interesting difference between these two types of compounds acting on the receptor: the effect of the inhibitors may be completely reversed, even after relatively prolonged times of exposure. This is, for instance, the case with eserine, procaine, and *d*-tubocurarine. The effect of activators cannot be reversed completely, and after a certain time the electric activity decreases again. This difference may be explained in the following way: keeping the barrier open to ion movements may result more readily in irreversible damage than keeping it closed. Moreover, the depolarization reflects structural changes which take place for rather long period of time compared with the milliseconds of the physiological process. This may lead to irreversible changes of the receptor protein or of other secondary factors. It is possible that in the improved preparation in which the concentrations required are markedly lower, even the effects of receptor activators will be completely reversible. This question has not yet been studied.

An important new fact was the demonstration that the quaternary nitrogen derivatives act exclusively upon the synaptic junction. Curare in low concentrations, about 0.1 μmole per milliliter or less, blocks the response to indirect, but not to direct, stimulation. Even in very high concentrations (18 μmoles/ml.) the propagated spike is not affected. As is known from nerve-muscle preparation, curare obviously affects exclusively the junction. In presence of curare in relatively low concentrations (about 0.075 μmole/ml.) acetylcholine, Prostigmine, carbamylcholine, and other receptor activators have no effect on the response to direct stimulation, even when the compounds are used in rather high concentrations (Table XVII). In previous experiments the depolarization of the membrane was erroneously interpreted to be a direct action of quaternary nitrogen derivatives on both synaptic and conducting membranes. The observations with low concentrations of curare have now shown that the conducting membrane of the electroplax is just as impervious to quaternary ammonium ions as are the axons or the muscle fiber, as will be discussed in greater detail in Chapter XIV. Tertiary analogs, on the other hand, known to act specifically on the acetylcholine system, such as eserine, procaine, and the tertiary analog of Prostigmine, block the response of the electroplax to direct stimulation in a similar way as has been observed with other conducting membranes. The question arises how the depolarization of the whole membrane may be explained if the quaternary compounds act only upon the synaptic junction. The innervated membrane does not have only one junction, as have most muscles, but as has been discussed before, a very large number of them. Each time the potential is changed across a small area of cell membrane, there is a gradient of potential between the depolarized area and the surrounding membrane. If the depolarized areas are as numerous as in the electroplax, their depolarization will short-circuit the conducting membrane and depolarize it.

C. Ion Flux

When an attempt was made to measure the flux of Na across the preparation, the values obtained were higher than those that could be expected on the basis of Na movements into and out of the cell. An analysis of the results led Schoffeniels (1957b) to the conclusion that most of the flux measured could be accounted for by the diffusion of Na into the extracellular space surrounding the cell. He, therefore, developed a new and improved preparation.

A compartment is formed by a membrane of connective tissue fibers surrounding the ground substance. One face is the innervated membrane. Thus there is a structure with a cleavage plane between the innervated membrane of the electroplax and the opposite face of the adjoining compartment (Fig. 54). Between these two structures the

TABLE XVII
PROTECTION WITH CURARE AGAINST DEPOLARIZING ACTION OF QUATERNARY AMMONIUM DERIVATIVES[a]

Compound	μmoles/ml.	Exposure (min.)	Direct spike in % of control	RP	*C* (mV.)
Acetylcholine	0.03	5	30	70	80
		7	0	65	
TbC + acetylcholine		10	100	82	82
		30	100	83	
		60	100	82	
Carbamylcholine	0.15	2	60	70	82
		5	10	65	
		6	0	60	
TbC + carbamylcholine		5	100	81	78
		60	100	82	
Prostigmine	0.26	6	0	73	78
		45	—	40	
TbC + Prostigmine		15	100	79	80
		25	100	80	
		45	100	79	
		55	100	81	
Decamethonium	0.11	1	0	—	75
		13	—	40	
TbC + decamethonium		15	100	75	74
		55	100	74	

[a] The electroplax is exposed to depolarizing compounds either with or without previous application of d-tubocurarine (TbC), 0.075 μmoles/ml. All compounds were applied to the solution bathing the innervated membrane. RP = resting potential after exposure, the control value (*C*) is added in the last column. In the experiment with acetylcholine, eserine was added at a final concentration of 0.07 μmoles/ml.

nerve fibers and blood vessels are located. In the original preparation, as illustrated in Fig. 55, the dissection was carried out through the ground substance separating each cell (arrows *1* and *2*), so that half of the connective tissue belonging to one compartment is left attached to the cell belonging to the next compartment. By dissecting the cell in a way in which the membrane of the adjacent compartment is removed (arrow *3* in Fig. 55), Schoffeniels achieved a complete separation of the innervated membrane from that of the contiguous compartment.

Using sulfate as a test ion, Schoffeniels found with the new preparation that it did not penetrate from one pool of fluid to the second. Thus the flux of Na and other ions measured in these conditions may be accepted as representing the exchange between the cell and the two compartments. The preliminary results obtained with this preparation

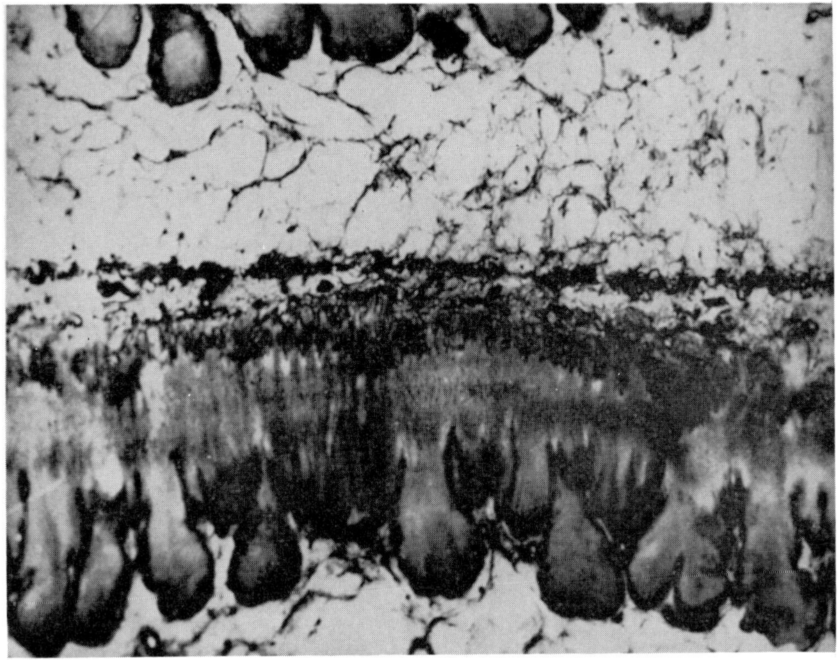

FIG. 54. Microscopic picture of a cross section of the electric organ (bundle of Sachs) of *Electrophorus electricus*.

The picture shows the space between the connective tissue of one compartment and the innervated membrane of the electroplax belonging to the next compartment, permitting a complete separation of the innervated membrane from most of the connective tissue belonging to the adjacent compartment.

under a variety of conditions fit—within the limits of experimental error—theoretical postulations and agree in general with those observed with other conducting cells (Schoffeniels, 1959). In view of the incomplete character of the data obtained so far, only one result, in which the concentration of intracellular Na was determined, may be described as an illustration.

The general procedure used by Schoffeniels was as follows: an electroplax mounted between two chambers is equilibrated in an oxygenated saline solution containing Na^{24}. After an adequate period of time, about 2–3 hours, the cell is mounted between two grids and either placed in front of a Geiger tube, the radioactive ions being washed out in a stream of inactive saline, or the cell remains in contact with the constant volume of inactive saline for fixed periods of time and the

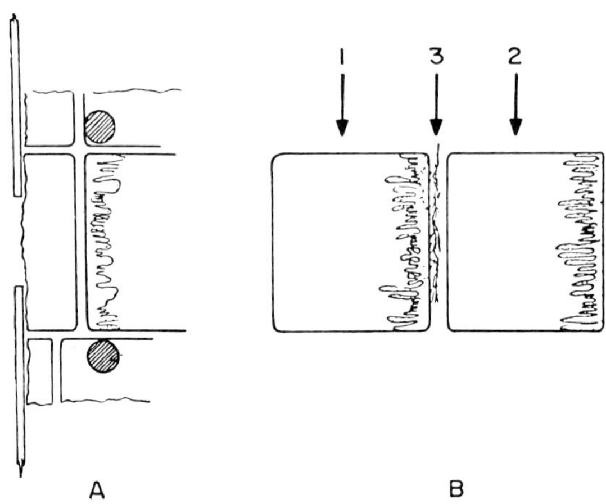

FIG. 55. Diagram illustrating the improved preparation of a single isolated electroplax.

Drawing A shows the original fixation of the electroplax in front of the window in the nylon sheet, the grid being seen in cross section. There is an interposition of connective tissue and ground substance between the innervated membrane and the window. Since this material is virtually incompressible, a certain space of ground substance is left between the cell and the sheet of nylon, and is apparently responsible for the escape of ions from one chamber to the other through this extracellular structure. Drawing B shows the new way of dissecting the electroplax. Previously the dissection took place at arrows *1* and *2*, in the improved preparation between *1* and *3*, removing all of the connective tissue and ground substance of the adjacent compartment. This permits fixation of the window directly on the innervated membrane.

activity of the solution is then measured. In the first arrangement the activity remaining in the cell is determined, in the second that coming out of the cell through either the innervated or noninnervated membrane or through both. Figure 56 shows the result of a typical experiment in which the activity remaining in the cell has been measured during the washing out. The data have been plotted on semilog paper. The curve can be divided into two main components, a fast and a slow one. The fast one, as is generally assumed and is also experimentally supported, represents the diffusion of Na in the extracellular space. The outflux is calculated using the intercept and the slope of the slow component. The extracellular space is calculated from the fast component, and after the necessary corrections the intracellular concentration can be estimated from the extrapolated value. The values obtained for the intracellular Na concentration with this method varied from 5 to 23 μmoles per milliliter intracellular fluid. Similar values for intracellular Na concentrations were obtained when they were measured by the second type of procedure, i.e., adding up the activity washed out and the activity remaining in the cell and correcting for extracellular space.

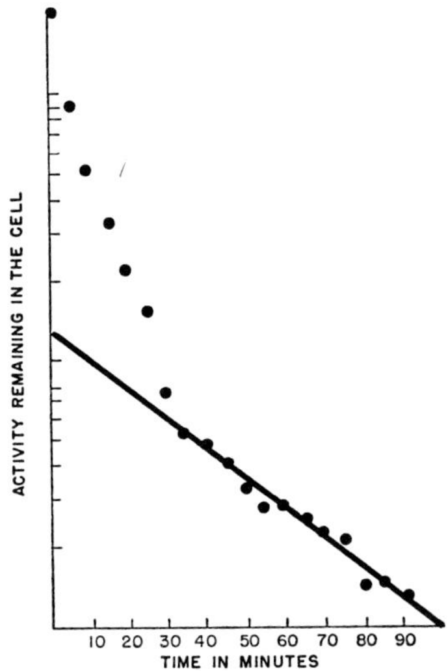

FIG. 56. Determination of Na^{24} in a single isolated electroplax.

In this particular experiment the cell was kept in oxygenated saline containing Na^{24}. After equilibrium was reached, the Na^{24} is washed out and the activity (c.p.m.) in the washing fluid is plotted on a semilog scale against time. For details see text.

Among the other results obtained in the exploratory experiments may be briefly mentioned that the influx of Na during activity seemed to increase across the innervated membrane. The important outcome of Schoffeniels' observations, although quite incomplete and preliminary, is the demonstration that the new preparation offers a valuable tool for the study of ion movements in rest, during activity and recovery, and of the physical and chemical factors controlling and affecting them. In addition, the three types of membranes in this preparation, the conducting, the postsynaptic, and the non-conducting, promise to provide information as to specific differences of mechanisms characteristic for the functional and structural peculiarities. In the new preparation, as might have been expected, the innervated membrane appears to be much more sensitive to chemical agents than the previous one. The higher sensitivity toward curare has been mentioned. A quantitative reevaluation of the effectiveness of the characteristic compounds appears therefore desirable.

In the current year (1959) R. Whittam and M. Guinnebault measured the efflux of K from the innervated membrane of the isolated single electroplax. The efflux during the resting state was about 1 mμmole K per square centimeter per second. During direct stimulation, both in the presence and the absence of curare, the efflux increased proportionally to the rate of stimulation: per square centimeter surface per impulse it was on the average about 6–8 μμmoles. This is comparable to the value obtained with squid giant axon. The significance of this result is clear: it shows that the loss of K during activity in a vertebrate conducting cell is comparable to that in the invertebrate nerve fiber, thus suggesting the general nature of K movements associated with conduction.

D. Isolation of the Receptor Protein in Solution

The experimental evidence for the existence of a receptor described in Chapter X raised the question of whether it is a separate protein (or conjugated protein) and whether its isolation is possible. Chagas and his co-workers have recently attempted to isolate the receptor (Chagas *et al.*, 1957, 1958; Chagas, 1959). Using a radioactive curarelike substance, triethiodide of gallamine, they found this compound bound to a component (or components) present in an extract of electric tissue of *Electrophorus*. However, the complex formation was reduced even by very dilute salt solution (0.02 *M*). This fact raises the question of whether the macromolecule, which forms the complex, is indeed the receptor, since curarization of intact electroplax is easily obtained with low concentrations of curare even in 0.18 *M* salt solution. Since curare has two cationic nitrogens, it is possible that unspecific Coulombic and van der Waals' forces led to complex formation with a number of macromolecules, which may be unspecific components and not identical with the receptor. Such a possibility is indeed suggested by the report that the component responsible for binding is not a protein.

The problem has been investigated by Ehrenpreis (1959a, b) with a procedure which differs from that used by Chagas and his associates in two respects. First, the tissue extracts have been subjected to ammonium sulfate fractionation and the resulting protein fractions have been examined for their ability to bind curare and other related substances known to react with the acetylcholine system and more specifically with the receptor. This procedure eliminated other components which were found by Ehrenpreis to react with curare, such as nucleic acid and chondroitin sulfuric acid. Second, the interaction was studied by equilibrium dialysis under controlled conditions of pH and ionic strength. Data thus obtained are less ambiguous to interpret than those in which binding has been determined after prolonged dialysis against large volumes of distilled water, the procedure used in Chagas' laboratory.

As a result of his studies Ehrenpreis was able to isolate a protein which has the following pertinent features: (1) There is a striking qualitative parallelism between the binding of curare and related compounds to this protein in solution and their affinity for the receptor as measured by their effects on electrical activity of the isolated electroplax (Schoffeniels and Nachmansohn, 1957; and unpublished observations of Guinnebault). The effectiveness in both types of studies decreases in the following order: curare, dimethylcurare, Prostigmine, eserine. None of these compounds is bound in comparable concentrations by bovine serum albumin. (2) The binding of curare, although salt sensitive, still occurred at a relatively high ionic strength ($\mu = 0.1$). As predicted from these results, a decrease in the ionic strength in the solution bathing the electroplax enhanced the effect of curare (Guinnebault, unpublished observations). (3) Decamethonium and Prostigmine are unable to displace curare from its complex with the protein in solution. The two compounds which depolarize the electroplax in low concentrations are without effect on the electroplax blocked by curare, even when applied in concentrations several times as high as curare (see Table XVII).

The observations strongly support the assumption that the protein isolated by Ehrenpreis may be the receptor. The protein has been purified to a considerable extent, to essentially one component in electrophoretic studies; the isolation is still being continued. Still more recently, Ehrenpreis tested binding of a whole series of choline derivatives to the protein which he had isolated. Strikingly, acetylcholine was by far the most strongly bound compound. The specificity was demonstrated by testing these compounds against other proteins (Ehrenpreis, unpublished data).

The procedure of identification, so easy in enzyme purification methods, offered a problem in the beginning, but the difficulty has been fortunately overcome by the availability of the isolated electroplax preparation which provides a highly sensitive and specific test for identifying the properties of the isolated protein with that of the receptor in the intact cell.

CHAPTER XII

Effects of Lipid-soluble Quaternary Ammonium Ions on Conduction

A. Depolarizing Action on Axons and Electroplax

The powerful effect of acetylcholine on synaptic junctions was one of the two essential observations on which the theory of neurohumoral transmission was based. Applied to axons, acetylcholine has no effect even in high concentrations. This failure of acetylcholine to affect electrical activity in axons has been for many years one of the main objections to the theory that acetylcholine plays an essential role in conduction. Acetylcholine is a methylated quaternary ammonium salt which is insoluble in lipids. It does not penetrate into the cell. This has been experimentally demonstrated and will be discussed in more detail in Chapter XIV.

The ability of PAM to reactivate acetylcholinesterase seemed to offer the possibility of an interesting test. If block of electrical activity in axons after exposure to alkyl phosphates is due to the inactivation of acetylcholinesterase, conduction should reappear when the enzyme is reactivated. In view of the specificity of the reactivating power of the oxime, it would be difficult to interpret the reappearance of electrical activity except by the reactivation of acetylcholinesterase. PAM itself is, however, insoluble in lipids. For the test desired, a lipid-soluble analog of PAM seemed to be a prerequisite. That was, in fact, the initial purpose of designing PAD: by replacing the methyl by a dodecyl group, the reactivating power decreased only slightly and the compound appeared, therefore, suitable for exploring whether a reversal of the effects of alkyl phosphates on the axon was possible.

When, however, PAD was applied, in control experiments, to axons without previous exposure to alkyl phosphate, the compound was, surprisingly, found to produce a rapid block of conduction in relatively low concentrations, about 10^{-4} *M* (Schoffeniels, Wilson, and Nachmansohn, 1958). This was the first demonstration that a quaternary nitrogen derivative, when made lipid soluble, blocks conduction. Other quaternary ammonium ions were therefore prepared, including a lipid-soluble analog of acetylcholine. When one methyl group on the nitrogen of the ester is replaced by a dodecyl group, the compound becomes very lipid soluble. It has been referred to as noracetylcholine 12. The lipid-soluble analogs of acetylcholine proved to be blocking agents when applied to crab and lobster nerve and to the isolated single electroplax. When the effects of the lipid-soluble quaternary ammonium ions were tested with intracellular electrodes on the latter preparation, depolarization was obtained even after complete curarization, consistent with the assumption that these analogs activate the acetylcholine receptor in the conducting membrane and thus mimic to some extent the physiological change of permeability.*

Effects similar to those obtained with noracetylcholine 12 on axons and on the electroplax of *Electrophorus* were obtained by Staempfli (1958) in experiments on Ranvier nodes of frog sciatic nerve. In 10^{-3} *M*, the compound produced a complete breakdown of the membrane potential. In concentrations of 3×10^{-5} *M* (equivalent to 10 μg./ml.) the compound increased the amplitude and duration of the action potential. The effects on the membrane take place within seconds; this is a rather high speed, only slightly less than that observed with K. The action is reversible if the compound is applied in low concentrations, but even in concentrations up to 10^{-4} *M*, it is at least partly reversible.

Similar and even slightly more potent effects on the conducting membrane of the desheathed frog tibialis nerve were obtained with PAD (Dettbarn, 1959a, b). This quaternary ammonium ion has a marked depolarizing action at 10^{-4} *M* and a small but definite effect at 10^{-5} *M*. In recent unpublished experiments with PAD on Ranvier nodes of frog sciatic nerve, Staempfli and Dettbarn obtained depolarizing effects in 10^{-6} *M* concentration (equivalent to 0.2 μg. of acetylcholine per milliliter). Effects on the electrical activity of Ranvier nodes were obtained by Dettbarn in 10^{-7} *M* concentration.

For the interpretation of these effects, the question is crucial whether or not they are produced by a specific action upon the acetylcholine system, postulated to be essential for the permeability change. The presence of a powerful acetylcholine system in all conducting tissues in or near the surface membrane suggests this mode of action. But if it were possible to demonstrate convincingly that this is the basis for the depolarizing action of the lipid-soluble quaternary ammonium ions, it would be rather conclusive evidence for the postulated role of the naturally occurring ester.

Evidence in this direction has been obtained (Dettbarn, Wilson, and Nachmansohn, 1958; Dettbarn, 1959b). If PAD depolarizes the membrane by reacting with the acetylcholine receptor, i.e., if it is a receptor activator, a receptor inhibitor should, by competition, protect the active surface of the protein against the quaternary ammonium ion. This is

* In our first observations all our records clearly showed an overshoot, but the findings could not later be confirmed.

indeed the case. Eserine has, as discussed before, a high affinity for the acetylcholine system: it is a strong inhibitor of the enzyme, the K_1 being 6×10^{-8}; it is also a receptor inhibitor. When eserine, in 7×10^{-3} *M*, is added to the desheathed frog sciatic nerve, addition of PAD in 10^{-4} *M* has absolutely no effect. Lower concentrations of eserine strongly counteract the effect of PAD, but do not abolish it. Higher concentrations of PAD may again overcome the protective action of eserine. If eserine is added after PAD, this decreases the depolarizing action. The antagonism between the two compounds is typical for a competitive action. The concentration of eserine must be higher since it is only the free base which is lipid soluble and penetrates into the cell. Moreover, at low concentrations it may act mainly as esterase inhibitor; for inhibiting the receptor higher concentrations are probably required.

Another problem essential for interpreting the significance of the action of the lipid-soluble quaternary ammonium ions was the question of whether the permeability change observed is consistent with the assumption that their action opens the barrier specifically for Na, as postulated by theory, or whether the compounds produce an indiscriminate increase in permeability. Since, during activity, Na conductance increases several hundredfold, while increase in K conductance follows later and is relatively small, a compound postulated to activate the acetylcholine receptor should specifically increase the permeability for Na.

This question has been tested by Dettbarn (1959a) in the following way: desheathed tibialis nerve fibers of frog were exposed to Ringer's solution containing the following lipid-soluble quaternary ammonium ions: noracetylcholine 12 and pyridinium dodecyl iodide in 10^{-4} *M* concentrations and pyridine aldoxime dodecyl iodide in 10^{-5} *M* concentration. At these concentrations no effect on the resting potential was observed. When the K concentration in these solutions was raised to 20 m*M*, the depolarization which is normally observed is slightly reduced both in respect to strength (about 20%) and in respect to speed of the response, indicating a small but definite decrease of permeability to K.

The change of permeability to Na effected by these compounds was tested by Dettbarn in the following way. The fibers were exposed to the same three lipid-soluble quaternary ammonium ions, but in concentrations five times as high as before, at which there is a pronounced depolarizing effect in normal Ringer's solution. When Na was replaced by choline chloride, the depolarization was strongly reduced. On again replacing choline by Na, the original strength of depolarization was restored. Thus the depolarization appears to be effected by a marked specific increase of permeability to Na.

The pertinent result of Dettbarn's experiments is the demonstration of an opening of the barrier in a way which distinguishes between Na and K, and the evidence that this discriminating increase in permeability is produced by a chemical compound analogous in structure to acetylcholine. The interpretation of the slight decrease in K permeability observed is, at present, difficult and must be postponed until further investigations have been carried out; the effect may depend on the concentrations used. The action of these compounds on the ion movements should be directly measured under various conditions. The single isolated electroplax offers a suitable material for these tests, and the experiments of Whittam and Guinnebault have established a good basis for further investigations.

The effects of the lipid-soluble acetylcholine analogs on conducting membranes have initiated a new and promising development. The investigations are still in an early phase. There are obviously differences between the effects described and the physiological events. This must be expected *a priori* since it is unlikely that these compounds, applied externally, should be able to mimic exactly the action of the ester, released *within* the conducting membrane and removed within a fraction of a millisecond. Some of the compounds, as for instance PAD and dodecylpyridinium, are, of course, not hydrolyzable at all; others, as for instance noracetylcholine 12, only at a very low rate. Studies of the interaction of these compounds with the receptor protein and with acetylcholinesterase are in progress. Most of them are rather strongly bound to the enzyme and are fairly good inhibitors. PAD, for instance, blocks, in 10^{-6} M concentration, about 40 per cent of the activity, in 2×10^{-6} M about 60 per cent (Hanna Greenberg, unpublished results).

According to theory, the proteins of the system are situated in close vicinity to each other within the conducting membrane and all compounds acting on one protein may also affect the others depending on the degree of affinity, as has been discussed in Chapter X. A proper evaluation of the detailed mechanism requires many more data than are available at present. Cholinesterase inhibition may contribute to some extent to the effects obtained. Most of the observations point, however, to a strong action on the receptor, especially the high speed with which the response is obtained. Even more potent esterase inhibitors do not produce such effects if they act predominantly on the enzyme, because acetylcholine accumulation due to enzyme inhibition is quite slow and it takes some time until a sufficiently high level is reached to produce biological effects. Another possibility to be considered is that of a reaction with the storage protein resulting in a release of acetylcholine. The question of specificity of the mode of action of these lipid-soluble quaternary ammonium ions would, of course, not be involved in the problem whether the receptor or the storage protein or the enzyme, or all three, are affected. The importance of the effects of the lipid-soluble acetylcholine analogs lies in their drastic and specific effects on electrical activity as postulated by theory.

B. Response of Striated Muscle

The block of conduction by lipid-soluble analogs of acetylcholine and their depolarizing action on axons and electroplax raises the pertinent question whether such compounds would evoke a functional response in muscle in which the end plates were blocked by curare. According to theory, lipid-soluble quaternary ammonium ions should be effective even under such conditions. This proved to be the case (Hinterbuchner, Wilson, and Schoffeniels, 1958; Hinterbuchner and Wilson, 1959a, b).

Two series of long-chain quaternary ammonium ions were tested. In one series 2-, 4-, 6-, 8-, 10, 12-, and 16-carbon chain derivatives of acetylcholine were used, in the second series 2-, 3-, 4-, 5-, 6-, 8-, 10-, 14-, and 16-carbon chain analogs of tetramethyl ammonium ions. In both series one methyl group on the nitrogen was substituted by the long chain. The distribution of the compounds between water and chloroform was determined as an indication of lipid solubility. A turning point in solubility was found with the 6-carbon compounds: those with less than six became more and more distributed in water, those with more than six were increasingly more concentrated in chloroform. Curarization was achieved by exposure to 50–100 μg./ml. of *d*-tubocurarine chloride, sufficient to block up to a hundred times the active concentration of acetylcholine.

In general, the muscle response to the compounds first decreases with the length of the chain; it is small with 8- and disappears with 10-member chain derivatives. However, it reappears with 12-, 14-, and 16-member chain derivatives. All compounds with a chain of less than ten carbons were blocked by curare, while those with a chain of more than ten carbons were not. The concentration of long-chain derivatives which gave a response similar to the reference contracture was about 10^{-3} *M*. The effective concentration was identical for curarized and noncurarized muscle. It is possible to produce at least ten successive contractures in the same muscle, whether curarized or not, without indication of impending loss of response. The fact that repeated contractures can be obtained from the same muscle indicates that whatever the other effects of these compounds may be, the response is a physiological one.

Since curare does not block the action of these compounds, the observations indicate that the site of action is not the end plate but the excitable membrane; the finding is consistent with the view that these compounds interact with the acetylcholine receptors in the conducting membrane of muscle.

Both noracetylcholine 12 and *N*-dodecyltrimethyl ammonium ions were also tested on whole sartorius and biceps muscles of frogs as well as on strips of sartorius and semimembranosus muscles (Hinterbuchner and Wilson, 1959b). These muscles respond by two twitches, a relatively fast one, reaching its peak in 10–15 sec., and lasting about 40–50 sec., and a slower one reaching its peak in 2 or 3 min. (Fig. 57). Six-

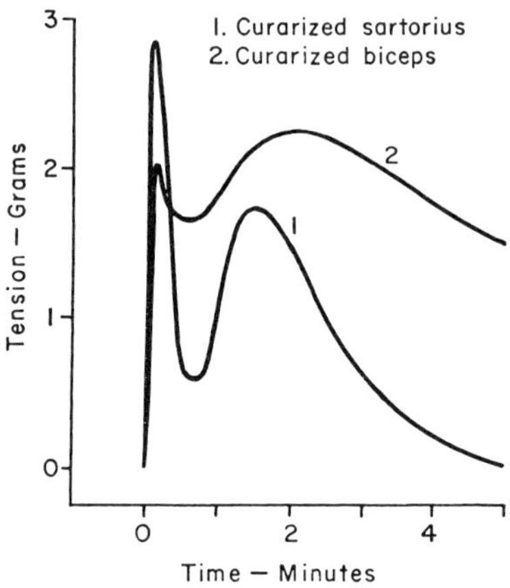

FIG. 57. Response of muscle to noracetylcholine 12 after complete curarization. Frog's sartorius muscle (*1*) was kept in Ringer's solution containing 20 µg./ml. of *d*-tubocurarine chloride for 60 min., biceps muscle (*2*) for 130 min. prior to exposure to the lipid-soluble anolog of acetylcholine (2.3×10^{-3} *M*). The record shows the double-peaked response.

and eight-carbon chain derivatives of acetylcholine and tetramethyl ammonium in comparable concentrations produce no response, nor did they prevent the usual response to long-chain derivatives on subsequent applications. In these muscles the effects of the lipid-soluble analogs were, however, not reversible. Exposure to *d*-tubocurarine in high concentrations, up to 1.0 mg./ml. for 40–70 min. failed to influence the response. The nature of the double twitch, frequently found in the literature under various conditions, remains to be elucidated. But the effects on striated muscle add further emphasis to the significance of the actions of the lipid-soluble analogs of acetylcholine, because they are able to reproduce the biological action postulated for the physiological role of acetylcholine.

CHAPTER XIII

The Complex Nature of the Permeability Change

The study of the molecular forces acting in the acetylcholine system and their relationships with electrical activity have been valuable for the understanding of some aspects of nerve activity. We are able to affect some physical events by specific chemical agents, acting on well-defined cellular proteins, in a way which may be at least partly predicted from our knowledge of the proteins in solution. However, whereas the essential role of the system in the permeability change of conducting membranes is well supported by a great variety of chemical and biochemical facts, the precise mechanism of its action still remains unknown. It is difficult to conceive that any single system should be the only factor regulating a biological function. The acetylcholine system is not located in a vacuum, but in a complex membrane formed by many constituents. Virtually nothing is known about the interaction between this system and other chemical components such as phospholipids, steroids, amino acids, other metabolites, ions such as calcium and magnesium, etc. So even if we had much more complete information about the acetylcholine system than is available at present, there still would be a large gap in our picture of all the factors controlling ion movements during activity.

Since the time when the membrane theory was developed at the turn of the century, much ingenious thought and many efforts have been devoted to the elucidation of the physicochemical properties of cellular membranes in general and specifically to those of the conducting membrane. Until recently, however, progress was slow. The situation is now changing owing to the advances in two fields. Studies with electron microscope and X-ray diffraction have brought new information concerning the ultrastructure of cells and their membranes. The existence of a conducting membrane surrounding the axon was, until recently, a purely hypothetical assumption. A special "plasma membrane" has been observed with the electron microscope: this is considered to be the conducting membrane, although the identity of structural and functional barriers has not been established as yet. The plasma membrane has a diameter of about 75 A. and is formed by three distinct layers of approximately equal widths. The outer and the inner layers seem to be essentially formed by protein, the middle layer by lipid believed to be arranged as a bimolecular leaflet (see, e.g., Robertson, 1956a, b, 1957a, b, 1959).

The second line of advance has been made possible by the availability of radioactive and stable isotopes. New information about functional aspects of cell membranes and barriers surrounding the cell has been obtained during the last decade. A really staggering amount of investigation has been devoted to the problems of ion transport across membranes. Several recent books and symposia deal with this special subject (see, e.g., Clarke and Nachmansohn, 1954; Symposium on Active Transport and Secretion, 1954; Discussions Faraday Soc., 1956; Harris, 1956; Murphy, 1957; Colloque de Biologie de Saclay, 1959).

There is general agreement that for an "active" transport of ions, i.e., transport against concentration gradients, metabolic energy is required. During nerve activity, however, Na and K ions move from higher to lower concentrations. For these movements only a trigger action for increasing permeability (or conductance) is required. This is a process specific for conducting cells; here a special chemical system with specific properties is postulated, and this monograph presents the evidence that the acetylcholine system is an essential part without which the process would not work. It is only on this special point that there is a difference of opinion with the Cambridge group.

Acetylcholine clearly has nothing to do with the Na or K pump, an erroneous interpretation sometimes encountered in the literature. The restoration of the concentration gradient, the source of EMF, takes place during recovery; the ratio of heat produced during activity over that in recovery indicates that the latter requires most of the energy released. But this process need not be, and probably is not, specific for conducting cells. It appears likely that similar forces regulate active ion transport in most cells. Although this aspect is outside the scope of the monograph, it is to some extent related and therefore a few general remarks may be appropriate.

There has been considerable discussion about the role of the redox pump (see, e.g., Conway, 1959). Respiration is the source of all energy and its importance for active transport will hardly be questioned. But this does not yet answer the problem of the specific mechanism involved. Active transport is possible even in anaerobiosis. The evidence of Whittam (1958), that ATP is the immediate source of energy in red blood cells, appears to be quite strong. Hodgkin and his associates tested the effects of ATP injection into squid giant axons on Na efflux, but the results, although interesting, require further elucidation (see, e.g., Caldwell, 1959). As far as the precise chemical mechanism is concerned, the situation is quite obscure.

The formation of complexes between ions and carriers has been proposed. Some observations suggest a distinction between Na and K carriers, others suggest that the same carrier may be used by the two ions. This touches upon one of the most challenging questions, namely how the membrane is able to distinguish between Na and K. Very little information exists about this point. Differences in some properties and behavior of certain Na and K salts have been reported. Some salts of polyphosphates show a specificity with respect to the degree of ionization (Von Wazer and Campanella, 1950). Organic compounds have been described by Schwarzenbach, Kampitsch, and Steiner (1945, 1946) which form complexes with Na; they have a weak affinity for Na but none for K. Lamm and Malmgren (1940) reported that polymers of metaphosphoric acid have a special affinity for Na. It has been proposed that acidic lipids may distinguish between Na and K, but at present there is no evidence for this view. Folch and his associates have studied the role of acidic lipids in the electrolyte balance of the nervous system of mammals. With all the lipids tested no preference was found as to their binding to either Na or K (Folch-Pi *et al.*, 1957). Polyphosphates, including nucleic acids and phospholipidic acids, have been suggested as possible carriers, the latter exchanging a cation with choline. These views are still admittedly a matter of conjecture.

Since ion movements during activity and recovery take place across the same membrane, some of the factors and chemical forces may play a role in both processes, as for instance in the distinction between Na and K. The acetylcholine system must act within a complex framework of structural arrangements and chemical forces. This, of course, cannot be construed as an argument against an essential role of the system. But our ignorance of interaction with other forces and of the structural arrangements makes it frequently impossible to make predictions as to the kind of effects which might be expected.

Even the effects of acetylcholine and its analogs when applied to conducting tissue, offer, as we have seen, many difficulties of interpretation as to their exact mode of action within the acetylcholine system, without taking into account additional factors. The limitations of our knowledge as to the properties of receptor activators and inhibitors have been discussed. Only in a few cases has it been possible to distinguish between the action on the receptor and that on the enzyme, namely, when the affinities of the compound to the two proteins are far apart; when they are close to each other, as is often the case, the distinction is difficult. The new preparation of Schoffeniels offers a useful tool for obtaining more information about this particular problem, but so far such studies are in an initial stage. Reports in the literature frequently consider merely one factor or one component in testing the action of a chemical; when the anticipated effects do not take place, this failure is used to discard the essential role of acetylcholine in conduction altogether. Although in an early stage of development this may have been understandable, such an oversimplification appears less justifiable at present after so much pertinent information has been accumulated.

As an illustration of the complexities and difficulties, the following point may be briefly discussed. Excitation and inhibition of the conducting process are frequently a kind of mirror image although there are differences in time relations. Acetylcholine may produce activation by depolarization, inhibition by hyperpolarization. Vagus stimulation produces hyperpolarization and has an inhibitory action. A role of acetylcholine in this process has long been postulated and has been recently confirmed with intracellular electrodes (Burgen and Terroux, 1953; del Castillo and Katz, 1955c). Hyperpolarization may be produced in the axon by anodal currents. A dual function of the specific operative substances in the elementary process may have appeared surprising a decade ago. But in the last few years we have learned about the dual function of ATP. From the observations of Bozler (1951), Bendall (1953), Hasselbach and Weber (1953), and others, we know that not only contraction, but relaxation requires polyphosphates and that ATP is the most efficient polyphosphate for relaxation. A factor discovered by Marsh (1951) and Bendall (1951) and referred to as Marsh-Bendall factor, probably a protein, has an essential function in directing the effect one way or the other. In its active state the factor inhibits both splitting of ATP and contraction. Magnesium ions in 10^{-3} M concentration are essential to keep the factor active. Decrease of Mg concentration may inactivate the factor. Calcium ions in 10^{-3} M concentration are inhibitors of the relaxing factor. As has been found with glycerol-extracted fibers, which is a greatly simplified system compared to the intact muscle fiber, the factor lowers the optimal concentration of ATP necessary to produce maximal tension. Thus, a concentration of ATP which may produce maximal tension in absence of the factor, becomes inhibitory in its presence. This situation is an excellent demonstration how, even in a relatively simplified biological system, the same elementary reaction may behave differently and be directed in a seemingly opposite way by a secondary factor. Exactly how the two opposite effects are produced by ATP is not yet understood. The work of Huxley and Hanson (1955, 1957; Hanson and Huxley, 1957) suggests that during contraction the actin and myosin filaments slide past each other. In this process the energy released by the hydrolysis of ATP is essential, although the mode of action is obscure. A tentative picture how the nucleotide may act has been presented on the basis of sequences of chemical reactions *in vitro,* but at present this is still a working hypothesis (Weber, 1958). Excitatory and inhibitory effects have long been known for many compounds to depend on concentration. Activation and inhibition of the enzyme activity are sometimes observed even in an isolated enzyme system as a function of substrate concentration, resulting in the so-called bell-shaped curve of activity-substrate concentration relationship (Haldane, 1930); even here, where only two components react with each other, the process is not yet well understood. At present we can only say that the ability of acetylcholine to act in opposite ways, i.e., to produce either depolarization or hyperpolarization, has been experimentally demonstrated. What accessory factors determine the direction, is not known. It is one step to find that a biochemical system has the prerequisites to be associated with an elementary cellular function, be it ATP actomyosin with contraction or the acetylcholine system with conduction. Much more knowledge is required to describe the exact mode of action of the system and to be able to explain physical manifestations in living cells of different types where a great number of accessory factors may modify the response.

Another illustration of the complexity stems from observations with quaternary ammonium ions on conducting tissue initiated by Lorente de Nó. He found that tetraethyl ammonium ions restore the excitability of B and C fibers of frog nerves which had been made inexcitable by Na deficiency (Lorente de Nó, 1949). The excitability of A fibers is restored by guanidinium and certain other quaternary ammonium ions (Larramendi, Lorente de Nó, and Vidal, 1956; Lorente de Nó, Vidal, and Larramendi, 1957). A possible explanation for the effect of guanidinium ions has been recently discussed by Paoloni (1959). Fatt and Katz (1953a) found that the electrical responses of crustacean muscle fibers are maintained or even intensified when Na in the external medium is replaced by choline, tetramethyl, tetratheyl, or tetrabutyl ammonium ions. Other fibers did not give an active response in complete absence of Na and its substitution by these quaternary ions. In fact, choline is frequently used as an inert replacement for Na. Hagiwara and Tasaki (1957) injected tetraethyl ammonium ions into squid axons. The action potentials became greatly prolonged, but were abolished by removal of Na. Recently, cell bodies in the spinal ganglion of frog were exposed to quaternary ammonium compounds (Koketsu, Cerf, and Nishi, 1958a, b). Electrical activity of the ganglion cells could be maintained when Na of the Ringer's solution was totally replaced by choline or tetraethyl ammonium ions. Although the nature of these action potentials is still an open question, the data indicate that choline and tetraethyl ammonium are not inert but have a definite effect on the excitatory process of these cells.

Although these observations do not invalidate the assumption of an important role of Na and K during nerve activity in general, they further demonstrate the intricate nature of the conducting process, since we are unable to account for bioelectrical potentials in simple terms of Na and K movements. They emphasize the necessity of a better knowledge of the chemical forces involved.

CHAPTER XIV

Synaptic Transmission

I. Reevaluation of the Original Neurohumoral Transmitter Theory

In the preceding chapters the evidence for the essential role of acetylcholine in the generation of bioelectric currents has been presented and the present knowledge of chemical and molecular forces controlling nerve activity has been discussed. It appears interesting to reevaluate in the light of these advances the early observations which formed the basis of the hypothesis of neurohumoral transmission. Much new information has been obtained during the last decade about the events taking place in synaptic and myoneural junctions, but some of the interpretations are still based on the original findings. Therefore, a critical review of these early findings appears desirable before we turn to the present state of the question whether there do exist differences between axonal conduction and synaptic transmission and if so, in what respect.

A. Permeability Barrier Surrounding the Axon

Let us first consider the two main lines of observations upon which the hypothesis of neurohumoral transmission was originally based: (1) Acetylcholine has a powerful action on the junction (Brown *et al.*, 1936). Applied to the fiber even in high concentrations, it is inactive. (2) If nerve fibers are stimulated, acetylcholine appears in the perfusion fluid of the corresponding junction, although only in the presence of eserine, whereas no acetylcholine leaks from the fiber into the surrounding fluid (Dale *et al.*, 1936, Dale, 1937). These facts are well established and have not been questioned. But what is their meaning? Do these findings really indicate that the physiological function of acetylcholine is limited to synaptic junctions and that the ester acts in the manner proposed in the hypothesis of neurohumoral transmission, or is there an alternative interpretation possible or preferable in the light of later developments?

The absence of a response of nerve and muscle fibers to externally applied acetylcholine was explained as being due to permeability barriers surrounding the axon and the muscle fiber which are impervious to acetylcholine. The ester is a methylated ammonium ion and, like other methylated quaternary nitrogen derivatives, it is lipid insoluble. One would, therefore, not expect it to penetrate into the interior of axons. Even in absence of myelin, a lipid membrane or simply a lipid layer may render the entrance difficult or impossible.

The existence of a permeability barrier was demonstrated in several ways (Bullock, Nachmansohn, and Rothenberg, 1946). As mentioned before, eserine blocks conduction. Eserine is a tertiary nitrogen derivative with a pK of 8.2; at neutral pH most molecules are charged but a small fraction is uncharged. The compound penetrates into the interior: its presence has been demonstrated in the axoplasm of the squid giant axon after it had been exposed to the drug. In contrast, Prostigmine is a quaternary nitrogen derivative; the nitrogen is always cationic. *In vitro* it is a powerful inhibitor of acetylcholinesterase; the K_I is close to that

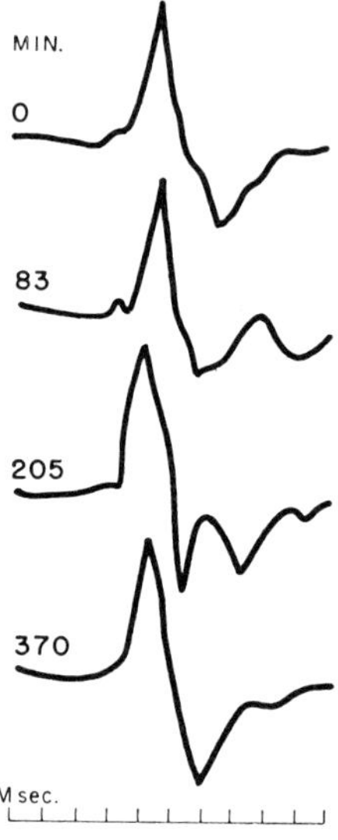

FIG. 58. The absence of any effect of Prostigmine on axonal conduction.

Fin nerve of squid was exposed to 0.01 *M* Prostigmine. Records before and after 83, 205, and 370 min. exposure are traced from enlarged photographs.

of eserine. But if nerve fibers, even so-called unmyelinated fibers, as, e.g., crab nerve fibers or the giant axon or fin nerve of squid, are exposed to Prostigmine in high concentrations, as high as 0.01 *M*, conduction is completely unaffected (Figs. 58 and 59). Even after long exposure of the axon to Prostigmine, the compound is not found in the axoplasm. In the experiments with intact crab nerve fibers mentioned above in which cholinesterase was determined on intact nerves with dimethylaminoethyl acetate as substrate, Prostigmine even in high concentration was found to block only about half of the enzyme; this fraction is apparently readily accessible even to lipid-insoluble compounds and is therefore referred to as "external" enzyme. But it did not depress the remaining enzyme activity, the "internal" enzyme, to the critical 20 per cent level and it did not affect conduction. When, however, eserine or the tertiary analog of Prostigmine was used, both the "internal" enzyme activity and the electrical activity were blocked although the inhibitory power of the tertiary analog is only 1 per cent of that of Prostigmine (Wilson and Cohen, 1953).

The chemical structure of Prostigmine is very closely related to that of acetylcholine; it is a typical competitive inhibitor of acetylcholinesterase. Nevertheless, it was questioned whether the permeability barrier shown to exist for Prostigmine prevents similarly the entrance of acetylcholine. Acetylcholine was therefore labeled with isotopic nitrogen (N^{15}). Giant axons of squid were exposed for 1 hour to 20 gm. per liter of

Fig. 59. The absence of any effect of Prostigmine on axonal conduction. Squid giant axon exposed to 0.01 *M* Prostigmine. Records before and after 45 min. exposure.

labeled acetylcholine. The axoplasm was then tested for excess N^{15}. The amount found was equivalent to about 1 μg. of acetylcholine per gram axoplasm; only traces had penetrated into the interior. In contrast, when the axon was exposed to a tertiary amine labeled with N^{15}, equilibrium between inside and outside concentration was reached within an hour (Rothenberg, Sprinson, and Nachmansohn, 1948). It appears unlikely that a biological permeability barrier is absolute. Traces may have penetrated, although it is difficult to exclude a slight contamination of the labeled acetylcholine with a tertiary amine as being responsible for the traces of excess N^{15} found in the axoplasm. But in any case, a very low degree of permeability may slow down the entrance to a rate which renders a compound biologically inactive. This demonstration of a permeability barrier has thus provided experimental support for the proposed explanation why no effect on conduction is observed when acetylcholine is applied externally. The inability of acetylcholine to act upon the axon once permeability difficulties are recognized, is no objection to the theory attributing an essential intracellular or intramembranous role to acetylcholine in the primary process of conduction.

Although it is certain that myelin offers an insurmountable barrier, it should not be considered as the only responsible factor. It has long been known that certain quaternary compounds, and specifically acetylcholine, act exclusively on the motor end plate and do not affect the muscle fiber elsewhere. Here again the question must be raised: do these findings mean that the acetylcholine system, or more specifically the receptor upon which these quaternary compounds act, is absent in the conducting membrane of the muscle fiber or does there also exist a permeability barrier protecting these cell constituents against the quaternary nitrogen compounds applied externally. Prostigmine does not affect the response to direct stimulation of the muscle fiber after complete block of the motor end plate has been obtained by *d*-tubocurarine in order to exclude any effect through this junction (Fig. 60). In contrast, eserine and DFP block this response (Couteaux *et al.*, 1946). We thus find in the response of the muscle fiber the same difference between the action of tertiary and that of quaternary nitrogen derivatives as described above with nerve fibers.

Finally the recent observations with lipid-soluble analogs of acetylcholine provide new confirmation: when applied to the conducting membrane of nerve or muscle fibers they rapidly and reversibly block, in low concentrations, electrical activity, as postulated by theory.

If externally applied acetylcholine cannot penetrate into the fiber owing to a structural barrier, the ester released internally will not be able to leak through the barrier to the outside. This accounts for the observation that, following stimulation of nerve fibers, acetylcholine appears only in the perfusion fluid of synaptic junctions. However, more than twenty years ago Calabro (1933) and Bergami and his associates reported that if a fiber is cut, acetylcholine is released during activity from the cut surface into the surrounding fluid (Bergami, 1936a, b; Bergami *et al.*, 1936). Their findings were repeatedly confirmed and extended to several types of nerves. Brecht and Corsten (1941) have demonstrated a release of acetylcholine even from cut sensory nerve fibers, although the quantity is smaller, as might be expected on the basis of the various observations reported previously. These results indicate that acetylcholine is released during activity at points along the axon as well as at the synapse, although it cannot diffuse to the outside as long as the structural barrier is intact.

B. Curare

The permeability barrier also explains why the action of curare is limited to the motor end plate. The structure of *d*-tubocurarine has been elucidated by the studies of King (1935) and Wintersteiner and Dutcher (1943). Bovet, Bovet-Nitti, and their associates have prepared a great number of curarelike substances ("curari di sintesi") and thoroughly investigated the relationship between structure and pharmacodynamic

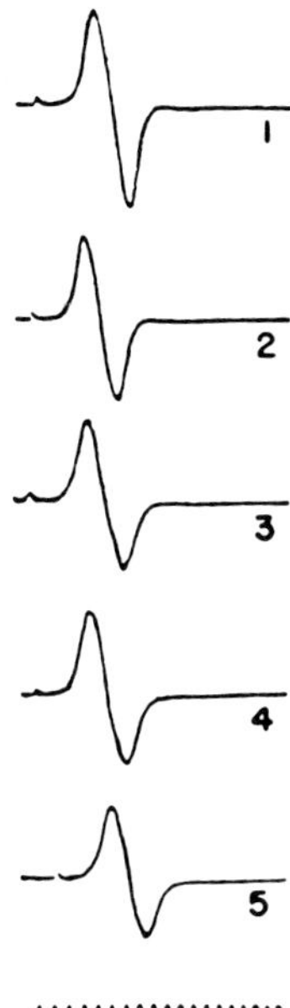

Fig. 60. Electrical activity of striated muscle exposed to Prostigmine; response to direct stimulation.

Frog sartorius muscle was kept in Ringer's solution containing 100 μg./ml. *d*-tubocurarine. After complete block of neuromuscular transmission the preparation was exposed to Ringer's solution containing Prostigmine (0.01 *M*) and *d*-tubocurarine. (*1*) Before, (*2–4*) after 10, 40, and 60 min. exposure. (*5*) Response after 20 min. recovery in Ringer's solution containing *d*-tubocurarine.

action (Bovet *et al.*, 1946, 1949a; Bovet and Bovet-Nitti, 1948). It has become apparent that the most important requirement for the action of curarelike substances is two quaternary nitrogen groups situated at a distance of about 13–15 A. There are two types of these compounds with seemingly different effects for which Bovet proposed the following classification: leptocurares (relatively small linear molecules) which may have a short depolarizing and stimulating action similar to that of acetylcholine itself, followed by a block, and pachycurares (relatively voluminous molecules like the curare itself) which antagonize the action of acetylcholine and exclusively block synaptic transmission without depolarization (Bovet, 1951, 1959; Bovet *et al.*, 1951; Bovet-Nitti, 1959). Of particular interest in the group of leptocurares turned out to be succinylcholine. Bovet and Bovet-Nitti and their colleagues (1949b) were the first to describe the curarizing action of this compound and to recognize the close structural similarity with acetylcholine: the molecule resembles two acetylcholine molecules linked together. The compound proved to be invaluble in surgery, being today widely used in anesthesia (Bovet and Bovet-Nitti, 1955).

Contrary to the views of other investigators, Bovet stressed the idea that both types of curarelike compounds must act by the same mechanism, viz., by competition with acetylcholine, in spite of apparent differences. His viewpoint has found strong support in the studies on the isolated electroplax described previously and the recent demonstration by Ehrenpreis (1959a, b) of the strong and specific binding of curarelike substances to the receptor protein in solution. Moreover, it seems that the leptocurares may be classified as receptor activators, whereas the pachycurares are receptor inhibitors, being too voluminous in their structure to permit the receptor protein to undergo the rearrangement effected by the activators.

Curarelike compounds are much more strongly bound to acetylcholinesterase than the corresponding monoquaternary salts. The binding increases by a factor of more than 100 (Bergmann, Wilson, and Nachmansohn, 1950b). A combination of increased Coulombic and van der Waals' forces may readily account for this increase in binding, if one assumes a second negatively charged group in the protein at the proper distance from the anionic site. The strong competitive action between curare and acetylcholine for the receptor suggests that this protein too forms stronger complexes with curare owing to a second negative charge at a suitable distance from that in the active site.

If monoquaternary ammonium salts are unable to penetrate the structural barriers surrounding nerve and muscle fibers, it is certainly to be expected that diquaternary salts will not enter. This is obviously the explanation for the observation of Claude Bernard, which have been so influential in promoting the idea of a special mechanism of synaptic transmission, basically different from that of axonal conduction. The absence of a response to curare of course does not indicate that the acetylcholine receptor upon which curare is generally assumed to act at the myoneural junction, is absent in the axon or does not function there; it only cannot be reached.

If this explanation is correct and the acetylcholine receptor is present in the axon and functioning there, a structure containing two cationic nitrogens at the proper distance, but in addition capable of penetrating the barrier, should have a curarelike action, i.e., it should block conduction. This is indeed the case as has been shown in experiments with stilbamidine. This compound is a diamidine derivative and has two cationic nitrogens at a distance of 12–15 A. Stilbamidine inhibits acetylcholinesterase in concentrations of the same order of magnitude as analogous curarelike diquaternary ammonium salts. It blocks, in low concentrations, transmission across the neuromuscular junction. In contrast to curare, a small fraction of the compound is uncharged at neutral pH. One could, therefore, expect that stilbamidine, in contrast to other diquaternary ammonium salts, would be able to penetrate into the interior of the axon and act on the active membrane. Since the fraction of the uncharged molecules is small, higher concentrations of the compound would be expected to be necessary to block axonal conduction than those used for the block of neuromuscular junction. Experiments with stilbamidine on conduction have borne out this expectation. Block of conduction has been readily obtained on axons (Bergmann, Wilson, and Nachmansohn, 1950b). The compound also blocks the directly evoked spike of the curarized single isolated electroplax preparation of *Electrophorus electricus* (Schoffeniels and Nachmansohn, 1957).

It appeared possible that quaternary compounds, once they are inside the axon, would affect conduction. To test this possibility, microinjection techniques worked out by Kao and Chambers were used to inject various quaternary ammonium salts into the giant axon of squid. Several compounds tested in a series of preliminary experiments seemed to block conduction in extremely low concentrations (Grundfest *et al.*, 1952). More recently, however, Hodgkin and Keynes (1956) repeated the microinjections with a more refined method. They did not observe any effect of curare in relatively low concentrations, about 10^{-5} *M*. It may be that the concentrations used were too low, but it also appears possible that the technique of microinjections used in the earlier experiments was inadequate and that the block observed was an artifact. The active membrane may be surrounded from both sides by a lipid layer. It would be interesting to test the effect of lipid-soluble inhibitors of acetylcholinesterase, such as eserine, DFP, Paraoxon, with the improved technique and to find out whether they block conduction, and if so, in what concentrations.

However, it must always be kept in mind that injection of compounds into a cell is a rather crude technique if one wants to mimic or affect specific intracellular processes. The extraordinarily complex structural organization of cells has always been suspected and is now well supported by electron microscope studies. The existence of a great many barriers, the evidence that the cellular processes take place in special, organized sequences, make it doubtful whether, and how efficiently, injection of a compound can affect a special structural and molecular arrangement in a way similar to that produced physiologically by the same compound released internally. Injection of ATP into the muscle fiber has, as is well known, no effect. Nobody will consider today this failure as an evidence against its role in the elementary process. Negative experiments cannot be used to exclude the functional importance of a particular compound. It was only after greatly simplified systems, the actomyosin threads and the glycerol-extracted fibers, were prepared that it became possible to demonstrate directly the specific ability of ATP to produce contraction and to study the interaction between ATP and actomyosin. The experience gained in several decades of muscle research should be a warning against too rapid and definite conclusions if one deals with such complex systems as an intact cell. The failure of del Castillo and Katz (1955a) to obtain an action of acetylcholine on the neuromuscular junction by intramuscular injections falls in the same category. Too many restrictions may prevent the molecule from reaching the protein with which it interacts. The special structural organization of cells permits only under unusually favorable and special conditions to reproduce and mimic a physiological event.

C. The Origin of Acetylcholine Found in Perfusion Fluids

The hypothesis of neurohumoral transmission assumed that acetylcholine is released from the nerve terminal and, crossing the nonconducting gap between nerve ending and muscle, acts as a mediator of the impulse on the effector cell. The significance of the appearance of acetylcholine in perfusion fluids in the presence of eserine will be discussed later. The original hypothesis would find at least some support if it could be demonstrated that the acetylcholine is released exclusively from the nerve terminal. It is, of course, exceedingly difficult to establish directly the site of origin in view of the inaccessibility of the synaptic junction for direct measurements. In 1936 Dale and his associates presented for the first time observations which seemed to suggest the nerve terminal as site of origin of acetylcholine found in the perfusion fluid. Motor nerves of cat's and dog's muscle were cut and the nerve endings permitted to degenerate; after direct stimulation of the muscle the authors failed to detect acetylcholine in the perfusate (Dale *et al.*, 1936). This failure has been interpreted as evidence that the acetylcholine found in the normal perfusate must originate from the nerve terminal, although this evidence is in the best case only very indirect and permits other interpretations.

These experiments were repeated by McIntyre and his associates (McIntyre *et al.*, 1950). They found in contradiction to the earlier findings that even after complete degeneration of the nerve terminals acetylcholine appears in the perfusion fluid of denervated muscle when fibrillations take place. At a recent symposimum on curare and curarelike substances at the Institute of Biophysics at Rio de Janeiro, McIntyre (1958, 1959) again described studies on the acetylcholine content in the eserinized perfusion fluid of denervated dog muscles and in that of the opposite, normally innervated leg muscles, before, during, and after application of stimuli to the sciatic nerve. In a number of experiments the postganglionic automatic nerves were extirpated at the time the sciatic nerve was cut. The acetylcholine content of perfusates from indirectly stimulated muscle and the perfusates from the opposite denervated muscle of the same animal were found to be of the same general order of magnitude, whereas the perfusates obtained from innervated muscle at rest contained only minute amounts of acetylcholine or none. In discussing the contradiction to the earlier findings and possible explanations for the previous negative results, the author points to several factors which may result in failure to detect acetylcholine in perfusates. The sample of perfusate collected was possibly not derived from perfusate that had really traversed the muscle capillaries. With the use of India ink in the perfusate and histological examination at the end of the experiment, considerable variations were found as to the distribution in the capillaries: when the ink was uniformly distributed in normal or denervated muscle, the perfusates contained acetylcholine; when, however, the ink was poorly distributed, little or no acetylcholine was found. The denervated preparations appear to be more variable as to the state of capillaries than the normal controls on the opposite side of the same animal. Since presence of eserine results in release of adrenaline, noradrenaline, and serotonin from the adrenal medulla, and the denervated muscles are hypersensitive to these compounds, he suggests the possibility that these factors may account for the previous failure to detect acetylcholine in denervated muscles. When all precautions were taken for obtaining accurate quantitative estimates in the light of these sources of error, acetylcholine was present in the perfusate of the denervated muscles. These data and a number of physiological tests, including effects of *d*-tubocurarine, lead McIntyre (1959) to the conclusion that the motor nerve is not necessary for the elaboration of acetylcholine. Thus, the only experimental evidence ever offered in support of the assumption that acetylcholine is released exclusively from the nerve terminal has been questioned and explained in a way contrary to the postulate of the hypothesis of neurohumoral transmission. It is obvious that this question should be reinvestigated and decided before the early findings are used as evidence in favor of neurohumoral transmission.

In summary, the early evidence in favor of neurohumoral transmission presented in the 1930's supports the assumption that acetylcholine must have an important role at synaptic and neuromuscular junctions. This was of great value because it has drawn the attention of many investigators to this compound and stimulated many studies of this phenomenon. Much new and valuable information has been obtained about many pharmacological aspects of the junctions. But the data are not adequate to answer satisfactorily the question whether acetylcholine acts as a neurohumoral transmitter in the original sense proposed or whether it acts in the membrane of the terminal in a way similar to that in the axonal membrane. The methods applied in the early investigations were too crude for elucidating a delicate cellular mechanism of an elementary process. For a proper analysis entirely different procedures were necessary.

A period of twenty years has elapsed. The huge amount of chemical and biochemical data and their relationship to function, accumulated since then, favor an essential role of acetylcholine in generating currents in the axon. We have now to consider other types of information about synaptic and neuromuscular junctions which have been obtained during the last two decades. Especially those findings will be discussed which appear pertinent for the interpretation of the mode of action of acetylcholine in synaptic and neuromuscular junctions, for the question whether the assumption of two different types of function of acetylcholine in the two phases of propagation is justified.

CHAPTER XV

Synaptic Transmission
II. Localization of Cholinesterase at Junctions

A. Chemical Determinations

Let us first consider the biochemical data concerning the synaptic junction, namely the concentration and distribution of cholinesterase at synaptic and neuromuscular junctions. If the action of acetylcholine is essential for the process of transmission, either in the way originally proposed or in the modified version, it is necessary to postulate that the compound must be rapidly inactivated by an enzyme. This was first recognized by Dale in 1914, and in 1936 he and his associates stated that one of the chief difficulties encountered by the theory of neurohumoral transmission across synapses and neuromuscular junctions was the necessity of a removal of the acetylcholine liberated with a "flash-like" suddenness. As mentioned before, it was the question whether or not the rate of inactivation of acetylcholine by cholinesterase at junctions is adequate to permit the assumption of an active role of acetylcholine in the transmitter process, which, in 1936, stimulated studies on the localization of cholinesterase. The results discussed in Chapter III had shown that acetylcholine may be metabolized at neuromuscular and synaptic junctions at a rate compatible with the assumption of a primary role in the process by which impulses are transmitted across these points. This fact does not, of course, reveal the precise mode of action of the ester.

When the picture of an intracellular or intramembranous action of acetylcholine emerged from the biochemical studies, a more precise localization of the enzyme at the junctions became of special interest for interpreting the mode of action of the ester at these junctions: is the enzyme localized at the nerve terminals or in the postsynaptic membrane or in both? If the mode of action in junctions is basically similar to that in axons, the enzyme should be present both in nerve terminals and in postsynaptic membranes, although the ratio of the concentrations at the two sites may vary according to structural and functional variations from synapse to synapse, from species to species, etc. If, on the other hand, acetylcholine does not act on a receptor in the presynaptic membrane but is only released there and acts on the opposite side, the presence of the enzyme in the terminal would be unnecessary and raise difficult questions as to its function and significance there. Why should there be a powerful mechanism of removal if the compound acts on the other side? Would the esterase in the terminal not interfere with the efficiency of the process by acting on the ester before it reaches its point of destination? On the other hand, its presence on both sides, although not proving the point, would support a similar function of the system at the two locations.

The investigations of the last two decades have supplied evidence that the enzyme is present on both sides of the junction. The findings of Couteaux and Nachmansohn (1940; Couteaux, 1942) that, following the section of preganglionic fibers, a rapid decrease of activity takes place in the superior cervical ganglion of cat during the first week, coinciding with the disappearance of the nerve terminals, have been discussed in Chapter III. They not only suggested that the enzyme is localized near the surface, an assumption borne out by the experiments on the squid giant axon, but present clear evidence for a high concentration of the enzyme in the nerve terminals inside the ganglion. The coincidence between the disappearance of the endings and the simultaneous rapid fall of the enzyme activity can most readily be interpreted in this way.

The high concentration of esterase in nerve terminals is also indicated by the experiments of Nachmansohn and Hoff (1944) on the spinal cord of cats. The enzyme concentration was determined after unilateral and bilateral deafferentation of dorsal and ventral roots, and a marked decrease was found in the four corresponding segments of the spinal cord coinciding with the disappearance of nerve terminals. It is of special interest that the decrease in the dorsal segments was higher than might have been expected from the concentrations of the enzyme in the dorsal roots; thus, the observations indicate that, as in the ganglion, the nerve terminals of sensory roots have a high concentration of enzyme per unit weight, higher than that found in the axons of dorsal roots; this again must be attributed to the endarborization.

In amphibian sartorius muscle, Feng and Ting (1938) reported a decrease of about 30 per cent of the muscle esterase within 4 weeks after section of motor nerves, which is the period of time required for the degeneration of nerve fibers in these muscles. During this time no change in activity had taken place in that part of the muscle which was free of nerve endings. The changes of cholinesterase concentration at motor end plates of guinea pig gastrocnemius after section of the motor nerve were studied by Couteaux and Nachmansohn (1940, Couteaux, 1942). The estimates in this case are complicated by the considerable loss of weight of these muscles so that the total concentration of the enzyme appears to have increased. Correcting for the factor of volume change, it was estimated that the amount of esterase which disappeared at the motor end plate within the first week after section of the nerve, i.e., the period of time during which the nerve terminals disappear, is about 10 per cent of the total enzyme concentration present at these junctions. Four weeks after section the enzyme concentration had decreased by about 50 per cent, but it was still high and remained high for prolonged periods of observation. The estimates in denervated muscles of guinea pig are complicated by several factors: there is a huge amount of "inactive" tissue in relation to the active points; the inactive mass changes greatly with the course of time; the amount of nerve terminals at the motor end plate of this particular muscle is rather small. The estimate of a decrease of 10 per cent during the disappearance of nerve terminals is, therefore, less precise and much more approximate than the figures obtained for the terminals of ganglia and spinal cord and those of other muscles. Although the results still point to the presence of the enzyme at nerve terminals, they must be considered in connection with the data obtained with more favorable material.

Stoerk and Morpeth (1944) followed the changes of acetylcholinesterase concentration in the gastrocnemius of rats after section of the sciatic nerve. These muscles are much more favorable for the study of this special problem, since there are no volume changes as in the corresponding muscle of guinea pigs. The authors found that within 2 days one-third of the enzyme disappeared; the remaining activity remained essentially the same over a period of 3 weeks. Their data indicate, in good agreement with the other biochemical results reported, a high concentration of cholinesterase both in the nerve terminals and in the postsynaptic structure. The ratio of the two fractions varies considerably, as might be expected from the variations in structure.

B. Histochemical Data

The introduction of new histochemical and staining techniques has brought valuable new information in many respects. The results will be discussed only insofar as they have a bearing on the discussion of the mode of action of acetylcholine in synaptic transmission. Using postvital staining with such stains as Janus green B, Couteaux (1947) described, in the superficial layer of the sarcoplasm of the motor end plate, a complex lamellar structure which he referred to as subneural apparatus. This structure is situated at the deep face of the nerve branches wherever they come in contact with the sarcoplasm. When Koelle and Friedenwald (1949) described their histochemical method for determining esterase activity, using acetylthiocholine as substrate for acetylcholinesterase, they found, in confirmation of the biochemical data, the existence of a high enzyme concentration at motor end plates. This technique was then applied by many investigators for more precise studies of localization of the enzyme at motor end plates as well as at many other sites. However, as is frequently the case with staining techniques, the method had to overcome many pitfalls. Several modifications of the technique became necessary before the results could be satisfactorily evaluated, and many of the early results did not stand after the necessary corrections and improvements. Detailed discussions may be found in the lucid presentation of Couteaux (1955; Couteaux and Taxi, 1952). The interpretation of the results obtained with the histochemical staining techniques offered some difficulties in regard to the problem of the presence of the enzyme in the nerve terminals. With muscle preparation of the mouse, staining of the nerve terminals appeared only after a treatment which may cause diffusion artifacts (Couteaux and Taxi, 1952). The authors consider it, therefore, possible that the picture obtained is the result of a diffusion of the reaction products; the question remained open as to the presence of the enzyme in the presynaptic membrane; in view of its vicinity to the subneural structure (see the electron microscope picture in the following section), the method did not permit a distinction. Coërs (1953), following with Koelle's method the changes of cholinesterase activity at the motor end plates of rat gastrocnemius muscle after denervation, did find, however, a weakening in the intensity of staining after degeneration of nerve terminals, indicating an important diminution of activity. Recently, Koelle and Koelle (1958) found definite evidence for the localization of acetylcholinesterase in axon terminals of various ganglia of cat by applying a new procedure in which staining techniques were used in combination with reversible and irreversible enzyme inhibitors (Koelle, 1957) and by comparing ganglia before and after denervation. The histochemical data thus are in agreement with the biochemical data: both indicate the presence of acetylcholinesterase at both pre- and postsynaptic structures.

The significance of the presence of the enzyme in the nerve terminal will be discussed in connection with other recent data on the synaptic junction.

CHAPTER XVI

Synaptic Transmission
III. Differences between Axon and Synapse

A. Structure

As stated so correctly by Couteaux (1955), the morphology and physiology of the neuromuscular junction have long been studied almost independently of each other. In recent years serious attempts have been made to bridge the gap. For a satisfactory interpretation of the mechanism of synaptic transmission, knowledge of structure and ultrastructure are indispensable.

In recent years interesting new information about the structure of myoneural junctions has been obtained mainly by Couteaux (1947, 1955; Couteaux and Taxi, 1952) in his studies with the light microscope in combination with various staining techniques. During the last few years extensive electron microscope studies, as mentioned before, have greatly advanced our knowledge of cell membranes, including those of nerve and muscle, and have contributed to a better picture of the ultrastructure of myoneural and synaptic junctions (Robertson, 1956a, b; 1957a, b; 1959; Reger, 1957; Palade and Paley, 1954; Palay, 1956; De Robertis and Bennett, 1955; Luft, 1956; Edwards *et al.*, 1958).

The electron microscope picture of the myoneural junction is illustrated by Fig. 61, which is an interpretive diagram of the myoneural junction of lizard by Robertson (1956a). The small branches of axon terminals lie in troughs in the surface of the muscle fibers. The surface membrane in these troughs is formed by complex branching and anastomosing folds; they are identical with the subneural apparatus of Couteaux. A compound membrane 500–700 A. thick separates axoplasm from sarcoplasm. This compound membrane, referred to by Robertson as synaptic membrane complex, consists of five distinct layers. The middle layer was for a time interpreted by Robertson to be the continuation of the outer membrane covering axonal and muscular fibers. Recently he is inclined to consider it as an artifact, resulting from a precipitation of the ground substance.

Several features are still interpretative and need further clarification. But looking at the electron microscope picture of a myoneural junction, it appears quite obvious that such a remarkably complex morphological structure must greatly affect the electrical events. It appears *a priori* likely that certain functional properties, and especially the electrical manifestations of such a structure, should show marked deviations in many respects when compared to those in a simple cylinder such as the axon. It may be recalled (Eccles, 1946) how electric fields and the response to them are dependent on the "geometry" when studied even with such simple structures as axons. No information exists about the effects of geometry on electrical events in structures of such complex nature.

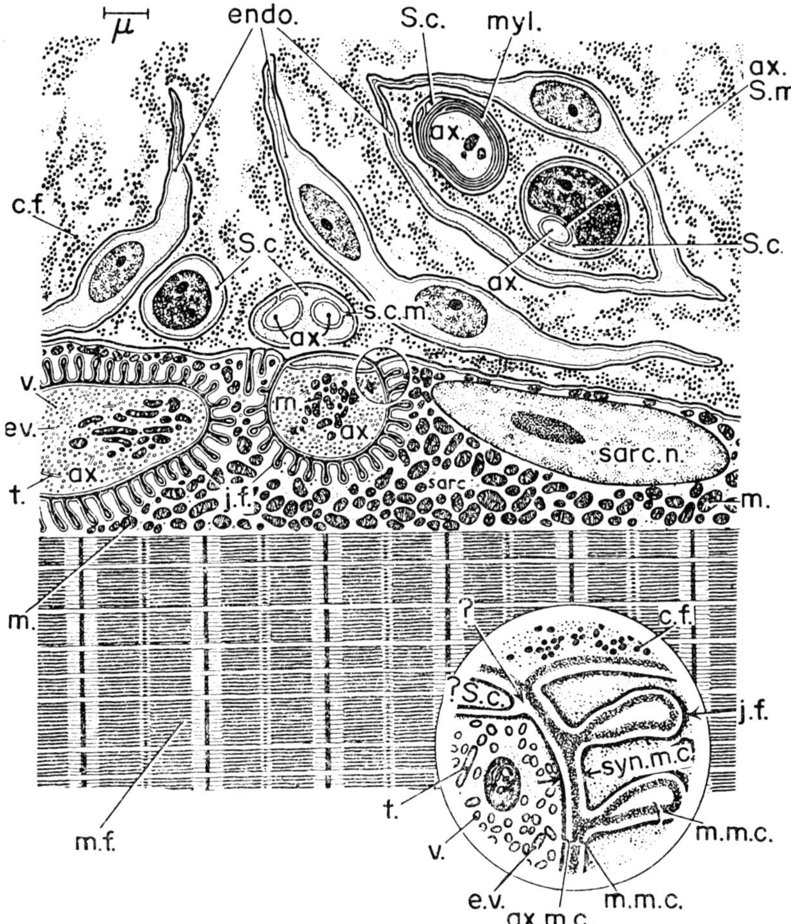

FIG. 61. An interpretative diagram of a myoneural junction in lizard, taken from Robertson (1956a).

The manner in which the muscle surface complex is thrown into the junctional folds of the neuronal apparatus is indicated. The region marked by the circle is enlarged to show more detail. The layers of the synaptic membrane complex are shown between the arrows (*syn. m.c.*).

Without this knowledge, interpretations of electrical signs and their meaning in regard to the mechanism of transmission can be only speculative and tentative.

An interesting finding of electron microscope studies are the vesicles (De Robertis and Bennett, 1955) present in great numbers in terminal portions of neurons. They have also been found in postsynaptic structures, in Schwann cells (Robertson, 1956b), and near Ranvier nodes. Recently "aposynaptic" granules have been described on the postsynaptic side of the junction in insects (Edwards *et al.*, 1958). A considerable amount of literature already exists about the chemical aspects of these vesicles. It has been proposed that they contain packages of acetylcholine. The number of acetylcholine molecules per vesicle has been estimated. Physicochemical mechanisms as to how these packages of acetylcholine are released from the vesicles through the terminals during activity have been proposed. The formation and the hydrolysis of acetylcholine by choline acetylase and by cholinesterase has been envisaged as being related to variations of these vesicles as observed with the electron microscope. Those who are accustomed to more orthodox methods of chemistry and biochemistry will find it difficult to accept theories as to biochemical events and physicochemical mechanisms based exclusively on electron microscope pictures.

B. Electrical Signs; the End Plate Potential

Göpfert and Schaefer (1938), discovered the existence of a special end plate potential. Its characteristics were further elaborated by Kuffler (1942a, b). Since then a considerable volume of studies has been done on the various features of this and other synaptic potentials and the effect of drugs upon these potentials. The technique of penetrating frog muscle fibers with microelectrodes, glass pipettes filled with KCl, having a tip less than 1 μ in external diameter, introduced by Ling and Gerard (1949) and further developed by Nastuk and Hodgkin (1950), greatly helped detailed investigation of the postsynaptic potentials. For these studies the microelectrodes are inserted in close vicinity of the end plate, the other electrode being outside the muscle fiber. The action of acetylcholine on the potentials was studied with this technique by Fatt and Katz (1951) and has been since then the subject of particularly numerous and extensive investigations. It is generally agreed today that the end plate potentials are produced by the action of acetylcholine and that the ester has a depolarizing action there. This idea was actually suggested a long time ago by Dubuisson and Monnier (1934) and Cowan (1936) and experimentally demonstrated by the depolarizing action found in 1939 on the *Torpedo,* as reported before. Further progress was achieved when Eccles and his associates (Brock *et al.*, 1952a, b) succeeded in inserting microelectrodes into motoneurons. A detailed discussion of the electrical characteristics of the postsynaptic potentials may be found in various reviews (e.g., Fatt, 1954; del Castillo and Katz, 1956) and in a recent book of Eccles (1957).

Nerve terminals are inaccessible to corresponding studies with inserted microelectrodes; but few electrophysiologists will be willing to assume that the negative waves sweeping down the axon stop before the nerve terminal and that there are no presynaptic action potentials. The technical difficulties leave us at present with the necessity to speculate about various aspects of these potentials. What is the basis for the synaptic delay which is of the order of magnitude of 0.5 msec.? Is it due to the slowdown of propagation in the fine terminal filaments or to the delay caused by the initiation of the potential in the second unit, whatever its nature, or to both? The delay cannot be attributed to a diffusion of a chemical substance from the pre- to the postsynaptic membrane. The high temperature coefficient of the latency period and the high energy of activation exclude diffusion as a major factor responsible for the synaptic delay.

The more recent electrical data do not offer the possibility of a definite decision between the two opposing views, the one accepting neurohumoral transmission in its original form and the other assuming that ion movements are the propagating agents and that release of acetylcholine takes place in both pre- and postsynaptic membranes. The failure to demonstrate experimentally electrotonic spread from the terminal has been used as an important argument in favor of the assumption of neurohumoral transmission. But in their review, del Castillo and Katz (1956) admit the weakness of such reasoning. Attenuation of subthreshold electric signals in the fine terminal filaments may be much more severe than in larger and more accessible parts of the system. Therefore, the signals might have faded out before reaching the junction. In fact, one may repeat here the pertinent question of Erlanger (1939) and include the new information about the actual distance involved: Since electric currents are known to be able to cross a nonconducting gap of more than 1 mm., is it reasonable to assume that they would not be able to cross the distance of 500–700 A. between pre- and postsynaptic membranes? Electrotonic spread is inadequate to propagate the impulse even along the axon. Forty years ago, Lucas (1917) and Adrian stated the necessity of locally supplied energy at successive points of the axon during propagation. If this is true for the axon, it must certainly apply to the junction. If the ionic concentration gradients are the local source of energy for currents propagating the impulse along axons, and this potential source is made available at the excited points by the action of acetylcholine, currents generated in the presynaptic terminal may be generated in a similar way. These currents, or ion movements, must reach the postsynaptic membrane where acetylcholine will be mobilized, thereby producing the postsynaptic potential. Action currents passing through fibers are able to stimulate adjacent fibers under appropriate conditions and suitable composition of the outer milieu. Such favorable conditions may prevail at the junction.

In this connection the interesting observations of Bullock and Hagiwara (1957) on the giant synapse of the stellate ganglion of squid may be mentioned. This synapse, because of the large size of the pre- and postsynaptic structures, permits recording of the presynaptic spike and of the postsynaptic response. There is a synaptic delay of 0.5 msec. at 22° C. and 2 msec. at 10° C. Only an insignificant amount of current (100–300 microvolts) can be detected passing between the two structures. These experiments raise interesting questions; the interpretation of the possible mode of transmission is, however, difficult as long as the ultrastructure is not well established. There is a possibility that the pre- and postsynaptic membranes are fused at some points at which there would then be no extracellular space. This would, of course, create very unusual and peculiar conditions since propagation across these areas could not be explained in terms of any known mechanisms and might instead proceed around these points.

Inhibitory Action on Nerve Cells: Hyperpolarization. Inhibitory synaptic activity has long been known. Various explanations have been given, based on indirect evidence such as the depression of a testing reflex discharge. Bernhard, Skoglund, and Therman (1947) demonstrated that conduction of impulses by certain spinal pathways may result in excitatory effects in some motoneurons, inhibitory in others. New additional information has been obtained with microtechniques which permitted intracellular recording. Studying the inhibitory actions on motoneurons with these techniques, Eccles and his associates observed a transient increase of the membrane potentials, i.e., a hyperpolarization (Brock *et al.*, 1952a, b). The potentials have been designated as inhibitory postsynaptic potentials in contrast to the excitatory postsynaptic potentials. Synaptic inhibitory action on crustacean stretch receptor was shown by Kuffler and Eyzaguirre (1955) to be attributable to a repolarizing action exerted on cells which had been depolarized by stretch. Several similar effects, or slight variations, of inhibitory mechanisms were described for other junctions with the use of microelectrodes. The inhibitory action of the vagus on the heart has long been known since Gaskell (1886) demonstrated increased demarcation potential of the cardiac muscle which accompanies stimulation of the vagus nerve (see also Chapter XIII). Hyperpolarization and decreased action potential were shown to be produced by acetylcholine or carbamylcholine applied to a strip of muscle from cat's auricle (Burgen and Terroux, 1953). These observations were confirmed by del Castillo and Katz (1955c) and by Hutter and Trautwein (1955). Inhibitory mechanisms at the neuromuscular junction of crustaceans were observed by Fatt and Katz (1953b). A large and prolonged postsynaptic hyperpolarization evoked by a presynaptic stimulus was observed by Laporte and Lorente de Nó (1950) in the deeply curarized cervical ganglion of turtle.

The phenomenon of inhibition by hyperpolarization greatly surprised and impressed Eccles and was apparently an important factor in his famous "conversion" to the neurohumoral transmitter theory. It is certainly good reasoning to attribute hyperpolarization to a chemical reaction. Hyperpolarization is unexplainable in purely physical terms. But so too is depolarization. Both phenomena require chemical processes in the active membrane regardless of whether they occur in axons or at synapses. However, the fact that the phenomena are produced by chemical reactions does not give a clue to the problem of how they are initiated. The crucial events may be produced by current or ion flow from the pre- to the postsynaptic membrane mobilizing chemical reactions there. The amperage of the currents generated by the presynaptic fiber would, according to Eccles (1957), fail by a factor of 50 to produce the current required for activating the postsynaptic membrane of motoneurons even if the presynaptic fiber could very effectively inject its currents across the postsynaptic membrane; in motor end plates of amphibian muscle the factor may be higher than 1000. This, of course, requires the mobilization of chemical energy. Again it does not reveal the mode of activation and the chain of events by which a small effect initiates a chemical reaction. This action may eventually have very powerful consequences. The energy of the electric wave propagated along the muscle fibers may, for example, release amounts of energy of contraction many thousand to a million times greater. Many other examples could be given.

Variations of the postsynaptic membrane potential by extrinsic current cause large changes and even a reversal of the inhibitory postsynaptic potentials. This observation may rather be used as argument against neurohumoral transmission than for it, because it indicates the possibility of intracellular or intramembranous chemical changes produced by extrinsic current as being responsible for variations of hyperpolarization. If that is the case, currents flowing from the pre- to the postsynaptic membrane should be able to account for initiating the chain of events leading to those changes. Special structural arrangements may make the small eddy currents at junctions quite efficient in initiating the chemical reactions responsible for the changes in the postsynaptic membrane for either hyper- or depolarization.

Of interest is Eccles' discussion of the observations on the electric response of deeply curarized ganglia, first reported by Laporte and Lorente de Nó (1950) and later studied by other investigators. Two phases were found: first a hyperpolarization, then a depolarization. Eccles admits that this finding presents a difficult and unresolved problem that defies simple explanations in the usual terms of neurohumoral transmission. He attributes the hyperpolarization, as usual, to some unknown transmitter. The finding that the depolarization is effectively depressed by all anticholinesterases, but not by *d*-tubocurarine and its analogs, is difficult to explain. Eccles suggests a new type of cholinergic transmission, wherein acetylcholine depolarizes the ganglion cells by acting on special receptor areas having a cholinesterase configuration. It must be envisaged, according to Eccles, that a single transmitter substance may act on two completely different receptor surfaces in the same subsynaptic membrane. In this instance Eccles apparently becomes aware of the complexity of chemical systems. It is, of course, easy to propose many alternative explanations. Whatever the interpretation may be, Eccles recognizes that the action of acetylcholine should not have the same effect under all conditions and proposes that in addition to the active ester other members of the system must be considered. Since, therefore, even according to Eccles, such reactions are unpredictable as long as so little is known about the other factors and components of the membrane involved, interpretations of the detailed mechanism of synaptic transmission based on electrical and pharmacological signs can be only speculative until more information becomes available as to chemical processes involved. It seems desirable to emphasize the tentative nature of our views concerning the process of transmission, rather than to present statements in a categorical form, frequently ignoring the well-established biochemical data, as if the picture were final and excluded the possibility of any modification.

C. The Site of Drug Action

The powerful effect of acetylcholine and analogous compounds on synaptic junctions was one of the two essential facts which formed the basis of the neurohumoral transmitter theory. It was assumed without much questioning that acetylcholine is released from the nerve terminal and acts on a receptor in the second cell. The crucial question as to the precise site of action was not even raised. Are these agents really acting only on the postsynaptic membrane? If an alternative mode of action is considered, i.e., the possibility of an *intra*cellular process taking place in pre- and postsynaptic membranes, the receptor should be present in both membranes and in this case both sites might be affected. They may not be reached with the same ease; a larger amount of receptor may be present in one or the other membrane, the sensitivity may differ for other reasons, but the crucial question remains whether or not the receptor is present in both membranes.

Nineteen years ago Masland and Wigton (1940) reported the important observation that Prostigmine applied to the motor end plate produced rapid bursts of electrical activity in the motor roots, in addition to its action upon the muscle fiber. Curare blocked both muscle and nerve response. The authors concluded that their observations provide evidence that acetylcholine and Prostigmine stimulate the motor nerve terminal as well as the muscle; the concept of a specific local effect of acetylcholine on the muscle at the myoneural junction is, as emphasized by the authors, insufficient to explain the findings. Most investigators, and the writer is one of them, did not recognize the startling implications of their findings, the evidence for an acetylcholine receptor in the terminals. Masland and Wigton's results found support in observations of Eccles, Katz, and Kuffler (1942). These authors state that there is no doubt that nerve discharges observed in presence of eserine are independent of, and prior to, the muscle impulses. In the experiments of Laporte and Lorente de Nó (1950) curarization was mainly caused by a modification of the presynaptic fibrils and resulted in block of conduction in the presynaptic arborization; this again points to the presence on that site of a receptor for acetylcholine with which curare competes.

The problem of the site of action of compounds applied externally and acting on the myoneural junction has recently been taken up by Riker and his associates (Riker *et al.*, 1957, 1959; Riker, 1959). They tested the effects of 3-hydroxyphenyl trimethyl ammonium ion and of various analogs such as phenylethyl dimethyl ammonium ion (Tensilon) on the motor end plates of cat muscle. Riker and associates observed that the compounds applied to the junction produced repetitive discharges in motor roots. They conclude that these compounds must definitely act on a receptor in the nerve terminals. They postulate that acetylcholine is released from a bound form in the motor nerve terminal and reacts there with a receptor to initiate synaptic transmission. This view is in agreement with that proposed by the writer many years ago.

Further problems of interpreting the mode of action of drugs arise from the complexity of the acetylcholine system. An intramembranous action of acetylcholine implies, as stressed before, that all four members of the system are located at close vicinity and structurally organized. A compound penetrating from the outside may affect either one protein or several of them, depending on the affinity constants. The powerful action of acetylcholine and other analogous quaternary ammonium ions has always been associated with their action on a receptor substance. This interpretation has also been widely accepted for diquaternary curarelike nitrogen derivatives; they may have not only a blocking, but also a stimulating, action just as acetylcholine has, and this action may even be a stronger one. Paton and Zaimis (1949), for instance, found a depolarizing action in a series of diquaternary ammonium ions where the two trimethylated nitrogen groups were separated by a varying number of methyl groups; the C12 compound was the most effective. Although these diquaternary compounds have higher affinities to acetylcholinesterase than their monoquaternary analogs—the *K*'s of the former range between 10^{-4} and 10^{-5}, those of the latter are closer to 10^{-3} or even 10^{-2} —the effects of these compounds are generally attributed to an action upon the receptor.

In spite of the evidence for the reaction of quaternary nitrogen derivatives with the receptor, effects of some of the quaternary compounds were for a long time interpreted exclusively in terms of acetylcholinesterase inhibition because of their high affinity to the enzyme as measured by their inhibitory action *in vitro*. During the last decade, however, much evidence has accumulated from pharmacological studies that suggests an action of this type of compound on both members of the system. Aeschlimann and Stempel (1946) and Lehmann (1946) reported as direct effects of Prostigmine the producing of contractures of frog rectus muscle. The assumption of a direct action upon the receptor was further supported by observations of Riker and Wescoe (1946), who obtained acetylcholine-like contraction with Prostigmine after the enzyme had been strongly inhibited by DFP. The difference in action of Prostigmine and eserine on the end plate potential described by Eccles and MacFarlane (1949) also suggests—as has been pointed out by Riker (1953) —an action of the quaternary compound which cannot be explained exclusively in terms of esterase inhibition. The block of conduction by Prostigmine was shown on the electroplax to occur before a critically low level of the enzyme had been reached (Altamirano, Schleyer, Coates, and Nachmansohn, 1955).

The extensive studies on the pharmacological action of 3-hydroxyphenyl trimethyl ammonium ion and its analogs have been the object of discussion in regard to the component affected. Nastuk and Alexander (1954) have emphasized the importance of the inhibition of cholinesterase in experiments in which a potentiation is observed in the presence of Tensilon. Tensilon is a potent inhibitor. The enzyme-inhibitor dissociation constant is 3×10^{-7}, which is close to that of Prostigmine, 1×10^{-7} (Wilson, 1955b). Wilson also found that the compound reacts very fast with the enzyme, within seconds. The rapidity of the onset of the potentiating action is, therefore, not in contradiction with the interpretation proposed. On the other hand, the rapid anticurare action of these compounds is more difficult to explain in terms of enzyme inhibition. The absence of a depolarizing action in concentrations which counteract curare permits several alternative interpretations. An action on the receptor of the nerve terminals, which has already been considered as a possibility by Nastuk and Alxander, has now been demonstrated by Riker and colleagues. A bimodal action, i.e., on the enzyme as well as on the receptor, appears more likely than an action limited to one component; it may depend on concentrations. A further discussion of the pharmacological aspects may be found in the papers cited, but the studies are useful illustrations of the complexity of the drug effects. The isolation of the receptor protein by Ehrenpreis has opened the possibility of obtaining quantitative data as to the difference in dissociation constants of various compounds and the receptor. Until now such estimates were dependent on effects observed on living cells, a procedure subject to many errors. The availability of the receptor in solution promises, for this and other reasons, to clarify some of these problems.

Still another complication is due to the dependence of the reaction of the receptor on the particular structural properties of compounds. As we have seen, they may be receptor activators or inhibitors, i.e., they may produce depolarization and have a stimulating action or merely block electrical activity without depolarization. Finally, quaternary compounds may act with varying strength on the two other protein members of the system, the storage protein and the choline acetylase. Little consideration has been given to this possibility.

The difficulty of interpreting cellular mechanisms only on the basis of effects of drugs injected into blood vessels or applied to complex preparations is also suggested by the discrepancies found between factors affecting the acetylcholine action on synaptic junctions and those affecting physiological transmission. In perfusion experiments of the superior cervical ganglion of cat, Kewitz and Reinert (1954) found that the absence of glucose and lack of potassium decrease the response to injected acetylcholine; the response to electrical stimulation was not affected. Calcium deficiency modified the two processes in an opposite way: acetylcholine action was greatly increased, the response to electric stimulation was abolished. It would be difficult and not very profitable to discuss these and other discrepancies reported and their possible causes; they are quoted as other illustrations of the contradictions encountered. They should caution against using the effects of drugs, metabolites, antimetabolites, or ions as a means of explaining their role in a cellular mechanism. The actions of compounds on living cells frequently raise interesting problems; although they may provide important information, they permit only limited conclusions as to the mode of action, and only in connection with other data. This has been often stressed by leading pharmacologists. In 1933 A. J. Clark wrote: "The physical chemist can reasonably hope to simplify his conditions and to reduce the number of variables, until he obtains a system that provides formal proof of the laws which he enunciates, but the pharmacologist is interested in the action of drugs on the living cell and any attempt to simplify this material results in death. Hence he cannot hope to obtain formal proof for his theories and must be content with intelligent guesses. Even in the most favorable cases where quantitative relations have been established for the action of drugs on cells, there probably remain dozens of unknown variables, and there is usually a considerable range of possible alternative explanations." Two decades ago scarcely any data were available on the biochemical and physicochemical aspects of the acetylcholine system. Its proteins had not been isolated and were unavailable for studies in solution. A tentative interpretation of the mode of action of acetylcholine, given at that time on the basis of its pharmacological effect, was justifiable. The rigid maintenance of the original interpretation, as well as the vigorous opposition and categorical rejection of any modification which tries to integrate the results obtained with a variety of methods appear less justifiable.

D. Ion Movements and Permeability Changes at Junctions

Fatt and Katz (1952b) suggested that the efflux of acetylcholine from nerve terminals may be analogous to that of K from the axon and that acetylcholine may leave the terminal by a specific exchange with Na which enters the terminal during activity. Later they realized that this hypothesis was untenable and withdrew their suggestion. They now assume that acetylcholine acts on the postsynaptic membrane by increasing permeability to ions. Such a function of the ester is essentially the same as that proposed twenty years ago for its role in conducting membranes in general, only with the modification that release and action are intracellular. However, while in the axonal membrane a specific transitory increase in Na permeability takes place, followed by a smaller and less rapid change in K permeability, the change of permeability in the postsynaptic membrane is assumed to be unspecific. The authors base their hypothesis on the following finding: when a muscle is immersed in isotonic K_2SO_4 solution, the membrane is completely depolarized; the preparation is inexcitable electrically and application of acetylcholine does not produce a potential change in either direction. When, however, the membrane potential is displaced from zero by the passage of current, renewed application of acetylcholine reduces the potential difference built up across the end plate membrane (del Castillo and Katz, 1955b). This displacement of the potential difference is interpreted as an increased flux of K. It is, therefore, concluded that acetylcholine increases, in the postsynaptic membrane, both Na and K conductance and possibly the permeability to other ions. Consequently, it is argued, acetylcholine cannot be responsible for the more specific permeability changes in the axon.

High K concentration drastically affects membrane properties. Passage of transverse current must also influence them. Effects of acetylcholine applied under such extreme and unphysiological conditions may be entirely different from those of acetylcholine released within the membrane under physiological conditions. The ion movements across the axonal membrane during activity have been established with elaborate techniques, with the aid of radioactive material, over a period of many years. Only with the use of delicate methods and experimental conditions which avoided drastic changes was it possible to demonstrate the difference of the change of Na and K conductance during electrical activity. With less sensitive procedures, an equal change would have been found, i.e., an increased Na influx and K efflux of the same order of magnitude, since electrical neutrality must be maintained inside the fiber. This would then have suggested an equally increased permeability to both ions.

On the other hand, even if the action of acetylcholine is essential to the changes in membrane permeability, the ester cannot be the only determining factor. The complexity of the permeability change has been discussed in Chapter XIII. There are many factors which may modify the response of the synaptic membrane when compared with that of the axon. We know that there is no permeability barrier for lipid-insoluble compounds. On the other hand, it has been found that removal of Na from the extracellular fluid blocks synaptic transmission as well as the effect of applied acetylcholine (Nastuk, 1954). This observation would point to important similarities of the membrane properties and the function of the ester at both synapse and axon. Nevertheless, it is entirely possible that as a result of other factors there are modifications in the effect of acetylcholine action on ion permeability in axonal and synaptic membranes. However, experiments of a completely different type would be necessary to substantiate such an assumption.

Accepting Fatt and Katz's evidence that acetylcholine increases the permeability of the postsynaptic membrane to all ions and in this way produces the excitatory postsynaptic potential, Eccles (1957) proposes the hypothesis that the inhibitory postsynaptic potential in motor neurons is associated with an increase for small ions only, of about the size of K^+ and Cl^-, Na^+ being excluded. He envisages the excitatory and inhibitory postsynaptic membrane as being momentarily converted into a sieve-like structure with two different sizes of pores. He does not suggest that the actual pores have these sizes, only that the effective size of the pores as far as the passage of ions is concerned differs. This interesting picture is at present purely hypothetical, based mostly on variations of potentials due to intracellular increase of ions from the micropipette under the influence of current pulses. In view of the hypothetical nature of this picture, the significance of these observations need not be discussed here until further explorations have been made. However, Eccles proposes that some hitherto unknown transmitter substance, different from acetylcholine, produces the opening of the small pores. Since we know that acetylcholine is able to produce both hyper- and depolarization, it appears unnecessary to postulate another unknown substance taking the place of acetylcholine. Additional chemical forces and structural arrangements discussed previously may determine the character of permeability change. The reported effects of strychnine (Bradley *et al.*, 1953) rather support than contradict the assumption that acetylcholine action is associated with permeability changes in inhibitory effects. Strychnine has a high affinity to acetylcholinesterase; the dissociation constant of the enzyme inhibitor complex is about 2 to 3×10^{-5} (unpublished observations). The alkaloid may have a still higher affinity for the receptor. It blocks conduction in the giant axon of squid (Bullock *et al.*, 1946). Recently it was also found to block, in low concentrations, the response of the single isolated electroplax to both neuronal and direct stimulation (unpublished experiments of Schoffeniels).

In view of our limited knowledge as to the molecular forces effecting the permeability changes, all these hypotheses are obviously of a purely speculative nature. It is, therefore, difficult to understand how speculations proposed as possibilities on one page are used, frequently only a few pages later, as conclusive evidence either against the role of acetylcholine in conduction or for neurohumoral transmission at synaptic junctions.

E. Significance of Acetylcholine Appearance in Junctional Perfusates

The demonstration of the appearance of acetylcholine in junctional perfusates following activity was considered, in the 1930's, as important evidence for its transmitter role in the original sense. When those studies were made, no alternative explanation of the role of acetylcholine was envisaged. The significance of the appearance of acetylcholine in junctional perfusates must be reevaluated in the light of all the data accumulated in the last two decades in favor of an intracellular role of acetylcholine.

In contrast to the axon, which is surrounded by special structural permeability barriers, both pre- and postsynaptic membranes of the junction react with, and are therefore pervious to, acetylcholine and other quaternary ammonium ion derivatives. If a compound applied externally readily reacts with a protein in a membrane, the same compound, if it is there a naturally occurring metabolite, should be able to pass into the surrounding fluid with equal ease. If it is enzymatically metabolized at extremely high speed, as is the case with acetylcholine, we would expect that at most a negligible fraction might escape into the outside fluid. But if the cellular enzyme responsible for the rapid removal is inactivated, then of course a relatively large fraction may appear on the outside and may be collected in the perfusate.

According to the view that acetylcholine is essential for the permeability changes during activity in both nerve terminal and postsynaptic membrane, it would be expected that under optimal physiological conditions either negligible amounts of acetylcholine or none at all would leak to the outside from both membranes during the transmission of the impulse. This leakage should, however, be greatly increased on addition of cholinesterase inhibitors. This is exactly what happens. Appearance of the ester under these conditions cannot, therefore, be used either as evidence for neurohumoral transmission or against its intramembranous functions.

The appearance of the ester in perfusates not only played an important role in the earlier history, but is still frequently discussed as to its significance. It may be recalled that Otto Loewi encountered difficulties in reproducing the appearance of the *Vagusstoff* with regularity and did not succeed in finding it when he used *Rana pipiens*. Ascher (1925) and other investigators reported that they could confirm Loewi's finding only when the heart became "hypodynamic." These apparent contradictions may well indicate that the diffusion of a detectable amount of acetylcholine is a sign of a slight damage of the membrane. This is readily conceivable on account of several factors, especially since the membrane is unprotected and is known to be particularly vulnerable to all kinds of irritations. Once the membrane is damaged, the inactivation of the ester by the enzyme may not take place with the same speed and precision as under physiological conditions. Dale and his associates were puzzled by Kibjakow's (1933) early report of the appearance of acetylcholine, without the addition of eserine, in the perfusion fluid of ganglia. Kibjakow's perfusion methods may have been less elegant and more damaging for the membrane than those of Dale and his associates. Similarly, Lorente de Nó (1938), studying the appearance of acetylcholine in the perfusate of the superior cervical ganglion of the sympathetic and of the ganglion nodosum of the vagus found the ester on stimulation of the ganglion nodosum and on antidromic stimulation of the sympathetic ganglia. He attributed this appearance to cellular damage and described histological observations supporting his views. Lorente de Nó was one of the first to reject the perfusion method in principle as being inadequate to decide the function of acetylcholine in the process of transmission. He considered his findings to be incompatible with the idea that acetylcholine metabolism is limited to synapses: "The acetylcholine metabolism is not a process which is specific for the synaptic junction." This conclusion, written in 1938, has been borne out by many biochemical and physiological data accumulated since then. The apparent inconsistencies and contradictions reported are readily explained in terms of the intracellular processes proposed.

Another aspect which is frequently discussed in the literature, is the quantity of ester collected in the perfusate in relation to that required to produce an action. The first quantitative evaluation made by Dale and his associates revealed a discrepancy, of a factor of about 100, in spite of the presence of eserine; this was considered by the investigators themselves as a difficulty for the proposed mode of action. Revised figures even increased the factor to 100,000. Recently, refined electrophoretic methods of application of acetylcholine to the motor end plate were developed (Nastuk, 1953; del Castillo and Katz, 1955a). With this efficient technique it was found that about 6×10^8 molecules of acetylcholine produce a response at one motor end plate of a frog sartorius muscle, although the time course of the response (15 msec.) was still not the same as that evoked by neural stimulation. This figure is of the same order of magnitude as that estimated twenty years ago on the basis of biochemical studies (Marnay and Nachmansohn, 1938; Nachmansohn, 1939a). The amount of ester which may be hydrolyzed at a single motor end plate of frog sartorius muscle per millisecond is 1.6×10^9 molecules. This figure gives only the potential rate and not the amount actually metabolized, which could be calculated only if the excess of enzyme over the minimum required were known. The excess of enzyme in axons is about fivefold, but it may be higher at synaptic junctions. Moreover, the period of time required for the removal of the ester released per impulse is not known; this period may be either slightly more or slightly less than 1 msec. This introduces another factor of uncertainty. But the order of magnitude is comparable to the minimal amounts found to produce a response. On the basis of autoradiography of end plates of a cat diaphragm treated with radioactive curare or decamethonium, Wazer and Lüthi (1956, 1957) calculated that 8×10^6 molecules were fixed per motor end plate. This figure is quite close to the values obtained with different methods and a different type of muscle for evaluating the amount of acetylcholine required to activate a single motor end plate.

The amounts of acetylcholine after activity in perfusates for which Ringer's solution containing eserine was used, have been reevaluated in very carefully controlled observations by Emmelin and MacIntosh (1956). The discrepancy between the amounts per stimulus found in perfusates and those required to evoke a response, is markedly smaller than in earlier reports. A really quantitative evaluation of the ratio found is extremely difficult. Let us assume the difference to be five- to tenfold, which is on the low side of the estimates of these investigators who support neurohumoral transmission. This may mean that in the presence of eserine 10–20 per cent of the acetylcholine required for activity would escape from the membranes. This is still a very small quantity. In a muscle contracting a few times, probably no extra lactic acid is formed. The ATP hydrolyzed may be restored from phosphocreatine and from the pool provided by the citric acid cycle. With increasing amounts of work performed, lactic acid is formed and begins to accumulate, and with more strenuous work it leaves the cell and appears in the outside fluid. A great number of stimuli is required for obtaining in the perfusion fluid an amount of acetylcholine sufficient for its detection. It appears not at all surprising that under these conditions a small fraction of the ester, which is released intracellularly, escapes from the membranes located at the surface, since the ester released is protected from the attack by the esterase. Obviously, the action should be much more efficient if the compound released combines with the receptor within the same membrane. The discovery of the appearance of acetylcholine in the perfusates was important as a sign that acetylcholine plays a role in transmission. However, even the much smaller discrepancy reported now, when compared with that described earlier, fails to meet the principal objections raised above against using these findings for deciding the mode of action of the ester.

In this connection the observations of miniature end plate potentials

(Fatt and Katz, 1952a) may be briefly mentioned. It appears likely that the action of acetylcholine is essential for these potentials. The fact that these potentials are blocked by curare and are increased by esterase inhibitors supports such an assumption. These effects do not, however, indicate whether the release of acetylcholine is the primary cause, nor do they reveal the site of origin, since the ester may act in the same way when released on either side of the junction. Nerve terminals probably exert a strong influence upon the surrounding membrane of the effector cell. The events taking place when a nerve ending makes a new contact with the muscle fiber or when it degenerates and disappears, may be considered an indication. Various kinds of stimuli may pass all the time from the terminal, as long as it is functionally intact, to the opposite cell membrane. There may be an intermittent release of acetylcholine either within the pre- or in the postsynaptic membrane, or in both. At present there is no evidence for any relationship with the physiological events in conduction. The significance of the potentials becomes even more obscure by the finding that they continue, although in modified form, even in isotonic K_2SO_4-treated muscle, a condition under which the tissue is completely depolarized, the excitability of nerve and muscle has been abolished, and no Na is present in the fluid (del Castillo and Katz, 1955b). With increased magnesium and lowered calcium concentrations, the activity may also continue at an undiminished rate although the nerve impulse is no longer capable of initiating a response. Since these miniature potentials persist under conditions which are extremely unphysiological, i.e., when excitation and transmission are completely abolished, it appears not justified to link them directly to the physiological events taking place in transmission. Chemical reactions may have a stronger action than bioelectric currents: they may produce effects when the currents fail to do so. This is especially to be expected in such complex structures as the neuromuscular junction where geometry must play an important role in the response to electrical stimulation. This type of observation cannot be used as an argument that the membrane in physiological conditions cannot be excited by ions or by currents coming from the nerve terminals. In the same category belongs the reasoning based on the finding that degenerated structures, such as the electroplax of *Torpedo,* fail to respond to electric stimuli, whereas they still react to chemical compounds. Marked alterations take place during degeneration of nerve terminals, as indicated by the extensive changes of structure. Failure of a response to electric stimulation does not indicate that these structures are, under physiological conditions, irresponsive to currents or ions coming from the nerve terminals.

Just as in the case of the appearance of acetylcholine in the perfusion fluid in activity, the existence of the miniature potentials does not justify one to consider this observation as evidence for neurohumoral transmission. Neither of these findings gives an indication, and to an even lesser extent does it permit a decision, as to the nature of the propagating agent in physiological activity.

F. Present State of the Problem

Let us summarize the present state of the problem regarding the role of acetylcholine in synaptic and neuromuscular transmission. There is general agreement that acetylcholine plays an essential role in this process; the difference of opinion concerns the precise mechanism of action, i.e., the question of whether acetylcholine acts in the pre- and postsynaptic membranes to permit ion flux as in conduction, or whether acetylcholine acts as a neurohumoral transmitter as originally proposed.

To refer to the difference between axonal conduction and synaptic transmission as to the one being "electrical," the other "chemical," is an unfortunate and misleading terminology; it obscures the real issue. Some physiologists (Hodgkin, 1957; Eccles, 1957) referring to conduction as being "electrical" actually mean a purely physical process, as opposed to transmission, where they now accept the necessity of intervening chemical reaction. This is in contrast to previous views, e.g., those of Eccles (1946), in which transmission also was for a long time conceived in "electrical," i.e., purely physical terms. According to these views the difference between conduction and transmission is, of course, basic in nature. The real issue under discussion may then be formulated as follows: Does even conduction require a chemical reaction for the generation of the electric currents and is acetylcholine essential for this process everywhere in conducting membranes, or is the activity of acetylcholine specifically limited to the synaptic and neuromuscular junction, conduction being a purely physical process? Between these two views the gap is very wide indeed. It can hardly be bridged. If, however, one accepts the essentiality of acetylcholine in both processes, the question remains whether acetylcholine acts at synapses as a neurohumoral transmitter or whether it acts, there too, intramembranously as in the axon. This would obviously be a much less "fundamental" difference; it would be a minor matter of detailed mechanism. The writer does not find any convincing evidence for the mode of action originally proposed. On the other hand, the detailed mechanism still offers a great many problems, at the junction as well as in the axonal membrane.

Nobody denies that the events at the synaptic junctions show marked differences in many respects from those observed in the axon. Whether these differences are "fundamental" is a matter of definition: fundamental with respect to what? In regard to macro and micro structures there are striking differences; in this instance the attribute "fundamental" may be appropriate. These differences in structure must profoundly affect electric fields and electrical manifestations, time relations, response to various chemical agents, etc. But is the specific biochemical system acting in the elementary process on the molecular level and responsible for the propagation of impulses, different in the axon and at the synapse? If one categorically rejects a role of acetylcholine or, more generally, a chemical reaction in the primary process of conduction but admits it in synaptic transmission, then of course it would be justifiable to speak again of a basic difference.

There are a number of general principles and notions resulting from advances in dynamic biochemistry which make it difficult, from a biochemical point of view, to accept the hypothesis of neurohumoral transmission. All reactions taking place in a living cell are chemically and energetically coupled. This coupling of enzyme-catalyzed reactions is one of the most distinctive biochemical attributes of living matter. In contrast to processes in a homogeneous system, a condition of thermodynamic equilibrium does not exist, but rather a steady state. The coupling is very different from that observed in homogeneous enzyme solutions. Many biochemical reactions in an intact cell are structurally organized. Interesting documentations are provided by the recent developments obtained by electron microscope studies in combination with enzyme chemistry (see, e.g., Palade, 1956; Green, 1956, 1959). Even a process like oxidation, which is remote from more elementary activities of the cell, is not only localized in mitochondria, but in the cristae within the mitochondria, and the electron transfer proceeds there apparently in a well-organized pattern. It is true that some glands produce hormones which act as regulatory factors in the metabolism of body cells in general; in this case a compound produced by cells of one organ acts on cells in another organ, sometimes at very long distances. The idea of a secretion of neurohumoral transmitters from nerve endings was derived from the notion of hormonal secretion, and hence the transmitters have been referred to, and still are, as "neurohormones." The vesicles, for example, are considered to be little glands secreting acetylcholine (del Castillo and Katz, 1956). In fact, the whole brain is pictured as a big gland, each nerve ending secreting some kind of neurohormone (Blaschko, 1957). Although the situation in regard to adrenaline is in many respects different and need not be discussed here, the dependence of the elementary process of a cellular function on the secretion of hormones produced in another cell is an assumption which raises serious questions in the light of what biochemistry and electron microscopy have revealed as to coupling and organization of chemical systems. It is hard to picture that the fastest and most precise function developed by nature, the conduction of nerve impulses, includes at certain points chemical processes which are not structurally organized, not intramembranous, and not even intracellular, but intercellular; that a substance forming part of a complex system is produced in one cell and acts upon a specific protein in another cell. In that case we would have to assume that an uncoupling takes place in the acetylcholine system, which is one of the fastest-acting systems known, and that part of the reaction is localized in one cell, viz., the formation and the release of acetylcholine, while the reaction with the receptor and the hydrolysis by the esterase occur in the other cell. Admittedly, everything is possible, since the living cell is a heretic, to use an expression of F. G. Hopkins. But to make the view probable, substantial evidence must be presented before one accepts such a split of the system between two cells. The assumption of an unorganized and split system has become all the more difficult by the experimental evidence that both acetylcholinesterase and receptors are present in nerve terminals and that consequently the complete system is present on both sides of the intercellular gap, in the membrane of the nerve terminal and in the postsynaptic membrane.

The notion of the biochemical unity of life has frequently promoted the development of biochemistry. Most important chemical cycles take place, with only minor differences, in primitive organisms as well as in all types of cells of higher animals. The interaction between ATP and actomyosin is today considered to be the basis of motility of all types of muscle throughout the animal kingdom. The physical characteristics of this process vary greatly, depending on structural organization, etc. Since bioelectric current and permeability changes have been shown to be inseparably associated with the acetylcholine system in the axon, which is a much more favorable material for such studies than synaptic junctions because of its greater simplicity and accessibility, the difference of manifestations at the two sites of the same cell should be attributed to a different behavior of the system only on the basis of really strong experimental evidence. At present, no such evidence is available. Nobody will deny that at the synapse there may be quite a few additional secondary factors of importance modifying the events at junctions. Very little is known about such factors. Even such vigorous supporters of the idea of neurohumoral transmission as del Castillo and Katz (1956) make the statement: "Inferences about the physical state and chemical properties of nerve endings have had to be made by indirect argument and from analogies; and are therefore likely to be at fault." If one agrees with that statement, as the writer does, it appears difficult to accept the conclusion that the hypothesis of neurohumoral transmission is already firmly established.

Dale, in his Harvey Lecture (1937), envisaged the possibility of a much wider role of acetylcholine than was proposed at that time: "If the liberation of a chemical mediator at a nerve ending should prove to be not a process peculiar and limited to that ending but merely a local intensification, to ensure transmission to a contiguous cell, of a process which actually figures in the propagation of the impulse along the nerve fiber, we should have to make yet a further revision of our existing conceptions. Some minds have undoubtedly felt difficulty in postulating a complete breach in the nature of the processes concerned in transmission, where the excitation passes from nerve ending to effector cell. This particular difficulty would then disappear but only at the cost of a more fundamental change of conception concerning the nature of the propagated wave of excitation than any which has yet been seriously considered." This intuitive statement is all the more remarkable since at that time there was no experimental basis and scarcely even any indication for such a bold and imaginative assumption. It is impressive that at a symposium on neurohumoral transmission in 1953, Dale, one of the great pioneers in that field, recalled his statement of 1937 and warned the audience not to consider the theory of neurohumoral transmission as being a final solution of the problem (Dale, 1954). This spirit and attitude is certainly in striking contrast to the categorical statements of some of the adherents of his theory. Dale is reluctant to accept the unified theory; he considers it as a simplification which might lead us astray in biology. Methods of approach to the problem must of necessity greatly affect our thinking. The remarkable unity of biochemical principles and systems throughout the great diversity of living cells must influence the reasoning of the investigator attacking the problem primarily by chemical analysis. Physiology and pharmacology, on the other hand, study manifestations of intact cells characterized by a nearly infinite variety as to organization and structure. It may be useful to recall the words of J. J. Thomson, in his book "The Corpuscular Theory of Matter," that "the object of a theory is to connect or coordinate apparently diverse phenomena." It is on the molecular level that morphology, physiology, and biochemistry must eventually meet. Considering the present state of knowledge of events in conduction and transmission, it seems to the writer that there is at present no compelling reason to attribute the differences to either a modification of the molecular organization or of the functional significance of the acetylcholine system; it seems to him rather more appropriate to adhere to one of the fundamental rules of scientific thinking: not to assume two different principles without necessity.

Concluding Remarks

The explosive character of the progress of science in so many directions, the elaboration of new methods and techniques of high precision on an unprecedented scale have created unforeseen possibilities. At the same time the unavoidable specialization has increased the difficulties of integrating the great variety of phenomena observed and the advances achieved in any one field, and has multiplied the problems of communication. It appeared, therefore, useful to present the information and insight gained in the efforts to elucidate the chemical and molecular basis of nerve activity. The summary may contribute to clarify certain aspects of the problem of nerve impulse conduction and may remove misconceptions and misunderstandings.

The investigations centered essentially around the specific chemical system inseparably associated with the generation of bioelectricity, the acetylcholine system, the specific operative substance in the definition of Meyerhof. Studies of the proteins of the system in particular have been helpful for the understanding of several aspects of nerve activity; in fact, they have promoted advances of more general biochemical problems. The isolation of acetylcholinesterase, in 1938, and its subsequent purification, opened the way for the analysis of molecular forces in the active site and of the mechanism of hydrolytic action; the explanation of nerve gas action became thereby possible, and led eventually to the design of PAM. The discovery of choline acetylase, in 1943, offered the first demonstration of an enzymatic acetylation *in vitro;* it led to the discovery of CoA and initiated many investigations on acetylation mechanisms in general. The third protein, the acetylcholine receptor, has just been isolated; the significance of this result will be apparent to all those interested in the problems of neurotropic agents, many of which must react with this specific protein. Relationships have been established between chemical reactions of the proteins of the system in solution and physical events recorded on the intact cell. But it has been stressed that even this system is still unexplored and that it forms only part of a complex membrane, that we have to deal with an important but limited aspect of a complex process. Clearly we are still in an initial stage of our knowledge of chemical forces and structural arrangements of the conducting membrane. The many loopholes in our knowledge have been emphasized as well as the advances achieved.

In the twenty-four years which have passed since the beginning of these studies, modifications and corrections have been necessary. This is only natural and a sign of progress. With further advances, other adjustments will be required. In his search for truth, a scientist should never be influenced by the vain and complacent attitude of having been right with a theory. A much greater satisfaction should be that of having contributed new information and of having stimulated new ideas and developments. The words of Goethe: "Was fruchtbar ist, allein ist wahr" (for what creative is, alone is true) apply to scientific efforts as much as to any other endeavor of man.

The fascinating question of the functioning of living cells has stimulated the creative minds of many generations. In his essay "What is Life?" Schrödinger (1944) raises the question whether we are able to explain events in a living cell taking place in a given space and in a given time, in terms of physics and chemistry. His answer is that it is not possible as yet; however, the progress achieved indicates that eventually it will be possible. In his definition of life Schrödinger limits, of course, the discussion of the problem to one special aspect, or in the terms of Kant, to one single category. There are other categories which we will not understand merely by explaining the events in a living cell in terms of physics and chemistry. The results in various fields of biology support the optimistic viewpoint of Schrödinger in the limited sense of his definition. The elementary processes in muscular contraction and nerve impulse conduction belong to the problems which have reached molecular levels. At a recent symposium in the new Institute for Nuclear Studies in Saclay near Paris, Francis Perrin, Professor of Physics and French High Commissioner for Atomic Energy, made an interesting statement. The first part of the century, he said, has seen the greatest revolutions and many miraculous achievements in physics. He ventured the prediction that in the second part of the century the most exciting advances will take place in life sciences.

References

Abbott, B. C., X. Aubert, and A. Fessard (1958). *J. physiol. (Paris)*, **50**, 99.

Abbott, B. C., A. V. Hill, and J. V. Howarth (1958). *Proc. Roy. Soc.*, **B148**, 149.

Abderhalden, E., and H. Paffrath (1925). *Fermentforschung*, **8**, 299.

Adams, D. H. (1949). *Biochem. et Biophys. Acta*, **3**, 1.

Adams, D. H., and V. P. Whittaker (1949). *Biochim. et Biophys. Acta*, **3**, 358.

Adams, D. H., and V. P. Whittaker (1950). *Biochim. et Biophys. Acta*, **4**, 543.

Adanson, M. (1757). "Histoire naturelle du Sénégal." Bauche, Paris.

Aeschlimann, J. A., and M. Reinert (1931). *J. Pharmacol. Exptl. Therap.*, **43**, 413.

Aeschlimann, J. A., and A. Stempel (1946). *Jubilee Vol. Dedicated to Emil Christoph Barell*, p. 313.

Albe-Fessard, D. (1950). *Arch. sci. physiol.*, **4**, 299.

Albe-Fessard, D., C. Chagas, and H. Martins-Ferreira (1951). *Anais. acad. brasil. cienc.*, **23**, 327.

Albe-Fessard, D., A. Fessard, and L. Sollero (1959). *In* "Curare and Curare-like Agents" (D. Bovet, F. Bovet-Nitti, and G. B. Marini-Bettolo, eds.), p. 346. Elsevier, Amsterdam.

Aldridge, W. N., and A. N. Davison (1953). *Biochem. J.*, **55**, 763.

Alles, G. A., and R. C. Hawes (1940). *J. Biol. Chem.*, **133**, 375.

Altamirano, M., C. W. Coates, H. Grundfest, and D. Nachmansohn (1953). *J. Gen. Physiol.*, **37**, 91.

Altamirano, M., W. L. Schleyer, C. W. Coates, and D. Nachmansohn (1955). *Biochim. et Biophys. Acta*, **16**, 268.

Ammon, R., and H. Kwiatkowski (1934). *Arch. ges. Physiol. Pflüger's*, **234**, 269.

Ascher, L. (1925). *Arch. ges. Physiol. Pflüger's*, **210**, 689.

Auger, D., and A. Fessard (1939). *In* "Livro Homnagem aos Professores Alvaro e Miguel Ozorio de Almeida," p. 25. Rio de Janeiro, Brazil.

Augustinsson, K. B. (1948). *Acta Physiol. Scand.*, **15**, Suppl. 52, 1.

Augustinsson, K. B. (1949). *Arch. Biochem.*, **23**, 111.

Augustinsson, K. B., and A. G. Johnels (1958). *J. Physiol. (London)*, **140**, 498.

Augustinsson, K. B., and D. Nachmansohn (1949a). *Science*, **110**, 98.

Augustinsson, K. B., and D. Nachmansohn (1949b). *J. Biol. Chem.*, **179**, 543.

Baddiley, J., and E. M. Thain (1951). *J. Chem. Soc.*, p. 2253.

Barcroft, J., and D. H. Barron (1939). *Ergeb. Physiol. biol. Chem. u. exptl. Pharmakol.*, **42**, 107.

Bear, R. S., F. O. Schmitt, and J. Z. Young (1937). *Proc. Roy. Soc.*, **B123**, 496.

Bendall, J. R. (1951). *J. Physiol. (London)* **144**, 71.

Bendall, J. R. (1953). *J. Physiol. (London)*, **121**, 232.

Bentley, R., and D. Rittenberg (1954). *J. Am. Chem. Soc.*, **76**, 4883.

Berg, P. (1956). *J. Biol. Chem.*, **222**, 991, 1015.

Bergami, G. (1936a). *Atti accad. nazl. Lincei, Mem. Classe sci. fis. mat. e nat.*, [6], **23**, 518.

Bergami, G. (1936b). *Boll. soc. ital. biol. sper.*, **11**, 275.

Bergami, G., G. Cantoni, and T. Gualtierotti (1936). *Arch. ist. biochim. ital.*, **8**, 267.

Bergmann, F., and A. Shimoni (1952). *Biochim. et Biophys. Acta*, **9**, 473.

Bergmann, F., I. B. Wilson, and D. Nachmansohn (1950a). *J. Biol. Chem.*, **186**, 693.

Bergmann, F., I. B. Wilson, and D. Nachmansohn (1950b). *Biochim. et Biophys. Acta,* **6**, 217.
Berman, R., I. B. Wilson, and D. Nachmansohn (1953). *Biochim. et Biophys. Acta,* **12**, 315.
Berman-Reisberg, R. (1954). *Biochim. et Biophys. Acta,* **14**, 442.
Berman-Reisberg, R. (1957). *Yale J. Biol. and Med.,* **29**, 403.
Bernhard, C. G., C. R. Skoglund, and O. Therman (1947). *Acta Physiol. Scand.,* **14**, Suppl. 47.
Bernhard, K. (1940). *Z. physiol. Chem. Hoppe-Seyler's,* **267**, 91.
Bernstein, J. (1902). *Arch. ges. Physiol. Pflüger's,* **92**, 521.
Bernstein, J., and A. Tschermak (1906). *Arch. ges. Physiol. Pflüger's,* **112**, 439.
Bethe, K., W. D. Erdmann, L. Lendle, and G. Schmidt (1957). *Arch. exptl. Pathol. Pharmakol. Naunyn-Schmiedeberg's,* **231**, 3.
Bing, H. I., and A. D. Skouby (1950). *Acta Physiol. Scand.,* **21**, 286.
Blaschko, H. (1957). *In* "Psychotropic Drugs" (S. Garattini and V. Ghetti, eds.), p. 3. Elsevier, Amsterdam.
Bloch, K., and D. Rittenberg (1945). *J. Biol. Chem.,* **159**, 45.
Boehm, R. (1908). *Arch. exptl. Pathol. Pharmakol. Naunyn-Schmiedeberg's,* **58**, 265.
Boell, E. J., and D. Nachmansohn (1940). *Science,* **92**, 513.
Bovet, D. (1951). *Ann. N.Y. Acad. Sci.,* **54**, 407.
Bovet, D. (1959). *In* "Curare and Curare-like Agents" (D. Bovet, F. Bovet-Nitti, and G. B. Marini-Bettolo, eds.), p. 252. Elsevier, Amsterdam.
Bovet, D., and F. Bovet-Nitti (1948). "Structure et Activité Pharmacodynamique des Médicaments du Système Nerveux Végétatif." Karger, Basel.
Bovet, D., and F. Bovet-Nitti (1955). *Sci. Med. Ital.* **3**, 484.
Bovet, D., S. Courvoisier, R. Ducrot, and R. Horclois (1946). *Compt. rend.* **223**, 597.
Bovet, D., F. Bovet-Nitti, *et al.* (1949a). *Rend. ist. super. sanità, Numero speciale sui curare di sintesi,* **12**, 1.
Bovet, D., F. Bovet-Nitti, S. Guarino, V. G. Longo, and M. Marotta (1949b). *Rend. ist. super. sanità,* **12**, 106.
Bovet, D., F. Bovet-Nitti, S. Guarino, V. G. Longo, and R. Fusco (1951). *Arch. intern. pharmacodynamie,* **88**, 1.
Bovet-Nitti, F. (1959). *In* "Curare and Curare-like Agents" (D. Bovet, F. Bovet-Nitti, and G. B. Marini-Bettolo, eds.), p. 230. Elsevier, Amsterdam.
Bozler, E. (1951). *Am. J. Physiol.,* **167**, 276.
Bradley, K., D. M. Easton, and J. C. Eccles (1953). *J. Physiol.* (*London*) **122**, 474.
Brauer, R. W., and M. A. Root (1945). *Federation Proc.* **4**, 113.
Brecht, K., and M. Corsten (1941). *Arch. ges. Physiol. Pflüger's,* **245**, 160.
Brink, F., D. W. Bronk, F. D. Carlson, and C. M. Connelly (1952). *Cold Spring Harbor Symposia Quant. Biol.,* **17**, 53.
Brock, L. G., J. S. Coombs, and J. C. Eccles (1952a). *J. Physiol.* (*London*), **117**, 431.
Brock, L. G., J. S. Coombs, and J. C. Eccles (1952b). *Proc. Roy. Soc.,* **B140**, 170.
Brock, L. G., R. M. Eccles, and R. D. Keynes (1953). *J. Physiol.* (*London*), **122**, 4P.
Brown, G. M., J. A. Craig, and E. E. Snell (1950). *Arch. Biochem.,* **27**, 473.
Brown, G. L., H. H. Dale, and W. Feldberg (1936). *J. Physiol.* (*London*), **87**, 394.
Buchthal, F. (1954). *Pharmacol. Revs.,* **6**, 97.
Bueding, E. (1952). *Brit. J. Pharmacol.* **7**, 563.
Bullock, T. H., and S. Hagiwara (1957). *J. Gen. Physiol.,* **40**, 565.
Bullock, T. H., D. Nachmansohn, and M. A. Rothenberg (1946). *J. Neurophysiol.,* **9**, 9.

Bullock, T. H., H. Grundfest, D. Nachmansohn, and M. A. Rothenberg (1947a). *J. Neurophysiol.*, **10**, 11.

Bullock, T. H., H. Grundfest, D. Nachmansohn, and M. A. Rothenberg (1947b). *J. Neurophysiol.*, **10**, 63.

Burgen, A. S. V., and K. G. Terroux (1953). *J. Physiol. (London)*, **120**, 449.

Burton, K. (1958). *Nature,* **181**, 1594.

Burton, K. (1959). *Biochem. J.*, **59**, 44.

Burton, K., and H. A. Krebs (1953). *Biochem. J.*, **54**, 94.

Calabro, Q. (1933). *Riv. biol.*, **15**, 299.

Caldwell, P. C. (1959). *In* "Colloque de Biologie de Saclay" (J. Coursaget, ed.), p. 88. Pergamon Press, London.

Cantoni, G. L., and O. Loewi (1944). *J. Pharmacol. Exptl. Therap.*, **81**, 67.

Casida, I. E., T. C. Allen, and M. A. Stahmann (1954). *J. Biol. Chem.*, **210**, 607.

Chagas, C. (1959). *In* "Curare and Curare-like Agents" (D. Bovet, F. Bovet-Nitti, and G. B. Marini-Bettolo, eds.), p. 327. Elsevier, Amsterdam.

Chagas, C., E. Penna-Franca, A. Hasson, C. Crocker, K. Nishie, and E. J. Garcia (1957). *Anais acad. brasil. cienc.*, **29**, 53.

Chagas, C., E. Penna-Franca, K. Nishie, and E. J. Garcia (1958). *Arch. Biochem. Biophys.*, **75**, 251.

Chance, B. (1956). *In* "Currents in Biochemical Research" (D. E. Green, ed.), p. 308. Interscience, New York.

Chang, H. C., W. M. Hsieh, L. Y. Lee, T. H. Li, and R. K. S. Lim (1939). *Chinese J. Physiol.*, **14**, 27.

Childs, A. F., D. R. Davies, A. L. Green, and J. P. Rutland (1955). *Brit. J. Pharmacol.*, **10**, 462.

Clark, A. J. (1933). "The Mode of Action of Drugs on Cells." Edward Arnold, London.

Clark, A. J. (1937). *In* "Handbuch der Experimentellen Pharmakologie," (W. Heubner and J. Schueller, eds.). Springer, Berlin.

Clarke, H. T., and D. Nachmansohn (1954). "Ion Transport across Membranes." Academic Press, New York.

Coërs, C. (1953). *Acad. roy. Belg. Classe sci. Mém.*, **39**, 447.

Cohen, M. (1956). *Arch. Biochem. Biophys.*, **60**, 284.

Cole, K. S. (1949). *Arch. sci. physiol.*, **3**, 253.

Cole, K. S. (1955). *In* "Electrochemistry in Biology and Medicine" (T. Shedlovsky, ed.), p. 121. Wiley, New York.

Cole, K. S., and H. J. Curtis (1939). *J. Gen. Physiol.*, **22**, 649.

Colloque de Biologie de Saclay (1959). "La méthode des indicateurs nucléaires dans l'étude des transports actifs d'ions" (J. Coursaget, ed.). Pergamon, London.

Conway, E. J. (1959). *In* "Colloque de Biologie de Saclay" (J. Coursaget, ed.), p. 1. Pergamon, London.

Couteaux, R. (1942). *Bull. biol. France et Belg.*, **76**, 14.

Couteaux, R. (1947). *Rev. can. biol.*, **6**, 563.

Couteaux, R. (1955). *Intern. Rev. Cytol.*, **4**, 335.

Couteaux, R., and D. Nachmansohn (1940). *Proc. Soc. Exptl. Biol. Med.*, **43**, 177.

Couteaux, R., and J. Taxi (1952). *Arch. anat. microscop. morphol. exptl.*, **41**, 352.

Couteaux, R., H. Grundfest, D. Nachmansohn, and M. A. Rothenberg (1946). *Science,* **104**, 317.

Cowan, S. (1936). *J. Physiol. (London)* **88**, 4P.

Cox, R. T., C. W. Coates, and M. V. Brown (1945). *J. Gen. Physiol.*, **28**, 187.

Cox, R. T., C. W. Coates, and M. V. Brown (1946). *Ann. N.Y. Acad. Sci.*, **47**, 487.
Crescitelli, F. N., G. B. Koelle, and A. Gilman (1946). *J. Neurophysiol.*, **9**, 241.
Curtis, H. J., and K. S. Cole (1942). *J. Cellular Comp. Physiol.*, **19**, 135.
Dakin, H. D. (1909). *J. Biol. Chem.*, **6**, 221.
Dale, H. H. (1914). *J. Pharmacol. Exptl. Therap.*, **6**, 147.
Dale, H. H. (1937). *Harvey Lectures*, **32**, 229.
Dale, H. H. (1954). *Pharmacol. Revs.*, **6**, 7.
Dale, H. H., and H. W. Dudley (1929). *J. Physiol. (London)*, **68**, 97.
Dale, H. H., W. Feldberg, and M. Vogt (1936). *J. Physiol. (London)* **86**, 353.
De Candole, C. A., W. W. Douglas, C. L. Evans, R. Holmes, K. E. V. Spencer, R. W. Towange, and K. M. Wilson (1953). *Brit. J. Pharmacol.*, **8**, 466.
del Castillo, J., and B. Katz (1955a). *J. Physiol. (London)*, **128**, 157.
del Castillo, J., and B. Katz (1955b). *J. Physiol. (London)*, **128**, 396.
del Castillo, J., and B. Katz (1955c). *J. Physiol. (London)*, **129**, 48P.
del Castillo, J., and B. Katz (1956). *Progr. in Biophys. and Biophys. Chem.*, **6**, 121.
De Robertis, E. D. P., and H. S. Bennett (1955). *J. Biophys. Biochem. Cytol.*, **1**, 47.
De Roetth, A. J., Jr. (1951). *J. Neurophysiol.*, **14**, 55.
Dettbarn, W. D. (1959a). *Nature*, **183**, 465.
Dettbarn, W. D. (1959b). *Biochim. Biophys. Acta.* **32**, 381.
Dettbarn, W. D. (1959c). *Federation Proc.*, **18**, 36.
Dettbarn, W. D., I. B. Wilson, and D. Nachmansohn (1958). *Science*, **128**, 1275.
Discussions Faraday Soc. (1956).
Dixon, M., and E. C. Webb (1958). "Enzymes." Academic Press; New York.
Dodt, E., A. P. Skouby, and Y. Zotterman (1953). *Acta Physiol. Scand.*, **28**, 101.
Du Bois, K. P., J. Doull and M. J. Coon (1950). *J. Pharmacol. Exptl. Therap.* **99**, 376.
Du Bois-Reymond, E. (1877). "Gesammelte Abhandlungen zur allgemeinen Muskel- und Nervenphysik," B. II. Veit, Leipzig.
Dubuisson, M. (1954). "Muscular Contraction." C. C Thomas, Springfield, Illinois.
Dubuisson, M., and A. M. Monnier (1934). *Arch. intern. physiol.*, **38**, 180.
Eccles, J. C. (1937). *Physiol. Revs.*, **17**, 538.
Eccles, J. C. (1946). *Ann. N.Y. Acad. Sci.*, **47**, 429.
Eccles, J. C. (1957). "The Physiology of Nerve Cells." Johns Hopkins Press, Baltimore, Maryland.
Eccles, J. C., B. Katz, and S. W. Kuffler (1942). *J. Neurophysiol.*, **5**, 211.
Eccles, J. C., and W. V. MacFarlane (1949). *J. Neurophysiol.*, **12**, 59.
Edwards, G. A., H. Ruska, and E. de Harven (1958). *J. Biophys. Biochem. Cytol.*, **4**, 107.
Ehrenpreis, S. (1959a). *Science*, **129**, 1613.
Ehrenpreis, S. (1959b). *Federation Proc.*, **18**, p. 220.
Eisenberg, M. A. (1953). *J. Biol. Chem.*, **203**, 815.
Eisenberg, M. A. (1955). *Biochim. et Biophys. Acta*, **16**, 58.
Elliott, T. R. (1905). *J. Physiol. (London)*, **32**, 401.
Emmelin, N. G., and F. C. MacIntosh (1956). *J. Physiol. (London)*, **131**, 477.
Engelhardt, W. A. (1942). *Yale J. Biol. and Med.*, **15**, 21.
Engelhardt, W. A., and M. N. Ljubimova (1939). *Nature*, **144**, 668.
Erdmann, W. D., and L. Lendle (1958). *Ergeb. inn. Med. u. Kinderheilk.*, [N.F.] **10**, 103.
Erdmann, W. D., F. Sakai, and F. Scheler (1958). *Deut. med. Wochschr.*, **32**, 1359.
Erlanger, J. (1939). *J. Neurophysiol.*, **2**, 370.
Fatt, P. (1954). *Physiol. Revs.*, **5**, 1.

Fatt, P., and B. Katz (1951). *J. Physiol. (London)*, **115**, 320.

Fatt, P., and B. Katz (1952a). *J. Physiol. (London)*, **117**, 109.

Fatt, P., and B. Katz (1952b). *J. Physiol. (London)*, **118**, 73.

Fatt, P., and B. Katz (1953a). *J. Physiol. (London)*, **120**, 171.

Fatt, P., and B. Katz (1953b). *J. Physiol. (London)*, **121**, 374.

Feigl, F. (1946). "Qualitative Analysis by Spot Tests." Elsevier, Amsterdam and New York.

Feigl, F., V. Auger, and O. Frehden (1934). *Microchemie*, **15**, 12.

Feldberg, W. (1943). *J. Physiol. (London)*, **101**, 432.

Feldberg, W., and T. Mann (1946). *J. Physiol. (London)*, **104**, 411.

Feldberg, W., A. Fessard, and D. Nachmansohn (1940). *J. Physiol. (London)*, **97**, 3P.

Feng, T. P. (1936). *Ergeb. Physiol. exptl. Pharmakol.*, **38**, 73.

Feng, T. P., and V. C. Ting (1938). *Chinese J. Physiol.*, **13**, 141.

Fenn, W. O. (1927). *J. Gen. Physiol.*, **10**, 767.

Fessard, A. (1946). *Ann. N.Y. Acad. Sci.*, **47**, 501.

Flückiger, E., and R. D. Keynes (1955). *J. Physiol. (London)*, **128**, 41P.

Folch-Pi, J., M. Lees, and G. H. Sloane-Stanley (1957). *In* "Metabolism of the Nervous System" (D. Richter, ed.), p. 174. Pergamon, New York.

Frawley, J. P., E. C. Hagan, and O. G. Fitzhugh (1952). *J. Pharmacol. Exptl. Therap.*, **105**, 156.

Fulton, J. F. (1938, 1943, 1949). "Physiology of the Nervous System." Oxford Univ. Press, London and New York.

Fulton, J. F. (1939). *Science*, **90**, 110.

Fulton, J. F., and D. Nachmansohn (1943). *Science*, **97**, 569.

Gaskell, W. H. (1886). *J. Physiol. (London)*, **7**, 451.

Gerard, R. W. (1950). *Recent Progr. in Hormone Research*, **5**, 37.

Gerard, R. W., and O. Meyerhof (1927). *Naturwissenschaften* **15**, 538.

Gilman, A. (1946). *Ann. N.Y. Acad. Sci.*, **47**, 549.

Glick, D. (1938). *J. Biol. Chem.*, **125**, 729.

Glick, D. (1939). *J. Biol. Chem.*, **130**, 527.

Glick, D. (1941). *J. Biol. Chem.*, **137**, 357.

Glick, D. (1945). *Science*, **102**, 100.

Göpfert, H., and H. Schaefer (1938). *Arch. ges. Physiol. Pflüger's*, **239**, 597.

Granit, R., S. Skoglund, and S. Thesleff (1953). *Acta Physiol. Scand.*, **28**, 134.

Green, D. E. (1941). *Advances in Enzymol.*, **1**, 177.

Green, D. E. (1956). *In* "Enzymes: Units of Biological Structure and Function" (O. H. Gaebler, ed.), p. 465. Academic Press, New York.

Green, D. E. (1959). *Perspectives in Biol. Med.*, **2**, 163.

Greig, M. E., and W. C. Holland (1949). *Arch. Biochem.*, **23**, 370.

Grundfest, H., D. Nachmansohn, C. Y. Kao, and R. Chambers (1952). *Nature*, **169**, 190.

Gunther, F. A., and R. C. Blinn (1955). "Analysis of Insecticides and Acaricides." Interscience, New York.

Hagiwara, S., and Tasaki, I. (1957). *J. Gen. Physiol.*, **40**, 859.

Haldane, J. B. S. (1930). "Enzymes." Longmans, Green, London.

Hammett, L. P. (1940). "Physical Organic Chemistry." McGraw-Hill, New York.

Hanson, J., and H. E. Huxley (1957). *Biochim. et Biophys. Acta*, **23**, 250.

Harris, E. J. (1956). "Transport and Accumulation in Biological Systems." Academic Press, New York.

Hasselbach, W., and H. H. Weber (1953). *Biochim. et Biophys. Acta,* **11**, 160.
Hestrin, S. (1949a). *J. Biol. Chem.,* **180**, 249.
Hestrin, S. (1949b). *J. Biol. Chem.,* **180**, 879.
Hestrin, S. (1950). *Biochim. et Biophys. Acta,* **4**, 310.
Hill, A. V. (1932a). *Proc. Roy. Soc.,* **B111**, 106.
Hill, A. V. (1932b). "Chemical Wave Transmission in Nerve." Cambridge Univ. Press, London and New York.
Hill, A. V. (1959). *In* "Symposium on Molecular Biology" (D. Nachmansohn, ed.). Academic Press, New York. (In press.)
Hinterbuchner, L. P., and I. B. Wilson (1959a). *Biochim. et Biophys. Acta,* **31**, 323.
Hinterbuchner, L. P., and I. B. Wilson (1959b). *Biochim. et Biophys. Acta,* **32**, 375.
Hinterbuchner, L. P., I. B. Wilson, and E. Schoffeniels (1958). *Federation Proc.,* **17**, 71.
Hodgkin, A. L. (1951). *Biol. Revs. Cambridge Phil. Soc.,* **26**, 338.
Hodgkin, A. L. (1957). *Proc. Roy. Soc.,* **B148**, 1.
Hodgkin, A. L., and A. F. Huxley (1945). *J. Physiol. (London)*, **104**, 176.
Hodgkin, A. L., and B. Katz (1950). *J. Physiol. (London)*, **108**, 33.
Hodgkin, A. L., and R. D. Keynes (1955). *J. Physiol. (London)*, **128**, 28.
Hodgkin, A. L., and R. D. Keynes (1956). *J. Physiol. (London)*, **131**, 592.
Hodgkin, A. L., and R. D. Keynes (1957). *J. Physiol. (London)*, **138**, 253.
Holland, W. C., and M. E. Greig (1950). *Am. J. Physiol.,* **162**, 610.
Holtz, P., and H. J. Schueman (1954). *Naturwissenschaften,* **41**, 306.
Hopkins, F. G. (1932). *Proc. Roy. Soc.,* **B112**, 159.
Hunt, R. (1918). *Am. J. Physiol.,* **45**, 197.
Hunt, R., and R. de M. Taveau (1906). *Brit. Med. J.,* **2**, 1788.
Hutter, O. F., and W. Trautwein (1955). *J. Physiol. (London)* **129**, 48P.
Huxley, A. F. (1954). *In* "Ion Transport Across Membranes" (H. T. Clarke and D. Nachmansohn, eds.), p. 23. Academic Press, New York.
Huxley, H. E., and J. Hanson (1955). *Symposia Soc. Exptl. Biol. No.* **9**, 228.
Huxley, H. E., and J. Hanson (1957). *Biochim. et Biophys. Acta,* **23**, 229.
Jansen, E. F., M. D. F. Nutting, and A. K. Balls (1949). *J. Biol. Chem.,* **179**, 201.
Jansen, E. F., M. D. F. Nutting, R. Jang, and A. K. Balls (1950). *J. Biol. Chem.,* **185**, 209.
Kewitz, H. (1957a). *Arch. Biochem. Biophys.,* **66**, 263.
Kewitz, H. (1957b). *Klin. Wochschr.,* **35**, 521.
Kewitz, H., and D. Nachmansohn (1957). *Arch. Biochem. Biophys.,* **66**, 271.
Kewitz, H., and H. Reinert (1954). *Arch. Exptl. Pathol. Pharmakol. Naunyn-Schmiedeberg's,* **222**, 315.
Kewitz, H., and I. B. Wilson (1956). *Arch. Biochem. Biophys.,* **60**, 261.
Kewitz, H., I. B. Wilson, and D. Nachmansohn (1956). *Arch. Biochem. Biophys.,* **64**, 456.
Keynes, R. D., and P. R. Lewis (1951). *J. Physiol. (London)*, **114**, 151.
Keynes, R. D. (1958). *Intern. Congr. Biochem., 4th Congr., Vienna.* (In press.)
Keynes, R. D., and H. Martins-Ferreira (1953). *J. Physiol. (London)*, **119**, 315.
Kibjakow, A. W. (1933). *Arch. ges. Physiol. Pflüger's,* **232**, 432.
King, H. (1935). *J. Chem. Soc.,* **2**, 1381.
Kisch, B. (1930). *Biochem. Z.,* **225**, 183.
Klein, J. R., and J. S. Harris (1938). *J. Biol. Chem.,* **124**, 613.
Knoop, F. (1905). *Beitr. chem. Physiol. u. Pathol.,* **6**, 150.
Koelle, G. B. (1957). *J. Pharmacol. Exptl. Therap.,* **120**, 488.

Koelle, G. B., and J. S. Friedenwald (1949). *Proc. Soc. Exptl. Biol. Med.*, **70**, 617.
Koelle, G. B., and A. Gilman (1949). *Pharmacol. Revs.*, **1**, 166.
Koelle, W. A., and G. B. Koelle (1958). *Federation Proc.*, **17**, 384.
Koketsu, K., J. A. Cerf, and S. Nishi (1958a). *Nature,* **181**, 703.
Koketsu, K., J. A. Cerf, and S. Nishi (1958b). *Nature,* **181**, 1798.
Korey, S. R., B. de Braganza, and D. Nachmansohn (1951). *J. Biol. Chem.*, **189**, 705.
Korkes, S., A. del Campillo, I. C. Gunsalus, and S. Ochoa (1951). *J. Biol. Chem.*, **193**, 721.
Korkes, S., A. del Campillo, S. R. Korey, J. R. Stern, D. Nachmansohn, and S. Ochoa (1952a). *J. Biol. Chem.*, **198**, 215.
Korkes, S., A. del Campillo, and S. Ochoa (1952b). *J. Biol. Chem.*, **195**, 541.
Koshland, D. E., Jr. (1954). *In* "The Mechanism of Enzyme Action" (W. D. McElroy and B. Glass, eds.), p. 608. Johns Hopkins Press, Baltimore, Maryland.
Kuffler, S. W. (1942a). *J. Neurophysiol.*, **5**, 18.
Kuffler, S. W. (1942b). *J. Neurophysiol.*, **5**, 309.
Kuffler, S. W., and C. Eyzaguirre (1955). *J. Gen. Physiol.*, **39**, 155.
Kwiatkowski, H. (1936). *Fermentforschung*, **15**, 138.
Lamm, O., and H. Malmgren (1940). *Z. anorg. u. allgem. Chem.*, **245**, 103.
Langley, T. N. (1907). *J. Physiol. (London)*, **36**, 347.
Lapicque, L. (1936). *Compt. rend. soc. biol.* **122**, 990.
Laporte, V., and R. Lorente de Nó (1950). *J. Cellular Comp. Physiol.*, **35**(2), 61.
Larramendi, L. M. H., R. Lorente de Nó, and F. Vidal (1956). *Nature,* **178**, 316.
Lawler, C. (1959). *J. Biol. Chem.*, **234**, 799.
Le Heux, J. W. (1919). *Arch. ges. Physiol. Pflüger's*, **173**, 8.
Le Heux, J. W. (1921). *Arch. ges. Physiol. Pflüger's*, **190**, 280.
Lehmann, G. (1946). *Jubilee Vol. Dedicated to Emil Christoph Barell*, p. 314.
Lindhard, J. (1931). *Ergeb. Physiol.*, **33**, 337.
Ling, G., and R. W. Gerard (1949). *J. Cellular Comp. Physiol.*, **34**, 383.
Lipmann, F. (1945). *J. Biol. Chem.*, **160**, 173.
Lipmann, F. (1950). *Harvey Lectures Ser.* **44**, 99.
Lipmann, F. (1957). *In* "Metabolism of the Nervous System" (D. Richter, ed.), p. 329. Pergamon Press, New York.
Lipmann, F., and N. O. Kaplan (1946). *J. Biol. Chem.* **162**, 743.
Lipmann, F., and L. C. Tuttle (1945). *J. Biol. Chem.*, **159**, 21.
Lipmann, F., N. O. Kaplan, G. D. Novelli, L. C. Tuttle, and B. M. Guirard (1947). *J. Biol. Chem.*, **167**, 869.
Lipton, M. A. (1946). *Federation Proc.*, **5**, 145.
Lissak, K., and J. Pasztor (1941). *Arch. ges. physiol. Pflüger's*, **244**, 120.
Loewenstein, W. R., and D. Molins (1958). *Science,* **128**, 1284.
Loewi, O. (1921). *Arch. ges. Physiol. Pflüger's*, **189**, 239.
Loewi, O., and H. Hellauer (1938). *J. Physiol. (London)*, **93**, 34P.
Loewi, O., and E. Navratil (1926). *Arch. ges. Physiol. Pflüger's*, **214**, 678.
Lorente de Nó, R. (1938). *Am. J. Physiol.*, **121**, 331.
Lorente de Nó, R. (1949). *J. Cellular Comp. Physiol.*, **33**, 1, 231.
Lorente de Nó, R., F. Vidal, and L. M. H. Larramendi (1957). *Nature,* **179**, 737.
Lucas, K. (1907a). *J. Physiol. (London)*, **36**, 113.
Lucas, K. (1907b). *J. Physiol. (London)*, **36**, 253.
Lucas, K. (1917). "The Conduction of the Nervous Impulse" (revised by E. D. Adrian), Longmans, London.
Luft, J. H. (1956). *J. Biophys. Biochem. Cytol.*, **2**, Suppl., 229.

Lundholm, L. (1949). *Acta Physiol. Scand.,* **16**, 345.
Lundsgaard, E. (1930). *Biochem. Z.,* **217**, 162.
Lynen, F., E. Reichert, and L. Rueff (1951). *Ann. Chem. Liebigs,* **574**, 1.
McIntyre, A. R. (1958). *In* "Simposio internacional solve o curarecas substancias curarisantes," Instituto de Biofisica, Universidade do Brasil, p. 89.
McIntyre, A. R. (1959). *In* "Curare and Curare-like Agents" (D. Bovet, F. Bovet-Nitti, and G. B. Marini-Bettolo, eds.), p. 211. Elsevier, Amsterdam.
McIntyre, A. R., F. M. Downing, A. L. Bennett, and A. L. Dunn (1950). *Proc. Soc. Exptl. Biol. Med.,* **74**, 180.
Mann, P. J. G., M. Tennenbaum, and J. H. Quastel (1938). *Biochem. J.,* **32**, 243.
Marnay, A. (1937). *Compt. rend. soc. biol.,* **126**, 573.
Marnay, A., and D. Nachmansohn (1937a). *Compt. rend. soc. biol.* **125**, 41.
Marnay, A., and D. Nachmansohn (1937b). *Compt. rend. soc. biol.,* **125**, 489.
Marnay, A., and D. Nachmansohn (1937c). *Compt. rend. soc. biol.,* **125**, 1005.
Marnay, A., and D. Nachmansohn (1938). *J. Physiol. (London),* **92**, 37.
Marsh, B. B. (1951). *Nature,* **167**, 1065.
Masland, R. L., and R. S. Wigton (1940). *J. Neurophysiol.,* **3**, 269.
Mendel, B., and H. Rudney (1943). *Biochem. J.,* **37**, 59.
Meyer, K. H. (1937). *Helv. Chim. Acta,* **20**, 634.
Meyerhof, O. (1913). "Zur Energetik der Zellvorgaenge," Vandenhoeck und Ruprecht, Göttingen.
Meyerhof, O. (1924). The Chemical Dynamics of Life Phenomena *in* "Monographs on Experimental Biology." Lippincott, New York.
Meyerhof, O. (1925). *In* "Handbuch der Physik," Vol. **9**, 238.
Meyerhof, O. (1937). *Ergeb. Physiol. biol. Chem. u. exptl. Pharmakol.,* **39**, 10.
Meyerhof, O. (1941). *Biol. Symposia,* **3**, 239.
Michel, H. O., and S. Krop (1951). *J. Biol. Chem.,* **190**, 119.
Monnier, A. M. (1936). *Cold Spring Harbor Symposia Quant. Biol.,* **4**, 111.
Murphy, Q. R., ed. (1957). "Metabolic Aspects of Transport across Cell Membranes." Univ. of Wisconsin Press, Madison.
Nachmansohn, D. (1937a). *Nature,* **140**, 427.
Nachmansohn, D. (1937b). *Compt. rend. soc. biol.,* **126**, 783.
Nachmansohn, D. (1938a). *Compt. rend. soc. biol.,* **127**, 670.
Nachmansohn, D. (1938b). *Compt. rend. soc. biol.,* **127**, 894.
Nachmansohn, D. (1938c). *Presse méd.,* **48**, 942.
Nachmansohn, D. (1938d). *Compt. rend. soc. biol.,* **128**, 516.
Nachmansohn, D. (1938e). *J. Physiol. (London),* **93**, 2 P.
Nachmansohn, D. (1938f). *Compt. rend. soc. biol.,* **129**, 830.
Nachmansohn, D. (1938g). *Compt. rend. soc. biol.,* **128**, 599.
Nachmansohn, D. (1939a). *J. Physiol. (London),* **95**, 29.
Nachmansohn, D. (1939b). *Bull. soc. chim. biol.,* **21**, 761.
Nachmansohn, D. (1940a). *J. Neurophysiol.,* **3**, 396.
Nachmansohn, D. (1940b). *Yale J. Biol. and Med.,* **12**, 565.
Nachmansohn, D. (1946a). *In* "Currents in Biochemical Research" (D. E. Green, ed.), p. 335. Interscience, New York.
Nachmansohn, D. (1946b). *Ann. N.Y. Acad. Sci.,* **47**, 395.
Nachmansohn, D. (1947). International Congress of Physiology, Oxford.
Nachmansohn, D. (1950a). *In* "The Hormones" (G. Pincus and K. V. Thimann, eds.), Vol. 2, p. 515. Academic Press, New York.

Nachmansohn, D. (1950b). *In* "Metabolism and Function" (D. Nachmansohn, ed.), *Biochim. et. Biophys. Acta,* **4**, 96.
Nachmansohn, D. (1951). *In* "Phosphorus Metabolism" (W. D. McElroy and B. Glass, eds.), p. 568. Johns Hopkins Press, Baltimore, Maryland.
Nachmansohn, D. (1952a). *In* "Modern Trends in Physiology and Biochemistry" (E. S. G. Barrón, ed.), p. 230. Academic Press, New York.
Nachmansohn, D. (1952b). *Bull. soc. chim. biol.,* **34**, 447.
Nachmansohn, D. (1952c). *Estratto Rend. ist. Super. sanità,* **15**, 1267.
Nachmansohn, D. (1954). *In* "Biochemistry of the Developing Nervous System," (H. Waelsch, ed.), p. 479. Academic Press, New York.
Nachmansohn, D. (1955a). *Harvey Lectures Ser.* **49**, 57.
Nachmansohn, D. (1955b). *Ergeb. der Physiol.,* **48**, 575.
Nachmansohn, D. (1955c). *In* "A Textbook of Physiology" (J. F. Fulton, ed.), p. 192. 17th ed., Saunders Co., Philadelphia.
Nachmansohn, D. (1955d). *Am. J. Phys. Med.,* **34**, 33.
Nachmansohn, D. (1955e). *Circulation Research* **3**, 429.
Nachmansohn, D. (1957). *Bull. soc. chim. biol.,* **39**, 1021.
Nachmansohn, D. (1959a). *In* "Structure and Function of Muscle" (G. Bourne, ed.) Academic Press, New York. (In press.)
Nachmansohn, D. (1959b). *In* "Colloque de Biologie de Saclay" (J. Coursaget, ed.), p. 630. Pergamon, London.
Nachmansohn, D., and M. Berman (1946). *J. Biol. Chem.,* **165**, 551.
Nachmansohn, D., and E. A. Feld (1947). *J. Biol. Chem.,* **171**, 715.
Nachmansohn, D., and E. C. Hoff (1944). *J. Neurophysiol.,* **7**, 27.
Nachmansohn, D., and H. M. John (1944). *Proc. Soc. Exp. Biol. Med.,* **57**, 361.
Nachmansohn, D., and H. M. John (1945a). *J. Biol. Chem.,* **158**, 157.
Nachmansohn, D., and H. M. John (1945b). *Science,* **102**, 250.
Nachmansohn, D., and E. Lederer (1939a). *Compt. rend. soc. biol.,* **130**, 321.
Nachmansohn, D., and E. Lederer (1939b). *Bull. soc. chim. biol.,* **21**, 797.
Nachmansohn, D., and A. L. Machado (1943). *J. Neurophysiol.,* **6**, 397.
Nachmansohn, D., and Bettina Meyerhof (1941). *J. Neurophysiol.,* **4**, 348.
Nachmansohn, D., and M. A. Rothenberg (1944). *Science,* **100**, 454.
Nachmansohn, D., and M. A. Rothenberg (1945). *J. Biol. Chem.,* **158**, 653.
Nachmansohn, D., and M. S. Weiss (1948). *J. Biol. Chem.,* **172**, 677.
Nachmansohn, D., and I. B. Wilson (1951). *Advances in Enzymol.,* **12**, 259.
Nachmansohn, D., and I. B. Wilson (1955). "Electrochemistry in Biology and Medicine" (T. Shedlovsky, ed.), p. 167. Wiley, New York.
Nachmansohn, D., and I. B. Wilson (1956). *In* "Currents in Biochemical Research" (D. E. Green, ed.), p. 628. Interscience, New York.
Nachmansohn, D., C. W. Coates, and R. T. Cox (1941). *J. Gen. Physiol.,* **25**, 75.
Nachmansohn, D., R. T. Cox, C. W. Coates, and A. L. Machado (1942) *J. Neurophysiol.,* **5**, 499.
Nachmansohn, D., R. T. Cox, and C. W. Coates (1943a). *Proc. Soc. Exptl. Biol. Med.,* **52**, 97.
Nachmansohn, D., R. T. Cox, C. W. Coates, and A. L. Machado (1943b). *J. Neurophysiol.,* **6**, 383.
Nachmansohn, D., H. M. John, and H. Waelsch (1943c). *J. Biol. Chem.,* **150**, 485.
Nachmansohn, D., C. W. Coates, and M. A. Rothenberg (1946a). *J. Biol. Chem.,* **163**, 39.

Nachmansohn, D., C. W. Coates, M. A. Rothenberg, and M. V. Brown (1946b). *J. Biol. Chem.*, **165**, 223.

Nachmansohn, D., H. M. John, and M. Berman (1946c). *J. Biol. Chem.*, **163**, 475.

Nachmansohn, D., I. B. Wilson, S. R. Korey, and R. Berman (1952). *J. Biol. Chem.*, **195**, 25.

Namba, T., and K. Hiraki (1958). *J. Am. Med. Assoc.*, **166**, 1834.

Nastuk, W. L. (1953). *Federation Proc.*, **12**, 102.

Nastuk, W. L. (1954). *Federation Proc.*, **13**, 104.

Nastuk, W. L., and J. T. Alexander (1954). *J. Pharmacol. Exptl. Therap.*, **111**, 302.

Nastuk, W. L., and A. L. Hodgkin (1950). *J. Cellular Comp. Physiol.*, **35**, 39.

Novelli, G. D. (1953). *Physiol. Revs.*, **33**, 525.

Novelli, G. D., and F. Lipmann (1950). *J. Biol. Chem.*, **182**, 213.

Novelli, D., F. J. Schmetz, Jr., and N. O. Kaplan (1954). *J. Biol. Chem.*, **206**, 533.

Ostwald, W. (1890). *Z. physik. Chem. (Leipzig)*, **6**, 71.

Overton, E. (1902). *Arch. ges. Physiol. Pflüger's*, **92**, 346.

Palade, G. E. (1956). *In* "Enzymes: Units of Biological Structure and Function" O. H. Gaebler, ed.), p. 185. Academic Press, New York.

Palade, G. E., and S. L. Palay (1954). *Anat. Record*, **118**, 335.

Palay, S. L. (1956). *J. Biophys. Biochem. Cytol.*, **2**, Suppl., 193.

Paolini, L. (1959). *In* "Symposium on Molecular Biology" (D. Nachmansohn, ed.). Academic Press, New York. (In press.)

Paton, W. D. M., and E. J. Zaimis (1949). *Brit. J. Pharmacol.*, **4**, 381.

Persky, H., and M. Gold (1948). *Biol. Bull.*, **95**, 278.

Pézard, A., and R. M. May (1937). *Compt. rend. soc. biol.*, **124**, 942.

Podolsky, R. J., and M. F. Morales (1956). *J. Biol. Chem.*, **218**, 945.

Poziomek, E. J., B. E. Hackley, and G. M. Steinberg (1958). *J. Org. Chem.*, **23**, 714.

Reger, J. F. (1957). *Exptl. Cell Research*, **12**, 662.

Richter, D., and P. G. Croft (1942). *Biochem. J.*, **36**, 746.

Riesser, O., and S. M. Neuschloss (1921). *Arch. Exptl. Pathol. Pharmakol. Naunyn-Schmiedeberg's*, **91**, 342.

Riker, W. F. (1953). *Pharmacol. Revs.*, **5**, 1.

Riker, W. F. (1959). "Symposium on Physico-chemical Mechanisms of Nerve Activity," *N.Y. Acad. Sci.* (In press.)

Riker, W. F., and W. C. Wescoe (1946). *J. Pharmacol. Exptl. Therap.*, **88**, 58.

Riker, W. F., J. Roberts, F. G. Standaert, and H. Fujimari (1957). *J. Pharmacol. Exptl. Therap.*, **121**, 286.

Riker, W. F., G. Werner, J. Roberts, and A. Kuperman (1959). *J. Pharmacol. Exptl. Therap.*, **125**, 150.

Rittenberg, D., and K. Bloch (1944). *J. Biol. Chem.*, **154**, 311.

Rittenberg, D., and K. Bloch (1945). *J. Biol. Chem.*, **160**, 417.

Robbins, E. A., and P. D. Boyer (1957). *J. Biol. Chem.*, **224**, 121.

Robertson, J. D. (1956a). *J. Biophys. Biochem. Cytol.*, **2**, 381.

Robertson, J. D. (1956b). *In* "Electron Microscopy" (F. S. Sjöstrand and J. Rhodin, eds.), p. 197. Academic Press, New York.

Robertson, J. D. (1957a). *J. Biophys. Biochem. Cytol.* **3**, 1043.

Robertson, J. D. (1957b). *J. Physiol. (London)*, **140**, 58.

Robertson, J. D. (1959). *In* "Symposium on Molecular Biology" (D. Nachmansohn, ed.). Academic Press, New York. (In press.)

Rona, P., and P. Neukirch (1912). *Arch. ges. Physiol. Pflüger's*, **146**, 371.

Rosenberg, H. (1928). *In* "Handbuch der Normalen und Pathologischen Physiologie," Vol. VIII/2, p. 876.
Rothenberg, M. A. (1949). *Trans. Am. Neurol. Assoc.*, p. 230.
Rothenberg, M. A. (1950). *Biochim. et Biophys. Acta,* **4**, 96.
Rothenberg, M. A., and E. A. Feld (1948). *J. Biol. Chem.*, **172**, 345.
Rothenberg, M. A., and D. Nachmansohn (1947). *J. Biol. Chem.*, **168**, 223.
Rothenberg, M. A., D. B. Sprinson, and D. Nachmansohn (1948). *J. Neurophysiol.*, **11**, 111.
Sawyer, C. H. (1943). *J. Exptl. Zool.*, **92**, 1.
Schleyer, W. L. (1955). *Biochim. et Biophys. Acta,* **16**, 396.
Schoenheimer, R. (1942). "The Dynamic State of Body Constituents." Harvard Univ. Press, Cambridge, Massachusetts.
Schoenheimer, R., and D. Rittenberg (1936). *J. Biol. Chem.*, **114**, 381.
Schoenheimer, R., and D. Rittenberg (1940). *Physiol. Revs.*, **20**, 218.
Schoffeniels, E. (1957a). *Federation Proc.*, **16**, 497.
Schoffeniels, E. (1957b). *Biochim. et Biophys. Acta,* **26**, 585.
Schoffeniels, E. (1958a). *Nature,* **181**, 287.
Schoffeniels, E. (1958b). *Science,* **127**, 1117.
Schoffeniels, E. (1959). "Symposium on Physico-chemical Mechanisms of Nerve Activity," *N.Y. Acad. Sci.* (In press.).
Schoffeniels, E., and D. Nachmansohn (1957). *Biochim. et Biophys. Acta,* **26**, 1.
Schoffeniels, E., I. B. Wilson, and D. Nachmansohn (1958). *Biochim. et Biophys. Acta,* **27**, 629.
Schrader, G. (1952). "Die Entwicklung neuer Insektizide auf Grundlage organischer Fluor- und Phosphor-Verbindungen," Verlag Chemie, Weinheim, Germany.
Schrödinger, E. (1944). "What is life?" Cambridge Univ. Press, New York and London.
Schwarzenbach, G. (1936). *Z. physik. Chem. (Leipzig) Abt. A.*, **176**, 133.
Schwarzenbach, G., E. Kampitsch, and R. Steiner (1945). *Helv. Chim. Acta,* **28**, 828.
Schwarzenbach, G., E. Kampitsch, and R. Steiner (1946). *Helv. Chim. Acta,* **29**, 364.
Seaman, G. R. (1951). *Proc. Soc. Exptl. Biol. Med.*, **76**, 169.
Seaman, G. R., and R. K. Houlihan (1951). *J. Cellular Comp. Physiol.*, **37**, 309.
Shuster, L., and N. O. Kaplan (1953). *J. Biol. Chem.*, **201**, 535.
Simon, E. J., and D. Shemin (1953). *J. Am. Chem. Soc.*, **75**, 2520.
Skouby, A. P. (1951). *Acta Physiol. Scand.*, **24**, 174.
Snell, E. E., G. M. Brown, V. J. Peters, J. A. Craig, E. L. Wittle, J. A. Moore, V. M. McGlohon, and O. D. Bird (1950). *J. Am. Chem. Soc.*, **72**, 5349.
Sprinson, D. B., and D. Rittenberg (1951). *Nature,* **167**, 484.
Stadtman, E. R. (1952a). *J. Biol. Chem.*, **196**, 527.
Stadtman, E. R. (1952b). *J. Biol. Chem.*, **196**, 535.
Stadtman, E. R., G. D. Novelli, and F. Lipmann (1951). *J. Biol. Chem.* **191**, 365.
Staempfli, R. (1958). *Helv. Physiol. et Pharmacol. Acta,* **16**, C32.
Stedman, E., and G. Barger (1925). *J. Chem. Soc.*, **127**, 247.
Stedman, E., E. Stedman, and L. H. Easson (1932). *Biochem. J.*, **26**, 2056.
Stein, S. S., and D. E. Koshland (1953). *Arch. Biochem. Biophys.*, **45**, 467.
Stern, J. R., and S. Ochoa (1949). *J. Biol. Chem.*, **179**, 491.
Stern, J. R., and S. Ochoa (1951). *J. Biol. Chem.*, **191**, 161.
Stern, J. R., B. Shapiro, E. R. Stadtman, and S. Ochoa (1951). *J. Biol. Chem.*, **193**, 703.
Stern, J. R., S. Ochoa, and F. Lynen (1952). *J. Biol. Chem.*, **198**, 313.

Stoerk, H. C., and E. Morpeth (1944). *Proc. Soc. Exptl. Biol. Med.*, **57**, 154.
Symposium on Physico-chemical Mechanism of Nerve Activity (1946). *Ann. N.Y. Acad. Sci.*, **47**, 375.
Symposium on Active Transport and Secretion (1954). *Symposia Soc. Exptl. Biol. No.* **8**.
Taylor, I. M., J. M. Weller, and A. B. Hastings (1952). *Am. J. Physiol.*, **168**, 658.
Teorell, T. (1951). *Z. Elektrochem.*, **55**, 460.
Teorell, T. (1953). *Progr. in Biophys. and Biophys. Chem.*, **3**, 305.
Toman, J. E. P., J. W. Woodbury, and L. A. Woodbury (1947). *J. Neurophysiol.*, **10**, 429.
Ussing, H. H. (1954). *In* "Ion Transport Across Membranes" (H. T. Clarke and D. Nachmansohn, eds.), p. 30. Academic Press, New York.
Vahlquist, B. (1935). *Skand. Arch. Physiol.*, **72**, 133.
von Liebig, J. (1846). "Die Thier-Chemie oder die organische Chemie in ihrer Anwendung auf Physiologie und Pathologie." Vieweg, Braunschweig.
von Muralt, A. (1937). *Proc. Roy. Soc.*, **B123**, 399.
von Muralt, A. (1946). "Die Signalvermittlung in Nerven," Birkhauser, Basel.
von Muralt, A. (1954). *Ann. Rev. Physiol.*, **16**, 305.
Von Wazer, J. R., and D. A. Campanella (1950). *J. Am. Chem. Soc.*, **72**, 655.
Waelsch, H., ed. (1954). "Biochemistry of the Developing Nervous System," pp. 281-326. Academic Press, New York.
Walsh, J. (1773). *Phil. Trans. Roy. Soc. (London)*, **63**, 461.
Waser, P. G., and U. Lüthi (1956). *Nature*, **178**, 981.
Waser, P. G., and U. Lüthi (1957). *Arch. intern. pharmacodynamie*, **112**, 272.
Weber, H. H. (1958). "The Motility of Muscle and Cells," Harvard Univ. Press.
Weber, H. H., and H. Portzehl (1954). *Progr. in Biophysics and Biophys. Chem.* **4**, 60.
Weiland, W. (1912). *Arch. ges. Physiol. Pflüger's*, **147**, 171.
Whittam, R. (1958). *J. Physiol. (London)*, **140**, 479.
Whittam, R., W. Bartley, and G. Weber (1955). *Biochem. J.*, **59**, 590.
Williamson, H. (1775). *Phil. Trans. Roy. Soc. London*, **65**, 94.
Wills, J. H., A. M. Kunkel, R. V. Brown, and G. E. Groblewski (1957). *Science*, **125**, 743.
Wilson, I. B. (1951a). *J. Biol. Chem.*, **190**, 111.
Wilson, I. B. (1951b). *Biochim. et Biophys. Acta*, **7**, 466.
Wilson, I. B. (1951c). *Biochim. et Biophys. Acta*, **7**, 520.
Wilson, I. B. (1952a). *J. Biol. Chem.*, **197**, 215.
Wilson, I. B. (1952b). *J. Biol. Chem.*, **199**, 113.
Wilson, I. B. (1952c). *J. Am. Chem. Soc.*, **74**, 3205.
Wilson, I. B. (1954a). *J. Biol. Chem.*, **208**, 123.
Wilson, I. B. (1954b). *In* "The Mechanism of Enzyme Action" (W. D. McElroy and B. Glass, eds.), p. 642. Johns Hopkins Press, Baltimore, Maryland.
Wilson, I. B. (1955a). *Discussions Faraday Soc. No.* **20**, 119.
Wilson, I. B. (1955b). *Arch. intern. pharmacodynamie*, **114**, 204.
Wilson, I. B. (1958). *Biochim. et Biophys. Acta* **27**, 196.
Wilson, I. B. (1959). Symposium on "Physico-chemical Mechanisms of Nerve Activity," *N.Y. Acad. Sci.* (In press.)
Wilson, I. B., and F. Bergmann (1950a). *J. Biol. Chem.*, **185**, 479.
Wilson, I. B., and F. Bergmann (1950b). *J. Biol. Chem.*, **186**, 683.
Wilson, I. B., and E. Cabib (1956). *J. Am. Chem. Soc.*, **78**, 202.

Wilson, I. B., and M. Cohen (1953). *Biochim. et Biophys. Acta,* **11**, 147.
Wilson, I. B., and S. Ginsburg (1955a). *Arch. Biochem. Biophys.,* **54**, 569.
Wilson, I. B., and S. Ginsburg (1955b). *Biochim. et Biophys. Acta,* **18**, 168.
Wilson, I. B., and S. Ginsburg (1958). *Biochem. Pharmacol.,* **1**, 200.
Wilson, I. B., and E. K. Meislich (1953). *J. Am. Chem. Soc.,* **75**, 4628.
Wilson, I. B., and D. Nachmansohn (1954). *In* "Ion Transport Across Membranes" (H. T. Clarke and D. Nachmansohn, eds.), p. 35. Academic Press, New York.
Wilson, I. B., and C. Quan (1958). *Arch. Biochem. Biophys.,* **73**, 131.
Wilson, I. B., and F. Sondheimer (1957). *Arch. Biochem. Biophys.,* **69**, 468.
Wilson, I. B., F. Bergmann, and D. Nachmansohn (1950). *J. Biol. Chem.,* **186**, 781.
Wilson, I. B., S. Ginsburg, and E. K. Meislich (1955). *J. Am. Chem. Soc.,* **77**, 4286.
Wilson, I. B., S. Ginsburg, and C. Quan (1958). *Arch. Biochem. Biophys.,* **77**, 286.
Wintersteiner, O., and J. D. Dutcher (1943). *Science,* **97**, 467.
Wolfgram, T. J. (1954). *Am. J. Physiol.,* **176**, 505.
Young, J. Z. (1936a). *Quart. J. Microscop. Sci.,* **78**, 367.
Young, J. Z. (1936b). *Cold Spring Harbor Symposia Quant. Biol.,* **4**, 1.
Young, J. Z. (1952). "Doubt and Certainty in Science." Oxford Univ. Press, London and New York.
Zeller, E. A., and A. Bissegger (1943). *Helv. Chim. Acta,* **26**, 1619.
Zotterman, Y. (1953). *In* "Transactions 4th Conference on Nerve Impulse" (D. Nachmansohn, ed.), p. 140. Josiah Macy, Jr. Foundation, New York.

SUPPLEMENT I

Properties and Function of the Proteins of the Acetylcholine Cycle in Excitable Membranes

I. Cell Membranes

A. General Properties

Among the many dramatic developments in biological sciences one of the most important advances has been the spectacular progress in the understanding of the function and properties of cell membranes, achieved essentially during the last decade. Knowledge of membrane behavior has particular significance for the understanding of nerve function. Electrical activity during nerve impulse conduction has been assumed, since the turn of the century, to be associated with excitable membranes and with the ion movements across these membranes (see page 2). However, until the rise of electron microscopy the evidence for the existence of cell membranes was only indirect, no matter how suggestive and impressive the indications were. Because of examination by electron microscopy, combined with biochemical and biophysical analyses, a vast amount of information has accumulated and progress is continuing at an ever increasing rate. Before discussing the special features of excitable membranes it appears necessary to recall a few basic notions and characteristics of cell membranes in general.

It is now generally recognized that cell membranes are the site of most vital functions, such as energy supply, active transport, neural function, vision, excitation-contraction coupling, and photosynthesis. Moreover, recent investigations, in which membranes played an essential role, have revealed the fundamental importance of structure and organization for the chemical reactions taking place in living cells. Whereas *classical* biochemistry devoted most efforts to the isolation and characterization of cell constituents and their behavior in solution, modern biophysical chemistry has recognized that essential characteristics of reactions occurring in solution cannot be simply extrapolated to that of the same chemical processes occurring in the cellular organization. It has become increasingly apparent that rates and extents of reactions are profoundly influenced by structural factors such as microenvironment, for instance, charged groups surrounding the active site, cooperativity, allosteric effects, allotopy, regulatory factors, protein–protein and protein–lipid interactions (see, e.g., Loewenstein, 1966; Racker, 1970;

Manson, 1971; Rothfield, 1971). Therefore, modern biochemistry aims at analyzing the reaction behavior within cellular structures, in addition to that in solution and on molecular level.

One of the most significant developments in the field of membranes has been a conceptual change. At the turn of the century, cell membranes were considered mainly as passive barriers in which lipids were essential in preventing easy passage of cell constituents and metabolites. Today it is well established that cell membranes contain many proteins including enzymes. In some membranes 30–50 different proteins have been actually isolated and many of them identified. Biomembranes have been recognized as extremely dynamic structures, the site of a great variety of chemical processes; Aharon Katchalsky once referred to biomembranes as being powerful biochemical factories.

In many biological membranes more than two-thirds of the mass are proteins, about one-third are phospholipids. In a given membrane type, the phospholipids and their chemical properties show a great diversity. This diversity increases when different types of membranes are compared. In addition, there are oligosaccharides and a great variety of small molecules including different metal ions. However, the remarkable specificity, the diversity, and the efficiency of membrane functions seem to be more readily accounted for when we attribute the dominant role to the membrane proteins rather than to the lipid phase.

In spite of the large amount of information on the chemical composition of membranes, we are still very far from a real knowledge of the molecular organization of the various constituents in the intact membrane. The notion of "unit membrane," proposed by Robertson (1960a) and based essentially on the Danielli–Davson model, assumed a uniform structure of all cell membranes 80 Å thick and formed by a bimolecular leaflet of phospholipids to which proteins are attached on the inside and outside by ionic forces. While this notion first appeared attractive to many investigators, it proved to be inadequate to account for the great variety of biochemical, biophysical, and electron microscopic examinations (see, e.g., Sjöstrand, 1963; Elbers, 1964; Green and Perdue, 1966; Korn, 1966; Sjöstrand and Barajas, 1968; Green and MacLennan, 1969). In a very lucid and profound analysis Singer (1971), in discussing various thermodynamic problems due to the complexity of membrane structure, has pointed out the difficulties of reconciling the Robertson model with well-known thermodynamic facts.

In the last few years more than ten different membrane models have been proposed in an effort to integrate the available information (see, e.g., Lenard and Singer, 1966; Singer, 1971; Vanderkooi and Green, 1971; Benson, 1968; Blasie and Worthington, 1969). Without discussing

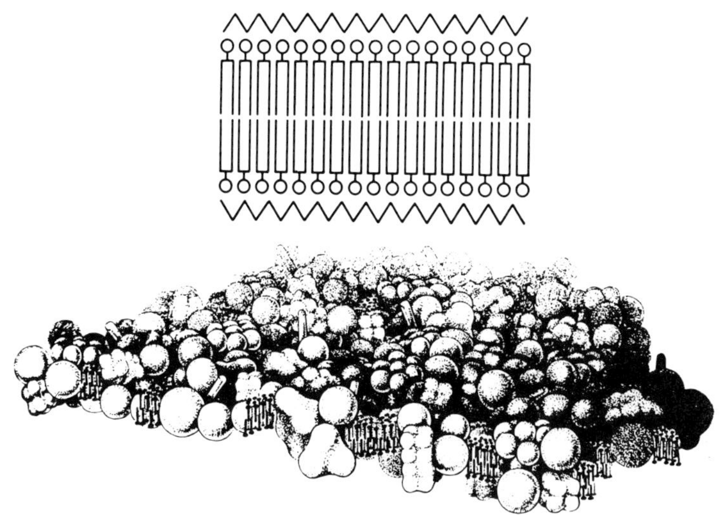

Fig. 1. *Top:* Model of Robertson (1960) of the "unit membrane," proposed to be 80 Å thick and formed by a bimolecular layer of phospholipids surrounded on the inside and outside by proteins attached to the phospholipids by coulombic forces. *Bottom:* Model of Sjöstrand and Barajas (1970) of the inner mitochondrial membrane. The model tries to integrate the enzymes and coenzymes with their subunits biochemically established to be located in this membrane. The phospholipids of this membrane are known to form only a relatively small fraction of this membrane, as indicated in the picture. Thickness about 150–200 Å.

the merits of these and various other models proposed, we would like to show two models as an illustration of the remarkable development of the last decade: the model of Robertson (1960a) of the "unit membrane" and the model of Sjöstrand and Barajas (1970) representing the inner membrane of mitochondria (Fig. 1). The latter model incorporates the many enzymes showing their subunits and coenzymes well established to be present in this membrane. It also accounts for the proper relationships between the amounts of proteins and phospholipids in this particular membrane. This model is rather attractive because it integrates the great number of known constituents.

Recognizing the importance of the proteins and the necessity of preserving them in their native conformation, Sjöstrand and Barajas have introduced new techniques for the preparation and fixation of the specimen for electron microscopy; they avoided the usual standard techniques which almost certainly denature the proteins. With the new techniques membranes have a thickness of 150–200 Å as compared to the 80–100 Å obtained with the standard procedures. A striking feature of their electron micrographs is the indication of many globular formations within the membrane (Sjöstrand and Barajas, 1968).

B. Excitable Membranes

Excitable membranes have the special ability of transiently changing their permeability in a controlled way to those ions which are the carriers of the electrical exchange currents accompanying membrane potential changes such as the action currents propagating nerve impulses. For more than two decades leading electrophysiologists strongly supported the view that electrical activity is a simple diffusion process following the concentration gradient (see, e.g., Hodgkin, 1951, 1964). Even at present in most textbooks of physiology this view is accepted as an untouchable dogma. The assumption of chemical reactions controlling the ion movements during electrical activity was vigorously rejected (see, e.g., Baker *et al.*, 1962; Keynes and Aubert, 1964); the metabolic reactions were only accepted for the restoration of the concentration gradients during recovery which require energy to extrude the entered Na^+ ions against their concentration gradients (the so-called "Na-pump").

This view has become untenable in the light of many recent developments. A strong heat production and absorption has been found in recent measurements by Hill and his associates to roughly coincide with electrical activity (Abbott *et al.*, 1958); the authors find it hard to believe that the drastic and precisely controlled changes of permeability could occur without chemical reaction (see pp. 4–8). Little attention was paid to the statement of A. V. Hill (1960) that there appears no alternative to the assumption that "*the early production and absorption of heat after a stimulus are largely due to chemical reactions associated with, and following the permeability cycle.*" In still more recent measurements the heat production and absorption were found to be strictly coinciding with the action potential; heat is produced during the rising phase and absorbed during the falling phase (Howarth *et al.*, 1968). The authors reject the assumption of chemical reactions as responsible for these heat changes. Their interpretation is open to question. The heat changes during nerve activity are discussed by the author (Nachmansohn, 1973) while some more fundamental aspects are discussed by Neumann in Supplement II. Two points may be stressed in this connection. First, A. V. Hill and his associates had pointed out more than 15 years ago, that the heat changes measured must not be referred to gram nerve, as had been done in the 1920's, but to a membrane of about 100 Å thickness, since the events associated with the heat are restricted to that small fraction of nerve fibers. As pointed out by Neumann (see Supplement II), even this limitation does not yet give a correct value of the real magnitude of the heat changes, since only a small fraction, 1% or less, of the axonal membrane is active during conduction. Second, a specific attribution of measured thermodynamic parameters is at present impossible. Even when the thermodynamic measurements of a chemical reaction are made with a single protein in solution, for instance, such as the reaction of oxygen with hemoglobin, a protein with a well established tridimensional structure, there are theoretical difficulties preventing precise separation in individual processes (Rossi-Fanelli *et al.*, 1964; Wyman, 1964). Heat changes in highly complex structures, such as muscle or nerve cell, are the result of a large number of chemical reactions taking place during the measurements. It is simply impossible to try to specify the role of one of the very many reactions involved (say, e.g., the enthalpy change associated with AcCh hydrolysis, as has been repeatedly attempted).

Various other biochemical and biophysical observations are accumulating which are incompatible with the assumption of a simple electrodiffusion process as an explanation for the generation of the action potential. For instance, drastic modifications of ion composition both in the interior of the axon and in its outer environment have for a considerable length of time no effect on the electrical parameters, contrary to the prediction of the Hodgkin–Huxley theory (Tasaki, 1968). Furthermore, according to this theory the prolongation of various phases of the action potential at lower temperature should correspond to larger ion movements. In contrast to this prediction, Landowne (1973) found a decrease of the amount of ions transported during excitation with decreasing temperature. It is pertinent to recall in this context the high Q_{10} of action potentials, as discussed on page 8. Particularly impressive are the data obtained by Schoffeniels (1958) who used the Arrhenius formalism to evaluate the temperature dependence of the duration of the action potential in the monocellular electroplax preparation: he found the apparent energy of activation associated with the duration of the action potential to be about 20,000 cal per mole. This rather high value is hard to reconcile with a simple electrodiffusion process.

Finally, another objection to the mathematical model of Hodgkin and Huxley as being the final answer to the problem of bioelectric currents is based on more fundamental considerations. Membranes are not only electrically but also chemically nonequilibrium systems. The Hodgkin–Huxley theory is based purely on electrodiffusion, using the Nernst equation extensively. As is observed experimentally and is discussed in Supplement II, there is a straightforward dependence of the membrane potential on the logarithm of ion concentration only in a limited concentration range, i.e., the Nernst equation is not generally obeyed, particularly not during electrical activity.

It is, therefore, not surprising that an increasing number of investigators reject the assumption of a simple electrodiffusion process as an explanation of electrical activity. Cole (1965), one of the pioneers in the experimental and theoretical analysis of the physical properties and the bioelectrical phenomena of the excitable membranes, discusses the model for passive ion flow especially in respect to the squid giant axons, in the light of theories on ion movements, as originally described by Nernst (1888, 1889) and further developed by Planck (1890a,b). In view of the penetrating inquiry performed in this article the final conclusions of the author may be quoted in full:

> A general explanation of the voltage-clamp characteristics in terms of the Planck several-ion model has not been found, and this model appears to be unable to account for all the relaxations required by the membrane. The sodium ion phenomena in the membrane seem to be almost entirely contrary to the properties of a single-ion model. The single-ion theory has long been an attractive explanation for the potassium behavior in the membrane but further consideration suggests that here also the model is so inadequate as to make it an improbable explanation.
>
> The contrasts between the various calculations and the array of experimental facts give a basis to conclude that the simple process of electrodiffusion is not a principal factor in the behavior of the squid axon membrane.

More recently, Agin (1967, 1972) discusses the limitations of unspecific mathematical models, such as the Hodgkin–Huxley model, and particularly criticizes the uncritical use of partially unrealistic assumptions made to describe the electrochemical behavior of excitable membranes. He concludes that a real understanding of electrical activity requires knowledge of the physicochemical events in the excitable membrane.

In view of the unsatisfactory description of nerve excitability offered by the Hodgkin–Huxley model and convinced of the necessity of the biochemical reactions involved in the control of ion permeabilities, Aharon Katchalsky suggested some 10 years ago that the author integrate the vast amount of biochemical and electrophysiological data for obtaining an insight into the molecular events underlying electrical activity. Many aspects were discussed over the years, but no attempt was made to present such an integration in a systematic and detailed form. It was only in the spring of 1972 that such an attempt was started in a series of discussions at the Weizmann Institute in which Eberhard Neumann participated. After the death of Katchalsky we decided to continue these efforts. The collaboration led to a tentative integral model of nerve excitability (Neumann *et al.*, 1973). During the last two years the integral model has been extended by Neumann (1974, 1975; for a condensed version of Supplements I and II, see Neumann and Nachmansohn, 1975). Since Neumann's work is presented in Supplement II, the author will only occasionally refer to some aspects of the model where it is desirable for the understanding of the biochemical data. The material presented in this Supplement will be limited to new biochemical information accumulated since 1959, the first printing of this monograph. The progress of the biochemical basis has been frequently summarized in a number of review articles and are listed in the Reviews of the Author (page 335).

It may be mentioned that Dubois and Schoffeniels (1974) have recently also presented a paper on the molecular basis of the action potential using the biochemical data of the author and postulating an AcCh and a Ca^{2+} ion cycle as being responsible for the action potential.

II. Role of the AcCh Cycle in Control of Ion Permeability

A chemical theory of the events associated with nerve excitability and the conduction of nerve impulses in nerve and muscle fibers has been elaborated on by the author during the last three decades. He was convinced that the increased ion fluxes during electrical activity and their precise control must result from complex chemical processes, just as is required by every other cellular mechanism. The elucidation of the generation of bioelectricity required consequently a knowledge of the biochemical reactions involved. The work was based essentially on the analysis of the properties and function of proteins processing acetylcholine (AcCh)*: (1) the enzyme AcCh-esterase, (2) the AcCh-receptor, the target protein of AcCh, (3) the enzyme choline *O*-acetyltransferase (choline acetylase) that synthesizes AcCh. These proteins have been isolated, purified, and characterized. AcCh-esterase has been crystallized. Additionally there is the postulate of a protein, the so-called storage protein to which AcCh is bound.

Since 1937, the use of electric organs of electric fish has been instrumental for progress (see pp. 70–82). These organs are specialized for their function, the generation of electricity, to a degree virtually without parallel in living organism. Dubois-Reymond, who had worked for 20 years with these organs, predicted in 1877 that this material would eventually permit to explain the mechanism of the electrical activity in nerve and muscle fibers (see p. 70). For the biochemist interested in the

* While in the first printing ACh was used as abbreviation of acetylcholine, in the two supplements AcCh is used, since the Editors of the *Proceedings of the National Academy of Sciences U.S.* consider this abbreviation to be more appropriate.

proteins processing AcCh, these organs offer a unique material; their paramount importance became immediately apparent to the author when it was found that the tissue contains fantastic amounts of AcCh-esterase: 1 kg of tissue may hydrolyze 3–4 kg of AcCh per hour, although the tissue is formed of 92% water and only 3% proteins. Even this figure does not yet give the real value of concentration which is 4 orders of magnitude higher, as will be outlined later. Obviously, several other factors were essential for the advances as outlined in the Preface to this revised edition.

A. Role of Acetylcholine in the Excitable Membrane; The Acetylcholine Cycle

Since the 1930's it was proposed that AcCh acts as a "neurohumoral" transmitter, i.e., that it is released from nerve endings and after crossing the nonconducting gap, stimulates the effector cell, nerve, or muscle. Although this view was vigorously contested by many eminent neurobiologists (see pp. 16–17), it is still widely accepted among physiologists. The results of the biochemical analyses of the protein processing AcCh indicated very soon that this hypothesis is untenable. Twenty years ago the following hypothesis was proposed by the author for the role of AcCh in the permeability changes of excitable membranes (Fig. 2). The picture is taken from his Harvey Lecture presented in 1953* (Nachmansohn 1955a,b,c). In this view AcCh is assumed to be released by stimulation from the storage protein and acts on the AcCh-receptor protein. (Both proteins were at that time a postulate.) It was suggested that AcCh induces a conformational change of the receptor initiating thereby the processes leading to an increased ion permeability. In the following step AcCh is hydrolyzed (in microseconds) by AcCh-esterase, permitting the receptor to return to its resting condition, thus reestablishing the barrier for ion movements. All these processes were assumed to take place within the membrane in a structurally organized form. In the recovery phase AcCh is resynthesized by choline acetylase (choline *O*-acetyltransferase) using ATP-hydrolysis as source of energy; ATP is provided by the glycolytic or citric acid cycle.

It is unnecessary to describe in detail all the modifications of the cycle during the last 20 years induced by the advances of our information. Figure 3 shows the present view of the mechanism by which AcCh acts in the membrane: the AcCh cycle worked out by Neumann and myself during the last year in connection with the integral model of nerve excitability. The figure shows the AcCh cycle controlling the open-

* Although the figure is shown on p. 101, it is reproduced here for the convenience of the reader wishing to compare the original AcCh cycle with its present form.

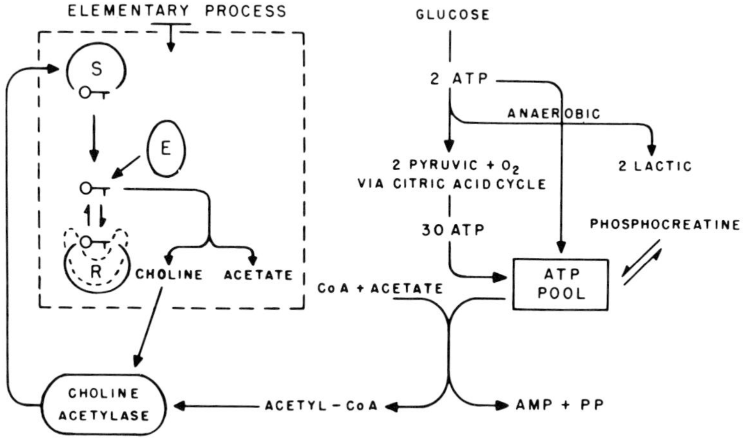

FIG. 2. Sequence of energy transformations associated with conduction, and integration of the AcCh cycle into the metabolic pathways of the nerve cell. The elementary process of conduction has been tentatively pictured as follows: (1) In resting condition AcCh (○‒┬) is bound, presumably to a storage protein (S). The membrane is polarized. (2) AcCh is released by stimulation. The free ester combines with the receptor (R), presumably a protein. (3) The receptor changes its configuration (dotted line). This process leads to increased Na ion permeability and permits its rapid influx. This is the trigger action by which the potential primary source of EMF, the ionic concentration gradient, becomes effective and by which the action current is generated. (4) The ester-receptor complex is in dynamic equilibrium with the free ester and the receptor; the free ester is open to attack by AcCh-esterase (E). (5) The hydrolysis of the ester permits the receptor to return to its original shape. The permeability decreases and the membrane is again in its original polarized condition. (Nachmansohn, 1953.)

ing and closing of the gateway, the permeation zone through which the ion flow is greatly increased during activity. A decrease of the resting potential by 15–20 mV (corresponding to an electric field strength of 15–20 kV/cm) is assumed to induce a conformational change of the storage protein. This view is derived from polyelectrolyte experiments of Neumann and Aharon Katchalsky (1972) and Neumann and Rosenheck (1972) and will be further discussed in more detail in Supplement II. The conformational change of the storage site leads to a realese of $AcCh^+$ ions translocated to the AcCh-receptor, where AcCh is suggested to induce a conformational change. As a result of this structural change Ca^{2+} ions are released from the receptor. Calcium ions are assumed to be involved in the change of the gateway state, possibly again through conformational changes of proteins and/or phospholipids (or lipo- or glycoproteins). Simultaneously AcCh is translocated to AcCh-esterase, which hydrolyzes the ester in choline, acetate, and protons. This is one

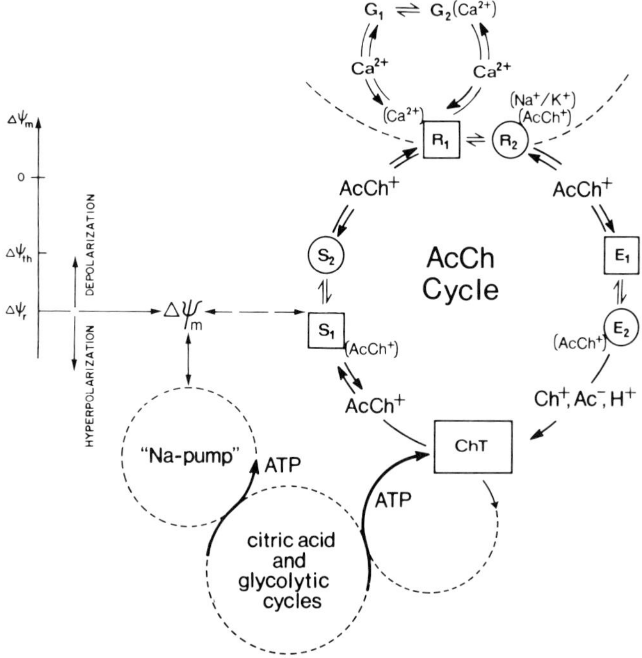

FIG. 3. AcCh cycle, for the cyclic chemical control of stationary membrane potentials and transient potential changes. The binding capacity of the storage site for AcCh is assumed to be dependent on the membrane potential, $\Delta\psi_M$, and is thereby coupled to the "Na^+/K^+ exchange-pump" (and the citric acid and glycolytic cycles). The control cycle for the gateway G_1 (Ca^{2+} binding and closed) and G_2 (open) comprises the SRE assemblies (see Fig. 16) and the choline *O*-acetyltransferase (Ch-T); Ch-T couples the AcCh synthesis cycle to the translocation pathway of AcCh through the SRE-assemblies. The continuous subthreshold flux of AcCh through such a subunit is maintained by the virtually irreversible hydrolysis of AcCh to choline (Ch^+), acetate (Ac^-), and protons (H^+) and by steady supply flux of AcCh to the storage from the synthesis cycle. In the resting stationary state, the membrane potential ($\Delta\psi_r$) reflects dynamic balance between active transport (and AcCh synthesis) and passive fluxes of AcCh through the control cycles surrounding the gateway and of the various ions unsymmetrically distributed across the membrane. Fluctuations in membrane potential (and exchange currents) are presumably amplified by fluctuations in the local AcCh concentrations maintained at a stationary level during the continuous translocation of AcCh through the cycle. (Neumann and Nachmansohn, 1975.)

of the virtually irreversible reactions in the cycle. For the resynthesis of AcCh energy is required, provided by ATP hydrolysis. Choline *O*-acetyltransferase acetylates choline and translocates it to the storage protein. The result of all these processes is the passage of 15–30,000 ions per molecule of AcCh activated across the membrane in both direc-

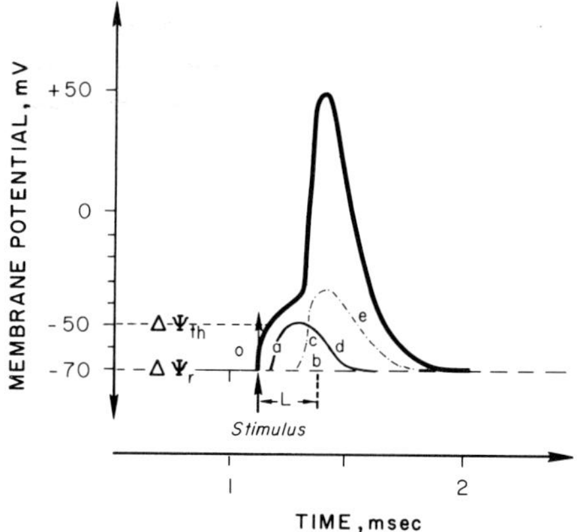

FIG. 4. Schematic representation of the time course of chemical reactions in relation to the action potential (o). Potential change owing the change of the membrane potential in rest ($\Delta\psi_r$) to the threshold potential ($\Delta\psi_{th}$). (a) and (d), time of the AcCh course of AcCh release and hydrolysis, respectively; (b) and (c), time course of Ca^{2+} release from receptor and of gateway opening; (e), time course of Ca^{2+} reuptake and conformational relaxation of the receptor, leading to closure of the gateway (Neumann *et al.,* 1973). This is a very simplified scheme indicating the relationship between the potential change during electrical activity and the assumed chemical processes causing the potential change.

tions. The processes of the cycle are, of course, taking place continuously, as are all events in living cells, but during activity the rate is greatly increased.

Although this figure is not yet 2 years old, a still more recent and more advanced version of the AcCh cycle is presented in Supplement II. A schematic presentation of the time course of the assumed chemical reactions in relation to the action potential is shown in Fig. 4.

B. *Macromolecular Conformation and Ca^{2+} Ions*

It may be useful to recall a few physicochemical aspects of macromolecular conformational changes in connection with Ca^{2+} ions. Structural changes of proteins and macromolecular organizations such as membranes are often cooperative in nature. One of the consequences of cooperativity is the possibility of far-reaching conformational changes by small local changes of environmental conditions. Moreover, conformational changes induced by binding of a ligand at one site may change the reactivity of other possibly even far remote sites of a macromolecular system (allosteric effects). Calcium ions are particularly effective in inducing large conformational changes as, for instance, in muscular contractions, and are particularly efficient in systems that contain regions of a relatively high negative surface charge. In such polyelectrolytic regions the osmotic coefficient for Ca^{2+} is in the order of 0.01, i.e., about 99% of Ca^{2+} counterions are bound (A. Katchalsky, 1964). The high-binding capacity is one of the reasons for the assumption that Ca^{2+} ions play an essential role in maintaining structural and functional integrity of protein and lipoprotein organization. For almost a century Ca^{2+} ions were assumed to play an essential role in nerve excitability (Brink, 1954). More recently, Tasaki (1968) has emphasized that Ca^{2+} ions are absolutely necessary for nerve excitability.

The question, however, arises how the action of Ca^{2+} on the permeability gateway is controlled. If one assumes that an increase in the release of AcCh ions is acting as a signal for initiating the processes leading to increased permeability, it appeared reasonable to assume that the "AcCh-signal" is recognized by a specific target protein; only proteins have the ability to recognize a given ligand with a high specificity. In this case the conformational change of the AcCh-receptor would control the release of Ca^{2+} ions which in the resting steady state may be bound to the carboxyl groups of the protein. It was suggested that several Ca^{2+} ions per molecule receptor activated may be released (see, e.g., Nachmansohn, 1972b). The test of this prediction requires, however, the availability of homogeneous receptor protein for a variety of experimental tests.

III. Proteins Processing Acetylcholine

A. *AcCh-Esterase*

1. Isolation and Purification; Special Features

As already mentioned, AcCh-esterase was first isolated, in highly active form, in 1938 from electric tissue. Until then unpurified horse serum esterase, a mixture of several types of esterases, was used, e.g., for studies of "cholinesterase" inhibitors and related problems. AcCh-esterase, although not particularly specific, has many special features by which it may be distinguished from other esterases (pp. 20–26). The enzyme was purified in the early 1940's virtually to homogeneity of the protein (Rothenberg and Nachmansohn, 1947).

2. First Proposal of Active Site, Mechanism of Hydrolysis; Organophosphate Action

The availability of a highly active preparation permitted the analysis of pertinent properties and reaction mechanisms with a variety of important compounds, such as carbamoyl derivatives (widely applied as drugs in medicine), organophosphates (some of which are used as insecticides and some of which are potential chemical warfare agents—"nerve gases"). The early data of these analyses, which led to the suggestion of some features of the active site, the catalytic mechanism, the idea of an acyl enzyme, and other information still of interest today, are described on pp. 105–118.

Of the various results described, one particular aspect may be briefly discussed in view of its general practical and theoretical interest: the development of the antidote pyridine-2-aldoxime methiodide (2-PAM) against organophosphate insecticide poisoning. At an early stage of his studies on DFP the author became convinced that the fatal actions of DFP (and other organophosphates) are due to a specific biochemical lesion, i.e., are due specifically to the irreversible block of AcCh-esterase. Although DFP is known to react with a large number of ester-splitting enzymes, only AcCh-esterase is essential for so many vital functions of the body. Therefore, its block appeared to be the most likely effect responsible for the high toxicity. For more than 10 years this view was vigorously rejected by a majority of electrophysiologists and pharmacologists. They considered, almost unanimously, the actions of DFP to be a *general* toxic effect. The author was convinced that once the understanding of the reaction mechanism was achieved, this would provide the basis for the development of an antidote by specifically repairing the biochemical lesion. Once it became established that organophosphates form a phosphorylated enzyme with a rather stable P–O bond between their P atom and the serine oxygen in the active site of the enzyme, it became probable that the phosphoryl group may be removed by a nucleophilic compound attacking the phosphorus atom. A potent nucleophilic agent reactivating specifically the phosphorylated AcCh-esterase, in an S_N2 reaction, may possibly act as an antidote. This concept formed the basis of the further developments. In view of Hestrin's (1950) observations on the reactions of hydroxylamine with AcCh-esterase, which had led to the first—admittedly tentative and speculative—assumption of an acylated enzyme, the author assumed that hydroxylamine may similarly act on the phosphorylated enzyme. When his suggestion was tested by Wilson (1951), a reversal of the block of the enzyme activity was obtained for the first time, but only with high concentrations of NH_2OH and after long incubation times. Wilson and Meislich (1953) and Wilson *et al.* (1953) and Wilson and Ginsburg (1955a,b) then developed a series of nucleophilic compounds in order to obtain a more potent reactivator. The most rapid and effective reactivator in the test tube turned out to be 2-PAM (see pp. 127–129). This potency permitted the assumption that it might act as an efficient antidote in the body although this was by no means a foregone conclusion. When tested by Kewitz on animals, 2-PAM acted as a surprisingly powerful antidote, especially in combination with atropine, a compound that protects the AcCh-receptor against excess AcCh and thereby prevents dangerous side effects (Kewitz *et al.*, 1956). Today, 2-PAM (although as chloride and not as iodide) is widely used as an antidote against organophosphate poisoning all over the world and has saved many lives.

When Kewitz presented his dramatic results at the Federation Meeting in Atlantic City, the author was asked whether he had an explanation as to why the Army Chemical Center, with many hunderds of investigators, unlimited funds and large facilities, did not succeed over a period of more than 10 years in developing an antidote, whereas the author with a few associates had found the solution. In his answer the author quoted Claude Bernard, who once stated that one must ask the right question in order to get the right answer. The investigators of the Center were firmly convinced, as was everybody else, that organophosphate action caused a general toxic effect, in contrast to the view held by the author. It was this hypothesis which prevented them from finding the right answer. This story is briefly described as an illustration of the basic credo of the author of the paramount importance of a sound working hypothesis for the approach to any scientific problem (see Einstein's remark in the Preface). Every hypothesis requires, of course, flexibility of mind and continuous modifications and readjustments to the experimental facts. In fact, the same situation prevails in electrophysiology; as discussed before, the refusal to seriously consider the possibility of chemical and molecular events in the membrane responsible for excitability has led investigators in that field to the dead end street in which they are today.

The remarkably high reactivating power of 2-PAM was attributed by Wilson and his associates to an extraordinarily high degree of molecular complementary fit of 2-PAM and the phosphoryl enzyme: the quaternary nitrogen being attracted to the anionic site and the oxygen of the aldoxime being just a bond length away from the phosphorus atom of the phosphoryl enzyme (see pp. 124–127). This explanation, however, turned out to be wrong (Poziomek *et al.*, 1961a,b; Hobbiger, 1963). Repeating the preparation and characterization of the monoquaternary pyridine aldoximes previously synthesized, to which a syn configuration has been assigned, Poziomek *et al.* (1961a) came to the conclusion that these compounds were not oximes but carbinolamines. Preparing syn and anti isomers of 4-PAM, the syn isomer was found to be 3 times more potent than the anti isomer in reactivating isopropyl methylphosphoryl AcCh-esterase (Poziomek *et al.*, 1961b). If the original interpretation were correct and bonding of the quaternary nitrogen group of isomeric PAM's to the anionic site of phosphorylated AcCh-esterase were to explain the high rate of activation, the syn isomer of 4-PAM should have a considerably lower reactivating power than the anti isomer. Since exactly the opposite is the case, it must be concluded that the cationic nitrogen group is bound to another negatively charged group in the protein. Only the two configurations of the 4-PAM are known, the configuration of 2- and 3-PAM are unknown. While the strong reactivating power of 2-PAM thus remains to be explained, the original interpretation cannot be maintained and that has been recognized by Wilson and his associates (Kitz *et al.*, 1965).

3. Affinity Chromatography; Molecular Weight, Structure, and Crystals

When this monograph was written, AcCh-esterase had been purified by the laborious and by now obsolete procedures using ammonium sulfate fractionation. The molecular groups in the active site and the mechanism of hydrolysis were proposed essentially on the basis of kinetic studies (see pp. 105–118). Since then, as mentioned in the Preface, protein chemistry has greatly advanced. For instance, the exploration of the tridimensional structure includes several ester-splitting enzymes, as will be discussed below. Entirely new types of enzyme purification were developed, culminating in the most elegant and efficient method of affinity chromatography. The availability of this rapid, easy, and efficient purification procedure opened the way to the analysis of many important enzymes. Finally, it was recognized that many enzymes are formed by several subunits with different functions.

The advances mentioned had a great impact on our knowledge of AcCh-esterase and have made many revisions necessary. While these particular aspects will be discussed in more detail in the monograph in preparation, it seems appropriate to mention briefly some of the latest developments in the context of this monograph. A critical evaluation of the recent developments may be found in the review of Rosenberry (1975).

Multistep procedures using ion exchange and exclusion chromatography were used for the purification of AcCh-esterase from electric eel tissue (Wilson and Kremzner, 1963). The specific activity was about 10 mmoles of AcCh split per mg protein/min, a value relatively close to the best obtained with ammonium sulfate fractionation in the 1940's (8 mmoles/mg protein/min). The same procedure, with some minor

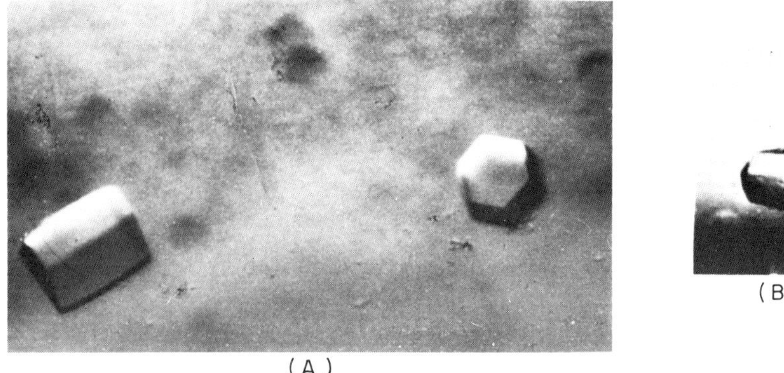

FIG. 5. Crystals of AcCh-esterase. (A) Regular hexagonal prisms; (B) possibly pyramidal termination. According to Dr. Barbara Low, the most common form of growth observed (A) is compatible with true hexagonal symmetry; (B) may imply, however, a crystal system of lower order. The crystals exhibit low birefringence so that it is not possible at present to determine whether they are uniaxial or biaxial. (Leuzinger *et al.*, 1968.)

modifications, was applied by Leuzinger and Baker (1967) to a large scale purification. The specific activity reported appeared to be even higher (12 mmoles/mg/hr), but these higher values have not been confirmed by Rosenberry *et al.* (1972), who discuss the various possibilities for the discrepancies. Leuzinger and Baker (1968) and Leuzinger *et al.* (1968) succeeded in crystallizing the enzyme. The crystals have the form of hexagonal prisms (Fig. 5). They were grown up to a size of about 150 μm length and 80 μm width. However, the size was not yet adequate to produce even preliminary X-ray pictures. Further efforts failed to increase the size and have been abandoned. Since the enzyme has four identical subunits of about 75,000 molecular weight with four active sites (Rosenberry, 1975) it may be more reasonable to try to obtain either active monomers or dimers and to postpone X-ray analysis until crystals of more manageable molecular weight may become available.

A new chapter in the procedure of enzyme (and proteins in general) purification was opened by the introduction of affinity chromatography (Cuatrecasas *et al.*, 1968; Cuatrecasas, 1970; Cuatrecasas and Anfinsen, 1971a,b). It avoids the multistep procedures with unavoidable losses at every step and yields much higher yields in very much simpler ways than the older methods. In view of this great efficiency, affinity chromatography has been generally accepted as the choice method for protein purification. It was first applied to the purification of AcCh-esterase by Kalderon *et al.* (1970), using toluene-treated electric tissue that was prepared in this laboratory in the usual way. In this first attempt phenyltrimethylammonium, a specific reversible inhibitor of the enzyme, was used as the affinity group to the resin matrix, Sepharose 4B. Affinity chromatography for purification of AcCh-esterase was also applied by Berman and Young (1971).

In view of the importance of the length of the spacer arm from the Sepharose matrix, emphasized by Cuatrecasas (1970) and Cuatrecasas and Anfinsen (1971a,b), a series of affinity ligands was prepared in this laboratory by Rosenberry *et al.* (1972). The most efficient affinity ligand was found to be di-(6-aminocaproyl)-phenyltrimethylammonium

$$H_3C\overset{\oplus}{-}N(CH_3)_2-C_6H_4-\left[NH-\overset{O}{\overset{\|}{C}}-(CH_2)_5\right]_2-\overset{+}{N}H_3 \quad 2\,Br^{\ominus}$$

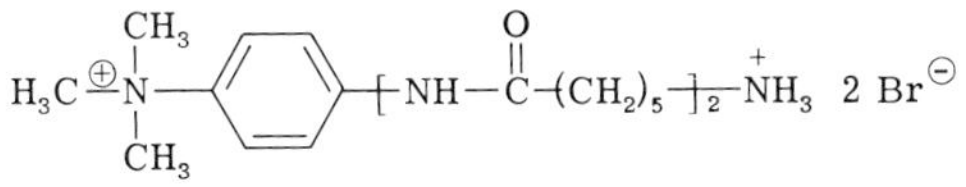

which was attached to the Sepharose (for a detailed description and evaluation, see Rosenberry *et al.*, 1972). Some further corrections and improvements by these authors permitted single-step column purification with yields of about 50% (Chen *et al.*, 1974). A critical evaluation of the results of affinity chromatography for the purification of AcCh-esterase for various tissues and by various authors may be found in the review of Rosenberry (1975).

Molecular Weight. Since 1940 all purifications performed in the author's laboratory used electric eel or *Torpedo* tissue which was stored in toluene for various periods of time, ranging from 2 months up to 2 years. In view of the difficulty of obtaining electric eels at that time, only small amounts of tissue were stored and kept for short periods of time; only when larger amounts of eel tissue became available for purification, was the material accumulated and kept for longer periods. A chance observation initially had indicated that toluene treatment both greatly facilitated purification with the method used at that time and prevented, moreover, bacterial infection. This toluene treatment was applied in most later extraction procedures including, initially, those involving affinity chromatography.

Estimates of the molecular weight (MW) vary considerably, probably owing not only to the many different methods and conditions used, but, as will be mentioned below, also to the various degrees of aggregation which arise from different treatments. In ultracentrifugation experiments in the early 1940's, the MW seemed to be in the neighborhood of 1,000,000, but there were some indications of aggregation and the experiments were discontinued due to the lack of material. In the 1960's several estimates were reported. Lawler (1961) estimated the MW to be 240,000. In a later paper (Lawler, 1963) in which she described studies on huge aggregates, she also estimated the MW of a smaller species to be 330,000. Kremzner and Wilson (1963) arrived at a value of 230,000 and Leuzinger *et al.* (1969) at a value of 260,000.

A most important new development started with the observation of Massoulié and Rieger (1969) that there are three different species of AcCh-esterase in fresh electric tissue extracts. Most homogeneous preparations obtained from toluene-treated tissue were characterized by a single band with a sedimentation coefficient of about 11 S. When, however, AcCh-esterase was extracted from fresh electric tissue, using high ionic strength (1 *M* NaCl), three molecular forms of enzyme were obtained in sucrose gradient centrifugation with sedimentation coefficients of 8 S, 14 S, and 18 S. The amount of enzyme extracted increased still further by addition of 0.5% Triton X-100. All three of these enzyme species were converted to the 11 S form either by treatment with trypsin, or other purified proteolytic enzymes, or on storage of crude enzyme extracts from electric tissue slices stored in toluene (Massoulié *et al.*, 1971; Rieger *et al.*, 1972a,b). Dudai *et al.* (1972) prepared as affinity ligand 1-methyl-9-aminoacridinium (attached to the resin matrix via the 9-amino group) and used fresh electric tissue extracted with high ionic strength, 1 *M* NaCl, for purification. They eluted the enzyme from the column with decamethonium. The highly purified enzyme, obtained in high yield, showed the same species differences as those described by Massoulié and his associates in crude extracts.

Dudai *et al.* (1973) measured the MW values for the 14 S and 18 S species by sedimentation equilibrium. They obtained MW's of 780,000 and 1,000,000, respectively, for the two species. Similar MW's of the 14 S and 18 S species were estimated by Bon *et al.* (1973) on the basis of measured values of the Stokes radius, the partial specific volume ($\bar{v}$), and apparent sedimentation coefficients. The value estimated for the 8 S species was 430,000. These authors propose that the 8 S, 14 S, and 18 S forms are aggregates which include varying numbers of 11 S tetrameric species. The molecular weight estimated for the 11 S species by Dudai and Silman (1973), was obtained by sedimentation equilibrium; it indicated a MW of 320,000 to 350,000. Based on studies of the subunit structures (see below) Rosenberry *et al.* (1974) estimate the MW of the 11 S species to be about 280,000 to 300,000 (for a detailed discussion, see Rosenberry, 1975). The MW of an 11 S form, extracted and purified from *Torpedo* electric tissue was found by Taylor *et al.* (1974) to be 335,000 either by sedimentation equilibrium or by analysis of sedimentation and diffusion coefficients.

It may be mentioned that some of the conflicting estimates of the specific activity, especially in earlier reports from various investigators, were based on differences between the extinction coefficients, $E_{280}^{1\%}$. The extinction coefficients reported for essentially homogeneous eel 11 S enzyme vary by 10 to 15%. A comparison of several methods of determination by Rosenberry *et al.* (1972) led to an average estimate of $E_{280}^{1\%} = 18.0 \pm 0.4$ for native AcCh-esterase. This value appears to be now generally accepted. Another source for reported discrepancies are apparently due to the variations of the protein determinations obtained with the Lowry method (Lowry *et al.*, 1951). The specific activity of purified 11 S enzyme from several laboratories is about 11 mmoles/mg protein/min (see Rosenberry, 1975).

Subunits. Many enzymes were found to be formed by varying numbers of subunits. These subunits may have different functions; there are regulatory and catalytic subunits; they exhibit allosteric activations and inhibition, positive and negative cooperativity, feedback inhibition, and other characteristics (see, e.g., Moyed and Umbarger, 1962; Monod *et al.*, 1963, 1965; Umbarger, 1964; Stadtman, 1966; Koshland, 1970). The remarkable complexity of some enzymes, formed by many subunits, required many efforts before satisfactory pictures of the functional significance of these subunits emerged. This complexity is illustrated by the well-known analyses on glutamine synthetase, formed by 12 subunits (see, e.g., Meister, 1968; Holzer, 1969; Stadtman, 1970a,b), and those on aspartate carbamylase (see, e.g., Gerhart and Schachman, 1965; Changeux *et al.*, 1968a; Schachman, 1972).

During the last few years when homogeneous proteins of AcCh-esterase became readily available, the subunit structure of the enzyme was studied in several laboratories (see, e.g., Froede and Wilson, 1970; Dudai and Silman, 1971; and for a comprehensive summary, see Rosenberry, 1975). Since many molecular forms with different sedimentation coefficients including the 8 S, 14 S, and 18 S species, appear to be clusters with varying numbers of 11 S forms attached, Rosenberry *et al.* (1974) analyzed extensively the subunit composition of this 11 S form into which the other forms are readily transformed by proteolytic enzymes or by autolysis in toluene-treated material. This species may be the common molecular form, although it is possible that it exists in membranes in various forms of aggregates.

The 11 S enzyme was purified by affinity chromatography from extracts of toluene-stored electric eel tissue as described by Rosenberry *et al.* (1972) and Chen *et al.* (1974). Polyacrylamide gel electrophoresis in 1% sodium dodecyl sulfate and gel exclusion chromatography in 6 *M* guanidine chloride were used. The data indicate that the active enzyme is a tetramer composed of 4 identical subunits. Each subunit was shown to contain one active site on the basis of analyses of ^{32}P phosphorylation by DFP and cyanogen bromide fragment compositions. The results are consistent with those of Rosenberry and Bernhard (1971), in which the active sites were shown to be kinetically homogeneous. After disulfide reduction by dithiothreitol and alkylation with *N*-ethylmaleimide, some cleavage of the subunit into two fragments, a major one of 50,000 molecular weight and a minor one of 20,000–22,000 molecular weight, is revealed. The native enzyme appears to exist as subunit dimers with a covalent intersubunit linkage that involves disulfide bonding. The data support a tetrameric structure with a dimer of dimers $(\alpha_2)_2$. The earlier suggestion by Leuzinger *et al.* (1969), that the enzyme has two different polypeptide chains and may be a dimeric hybrid $(\alpha\beta)_2$ is unequivocally ruled out.

4. Mechanism of Ester Hydrolysis by Enzymes

The molecular groups on the active site of AcCh-esterase were analyzed by kinetic studies some 25 years ago, using highly active enzyme solutions extracted from electric tissue. A possible mechanism of the ester hydrolysis was proposed by Wilson *et al.* (1950); it was suggested that a nucleophilic group in the so-called "esteratic" site attacks the carbon of the carbonyl group and that an imidazole group may play an active role in the catalytic process (for a summary, see Nachmansohn and Wilson, 1951). In the following two decades the mechanism of the hydrolysis of ester-splitting enzymes was analyzed by a combination of kinetic and chemical studies, using most frequently chymotrypsin and trypsin readily available commercially in crystal form. The nucleophilic group in the active site was identified to be the oxygen of a serine residue. The involvement of a histidine residue became probable through the work of several investigators (see, e.g., Neurath and Dixon, 1957; Westheimer, 1957; Dixon *et al.*, 1958; Koshland, 1959). Direct evidence for a histidine residue in the active site of chymotrypsin was provided by Schoellman and Shaw (1963). A large amount of information about the mechanism of hydrolysis by these enzymes was obtained by a great number of investigators, among them, in particular, Bender, Neurath, Westheimer, Koshland and their associates, but also many others (see, e.g., Spencer and Sturtevant, 1959; Koshland, 1959; Bruice, 1961; Bender, 1962; Koshland *et al.*, 1962; Jencks, 1963; Bender *et al.*, 1964; Bender and Kezdy, 1964, 1965; Neurath, 1964; Walsh and Neurath, 1964). Only a few publications have been quoted as examples among the great number of reports on the developments in the field.

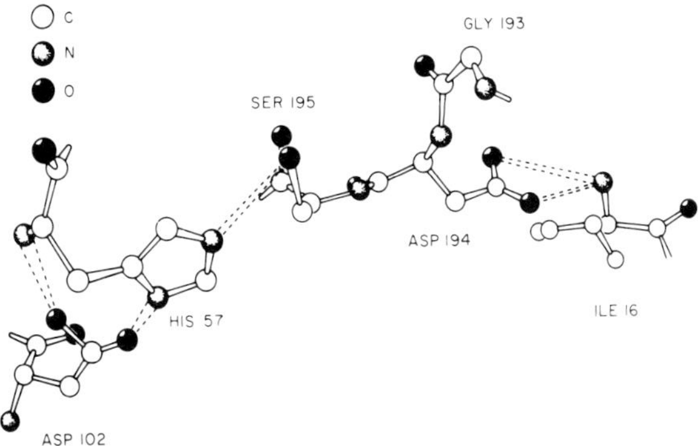

FIG. 6. The conformation of a few amino acids in the active center of α-chymotrypsin. The viewpoint is outside the surface of the molecule looking toward the interior. (Blow and Steitz, 1970.)

A new page was opened when the tridimensional structure of chymotrypsin and other ester-splitting enzymes was elucidated by X-ray diffraction measurements. The structure of bovine α-chymotrypsin was achieved by Blow and his associates at 2 Å resolution (for a summary of these developments, see Blow, 1971). Many of the notions developed in the previous studies were confirmed; others had to be modified, but the knowledge of the tridimensional structure has in a decisive way extended the understanding of the detailed mechanism. The amino acid sequence has been determined by Hartley (1964; Hartley and Kauffman, 1966). The conformations of some residues at the active center are shown in Fig. 6. Ser 195 and the imidazole of His 57 are the two main residues involved in the catalytic process. The α-amino group of Ile 16 is essential for maintenance of enzyme activity (Oppenheimer *et al.*, 1966; Hess, 1971): it forms an ion pair with a carboxyl group of Asp 194. The ion pair appears to be essential for maintaining the enzyme in an active conformation and substrate binding prevents the disruption of an ion pair. Both amino acids are internal. Blocking of the carboxyl group of Asp 194 with glycine methyl ester leads to a complete block of active sites of the enzyme, as was shown by Koshland and his associates (Carraway *et al.*, 1969). The carboxyl group of Asp 102 is buried in a hydrophobic environment, as was deduced by Blow and his associates (Blow, 1969; Blow *et al.*, 1969), in close enough contact with one N of the imidazole ring of His 57 to form either a hydrogen bond or an internal ion pair, depending on the ionization state of His 57 and Asp 102. Thus, the active center of chymotrypsin may be represented by a hydrogen-bonded network extended from Ser 195 through His 57 to Asp 102.

The significance of this arrangement for the catalytic process is emphasized by the fact that several serine proteases exhibit certain structural and chemical similarities as was already postulated by Bender and Kezdy (1965). The tridimensional structures of chymotrypsin, elastase (see, e.g., Hartley, 1970; Hartley and Shotton, 1971) and subtilisin (Wright *et al.*, 1969; Kraut, 1971) exhibit different substrate specificity and are structurally quite different at the subtrate binding site; however, the amino acid residues Asp, His, and Ser implicated in the catalytic reactions of chymotrypsin seem to bear a similar spatial relationship to each other in these enzymes (Hartley, 1974).

The ester hydrolysis by enzymes is now generally pictured as biphasic. When a reversible enzyme-substrate complex is formed, at first one split product is eliminated, then the second one. Sometimes acylation is very fast with rapid elimination of the alcohol leaving group, and deacylation is the rate-limiting reaction. In other cases, both steps may have about the same rate. The rate-limiting steps vary with the enzymes and substrates. For a full understanding of the hydrolytic mechanism it is, therefore, important to know the rates of the individual steps involved. Attempts were made to determine these rates involved in catalysis by steady-state kinetic experiments. However, this type of experiment does not permit evaluation of the speed of the various individual reaction parameters involving intermediates reactions since they yield only a combination of rate and equilibrium constants. The development of a great variety of methods for studying fast reactions by Eigen and his associates in the last 20 years has dramatically changed the situation for the analysis of enzyme-catalyzed reactions as well as in many other areas, as mentioned in the Preface. These methods permit precise determination of reactions taking place in the range from 1 to 10^{-8} sec (for summaries, see Eigen and DeMayer, 1963; Eigen and Hammes, 1963; Eigen, 1967b). Thanks to their great versatility the relaxation methods developed permit one not only to follow the overall rates of enzyme catalyzed reactions, but also to measure individual rate constants involving enzyme intermediates. In principle, these methods involve an external perturbation (e.g., of temperature, pressure, electric field, and density) which is rapidly applied to the reaction system and is thus able to shift the position of chemical equilibria. The rate at which the chemical equilibria relax to new values can be measured directly by spectrophotometric, fluorometric, polarometric, conductometric, and of other suitable techniques. These methods have permitted measurements of such diversified and extremely fast processes as activation of enzyme activities by metals, enzyme-substrate complex formation, conformation changes, proton and electron transfer, and many others. They have been applied to the study of intermediates in chymotrypsin-catalyzed reactions (see, e.g., Hess, 1971).

These various developments have been briefly described as an illustration of how rapidly our understanding of the mechanism of enzyme catalysis has moved in the last two decades. A more detailed description will be presented in the new monograph in preparation. It is obvious that the insight obtained into the mechanism of ester-splitting enzymes is of great interest to the study of the mechanism of hydrolysis by AcCh-esterase. At least some analogies must be anticipated, although there will certainly be important differences. In the case of AcCh-esterase we are not in the fortunate position to know the tridimensional structure. Nevertheless, a considerable amount of new information has accumulated. The catalytic mechanism of AcCh-esterase has been comprehensively and critically evaluated in the review of Rosenberry (1975). Therefore, the presentation will be limited to a few points only.

In addition to the now well-established oxygen serine and a histidine residue in the active site there is an acidic group of $pK = 9.3$ (Bergmann *et al.*, 1958). The studies of Krupka (1966a,b) suggest that the anionic site is a side chain carboxyl with a $pK = 4.0 - 4.5$. The presence of tryptophan has been demonstrated by Shinitzky *et al.* (1973). Krupka (1966a,b) has proposed a scheme for the ester hydrolysis on the basis of kinetic studies with a great number of substrates and inhibitors. He used purified bovine erythrocyte AcCh-esterase. On the basis of his results he suggested that acetylation is controlled by an ionizing group in the enzyme of $pK = 5.3$. The deacetylation is controlled by another ionizing group of $pK = 6.3$. He concluded that the two steps are catalyzed by different groups and proposed in his scheme that two different histidine residues are involved in the catalytic mechanism; one residue is located near the serine hydroxyl and catalyzes the first step, the acetylation. The second residue is located near the anionic site. As a result of a conformational change the acetyl residue is brought near the second imidazole which catalyzes the deacetylation. Although Krupka's observations have added many interesting aspects of the activity of AcCh-esterase, his scheme is open to some serious questions. The data on the hydrolytic mechanism as revealed by the work with α-chymotrypsin do not support the assumption of two histidine residues involved in acylation and deacylation. There are several questions raised in the review of Rosenberry (1975), in particular the difficulty of his interpretation assuming the group with the $pK = 5.3$ to be a histidine residue.

Recently, Rosenberry (1975) proposed an attractive scheme of the mechanism of hydrolysis by AcCh-esterase based on the "induced-fit" model of enzyme catalysis formulated by Koshland (1959; see also p.

268). In this model the initial binding of the substrate induces a conformational change of the enzyme to a highly active enzyme-substrate complex. The equilibrium formulation of induced fit would not account for the behavior of certain substrates in their hydrolysis by AcCh-esterase, as pointed out by Rosenberry, but the ability of the nonequilibrium induced-fit model to account for the pH dependence of hydrolysis of all substrates seems to be the best argument for its application to AcCh-esterase. Koshland and Neet (1968) have listed a number of phenomena all of which may be best explained by the induced-fit model and these phenomena have been observed with AcCh-esterase. Without going into the details of the discussion of Rosenberry (1975) in which he presents several observations in favor of this proposal, his scheme for the mechanism of hydrolysis is presented in the diagrammatic model of Fig. 7. The initial enzyme substrate is complex E·RX; the induced-fit complex (E·RX)′ and the acyl enzyme, ER. The enzyme-ligand complex with bisquaternary ligands like decamethonium ($E \cdot I_1$) involves the anionic site and a peripheral negative group P_1; the complexes formed by other multiquaternary ligands like Flaxedil ($E \cdot I_2$) with three quaternary

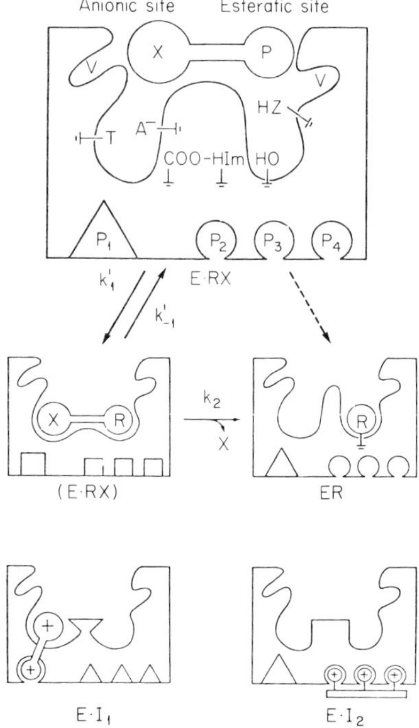

FIG. 7. Diagrammatic model of the acetylcholinesterase catalytic site based on the induced-fit mechanism. For explanation see text. (Rosenberry, 1975).

nitrogen groups involve ligand binding at other peripheral points. Identified residues at or near the catalytic site include the charge-relay complex COO^-—HIm—HO (Structure IV); an acidic group HZ; the anionic group A^- which defines the anionic site; a tryptophan residue T near the anionic site; and adjacent hydrophobic area V. For the exact interpretation of this scheme the reader is referred to the description of Rosenberry (1975). In the view of the author this is the most lucid model devised so far for the mechanism of hydrolysis by AcCh-esterase and best explains the great number of observations reported.

5. Distribution and Localization of AcCh-Esterase

Among the aims of the author, from the beginning of his studies on the proteins processing AcCh in 1936 (see pp. 26–37), has been the information about the distribution of the enzyme in a vast variety of types of nerve and muscle fibers, ranging from the lowest phyla to the highest forms of vertebrates. The enzyme was found to be present in *all* types of excitable tissues investigated, in so-called "cholinergic" and "adrenergic," in motor and sensory, in peripheral and central fibers, in invertebrates and vertebrates, and in all types of muscle fibers. The concentrations varied greatly. With the chemical methods used *not a single exception* was found. For details see pp. 26–30. Even in *Nitella,* an algae capable of electrical activity, Dettbarn (1962) demonstrated the presence of AcCh-esterase.

When histochemical staining techniques for detecting the presence of AcCh-esterase were developed by Koelle and Friedenwald (1949) and were applied to a variety of tissues, the enzyme appeared to be absent in many conducting fibers, and even in certain electric organs, examined in many laboratories over a period of a decade (for a summary, see Koelle, 1963b). These reports were based on the staining technique with the use of light microscopes. In the specimen slices used in these experiments, the thickness of membrane-covering tissue layers may have prevented the access of the added substrate (acetylthiocholine) to the enzyme. The absence of staining outside the neuromuscular junction was used for more than a decade as a strong evidence for the assumption that the enzyme is localized exclusively there and that, consequently, the function of AcCh is limited to the junction. In some tissues, in which the concentration was very high when determined by chemical methods, the hydrolysis products appeared to be hardly visible or were even absent. All these reports were in sharp contrast to the results obtained with chemical methods demonstrating the presence of AcCh-esterase in all types of fibers.

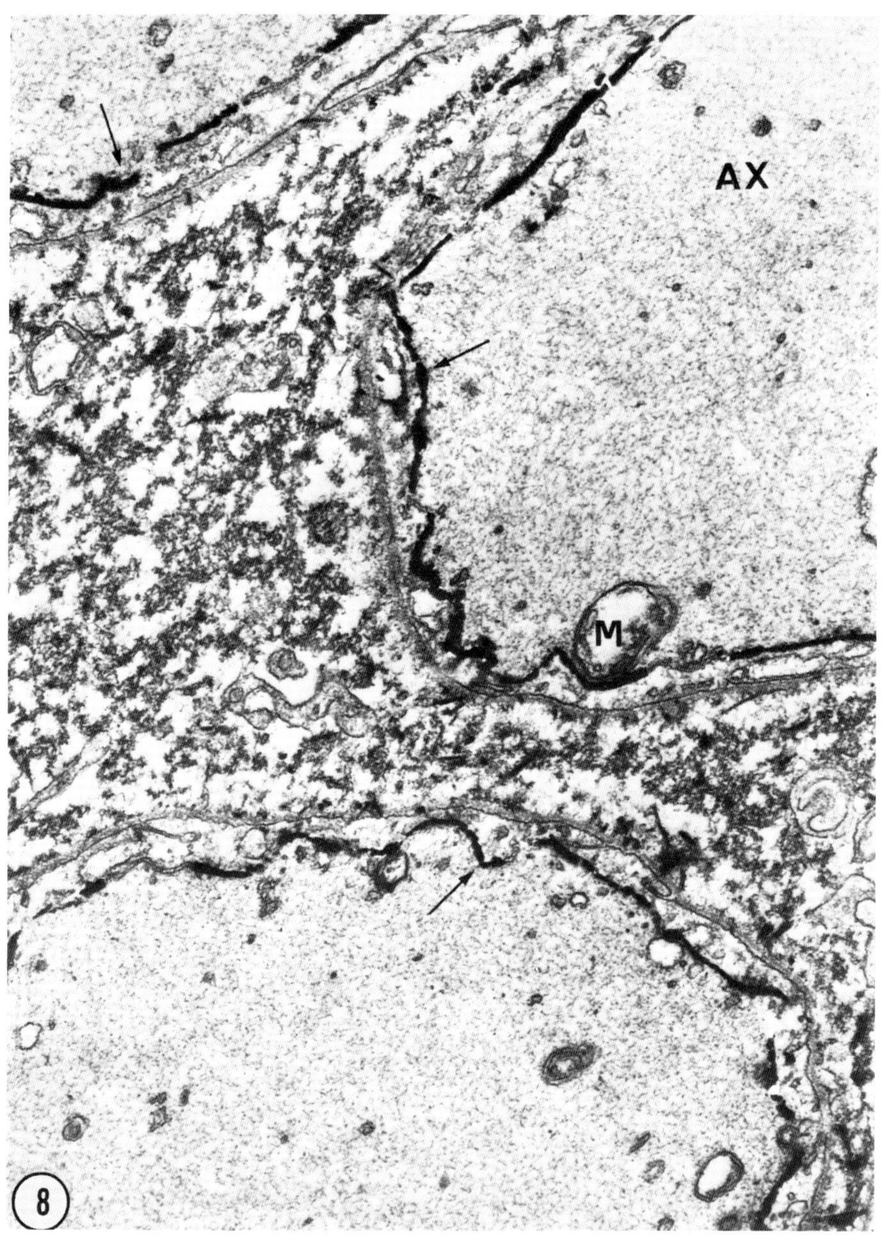

FIG. 8. Electron micrograph showing the exclusive localization of AcCh-esterase (arrows) in the plasma membrane of small nerve fibers (AX) in the stellar nerve of squid. A mitochondrion (M) is located close to the axolemmal surface. A basement membrane surrounds the Schwann cell, separating it from the fibrillar material of the intercellular matrix. Incubated in acetylthiocholine without inhibitors. (Courtesy of Dr. Virginia Tennyson.)

The situation changed drastically with the use of electron microscopy in combination with staining techniques, first applied by Barrnett (1962). Three decades ago (see Nachmansohn and Meyerhof, 1941), biochemical evidence began to accumulate that the enzyme is localized at or near the cell surface. The biochemical data were, however, much too indirect for claiming that the enzyme is localized in the membrane, although such an association was suspected. Electron microscopy combined with histochemical staining techniques has conclusively established that AcCh-esterase is associated with excitable membranes. This has been shown with a great variety of different types of fibers. A few examples are shown as an illustration in Figs. 8–13. Figure 8 shows the localization of AcCh-esterase in the surface membrane of small fibers in the stellar nerve of squid. The electron micrograph was prepared by Virginia Tennyson. Figure 9 shows the enzyme in the conducting membrane of a myelinated fiber of a ventral root axon of frog. It may be noted that several investigators, even using the electron microscope and histochemical staining techniques, failed to find AcCh-esterase in

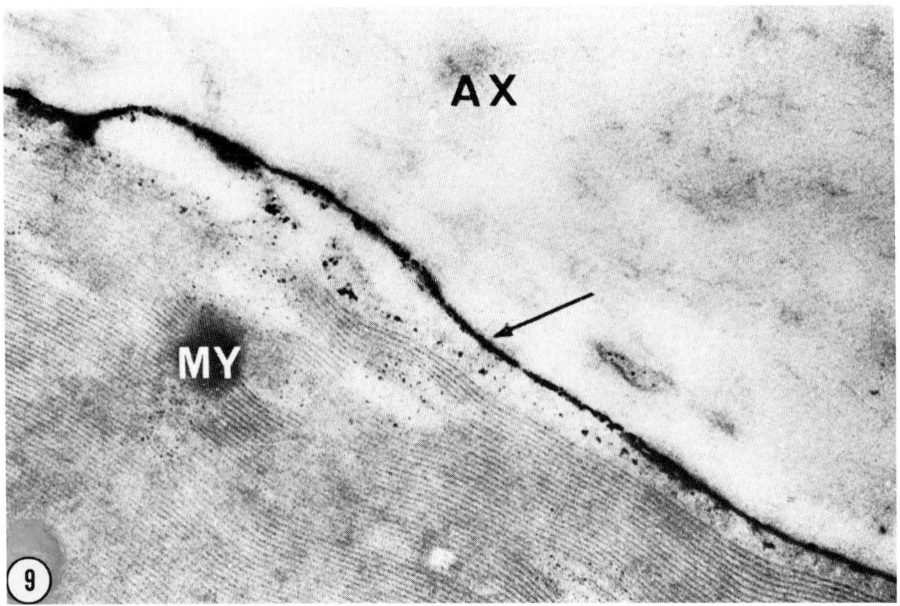

FIG. 9. Myelinated (MY) ventral root axon (AX) of frog. The slice was treated with Triton X-100 before the incubation for testing acetylcholinesterase activity with the standard procedure for histochemical staining of the enzyme (adding acetylthiocholine and copper sulfate). The hydrolytic product, thiocholine, forms a precipitate with copper sulfate. The dense end product is present in the axolemmal (plasma) membrane (arrow). (Brzin, 1966).

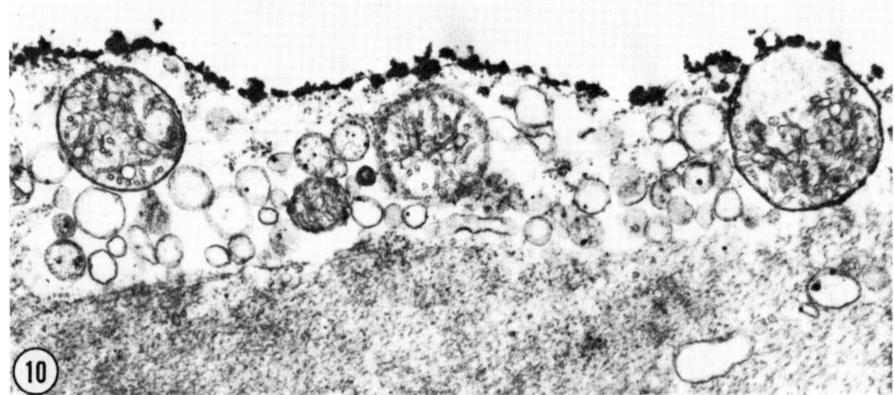

FIG. 10. AcCh-esterase activity located in the plasma membrane (sarcolemma) of a plaice muscle. Section of a cell kept at 0°C in isotonic sucrose solution showing black precipitates along the sarcolemma as a result of cholinesterase activity. Substrate: Acetylthiocholine for 45 min at 0°C. (Lundin and Hellström, 1968.)

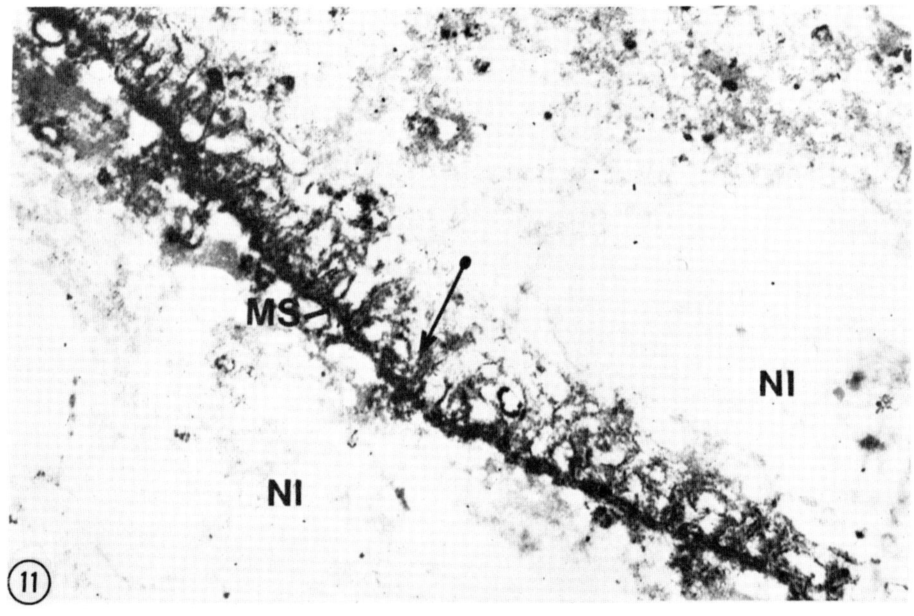

FIG. 11. Electron micrograph of an isolated fragment of excitable membrane from the electroplax of *Electrophorus*, tested for AcCh-esterase activity by standard procedures. The picture shows the uniformity of distribution of AcCh-esterase at the innervated membrane surface (MS). No staining was found in the noninnervated (NI) membrane. (Changeux *et al.*, 1969.)

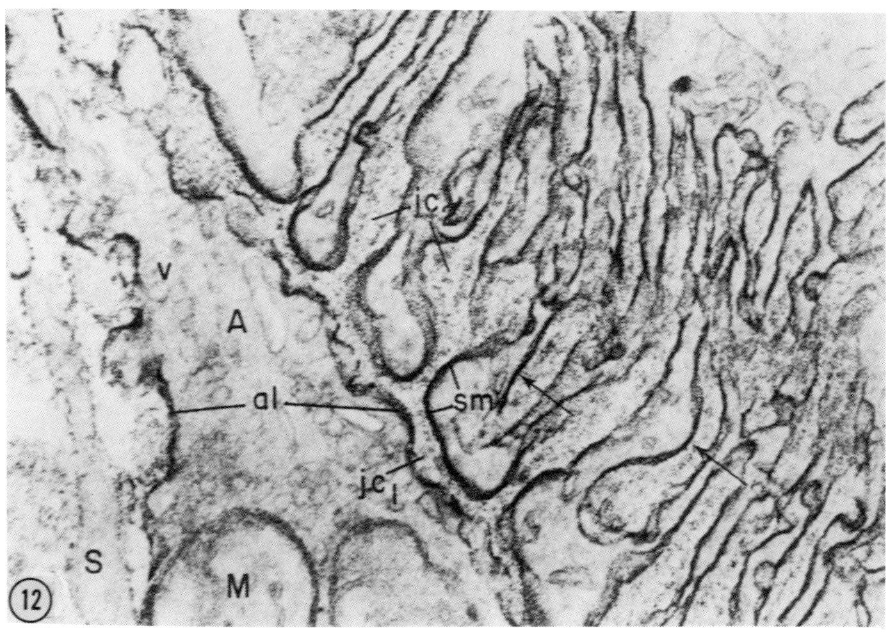

FIG. 12. Electron microscopic histochemical localization of acetylcholinesterase at the motor end plate of mouse intercostal muscle. A high magnification view of the junctional complex, showing the axonal terminal (A) containing mitochondria (M) and numerous synaptic vesicles (v), the junctional cleft (jc), and junctional folds of the sarcolemma (sm). The electron-dense granules, 40–50 Å in diameter, represent gold sulfide, the reaction product of the gold thioacetic acid method for the detection of acetylcholinesterase and nonspecific cholinesterase. The axolemma (al) exhibits marked enzymic activity both on the surface facing the primary junctional cleft (jc_1) and at the surface facing the teloglial Schwann cell sheath (S) (the axonal terminal is somewhat separated from the Schwann cell in this micrograph). Where the plane of section is perpendicular to the sarcolemma (arrows), the particles form a dense line about 120–140 Å thick. (Koelle, 1971.)

myelinated fibers. Suspecting that, in view of the large amount of lipid present in myelinated axons, the substrate may not be able to reach the membrane bound enzyme during incubation with the required speed (although here the slices are only about 500 Å thick), Brzin (1966) applied Triton X-100 and thus was able to demonstrate the presence of the enzyme in the conducting membrane. Figure 10 shows the presence of the enzyme in the muscle membrane, demonstrated by Lundin and Hellström (1968).

Figure 11 shows the enzymes in the isolated excitable membrane fragment of an electroplax of *Electrophorus,* a preparation obtained by

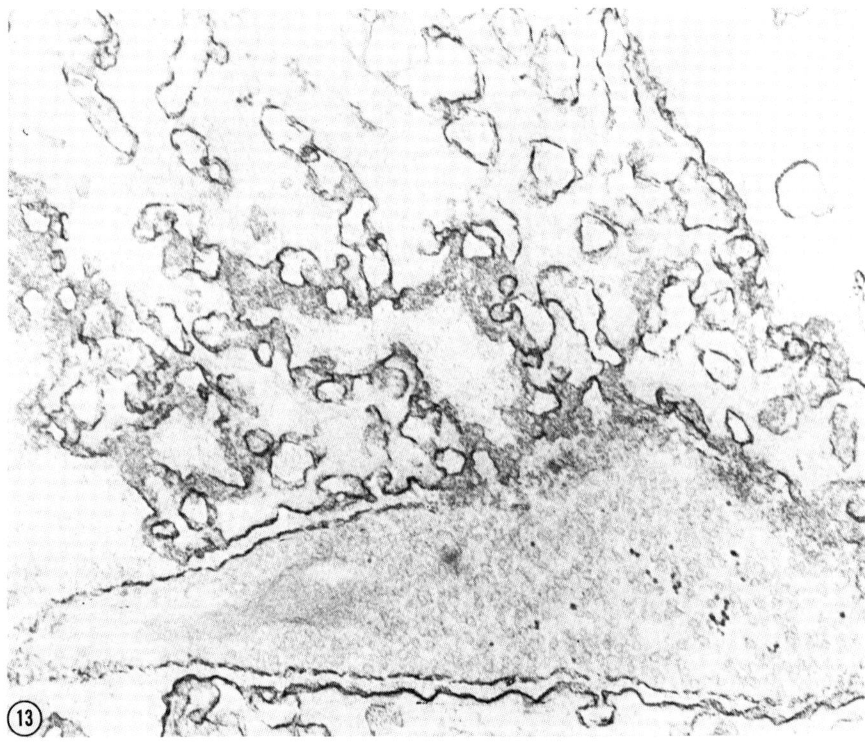

FIG. 13. Electron micrograph of the excitable membrane of *Electrophorus*. The presence of AcCh-esterase in the membrane is tested by the new method developed by Koelle *et al.* (1974). The enzyme is present in the nerve terminal, the postsynaptic and conducting part of the membrane. The circular formations seen in the picture and showing the stain are artifacts owing to the cutting in preparing the slices for examinaion; they are due to the deep invaginations all along the excitable membrane (Tomas *et al.*, 1975).

Changeux *et al.* (1969). No staining was found in the noninnervated membrane. Particular attention must be paid to the even distribution of the enzyme all along this electroplax membrane. It is known that only a few percent of this membrane is formed by synaptic parts, more than 95% by the conducting parts. Moreover, when, in 1937, the enzyme in the electric tissue was tested, it was found, as mentioned before, that 1 kg of tissue is capable of hydrolyzing 3–4 kg of AcCh per hour. Since it is now evident that most or all of the enzyme is localized in the membrane and since chemical determinations of the enzyme have shown that 10^{11} molecules of enzyme are present in the excitable membrane of a single electroplax of *Electrophorus,* a more correct value indicating the fantastic concentrations of AcCh-esterase in electric tissue is the ability of 1 gm excitable membrane to hydrolyze about 30 kg of AcCh per hour.

A further pertinent advance of this aspect is due to the development of a new and refined staining technique by Koelle and his associates (1974) using thiolacetic acid as the substrate and gold sulfide for precipitation. Figure 12 shows an electron micrograph of the histochemical localization of AcCh-esterase at the motor end plate of mouse intercostal muscle (Koelle, 1971). The physiological significance of this extremely important picture will be discussed in Section V. Another pertinent result obtained with this new staining method is the distribution of the enzyme in the excitable membrane of the electroplax of *Electrophorus* (Fig. 13). This electron micrograph shows even more conclusively than that of Fig. 11 the even distribution of the enzyme along the conducting part and in the pre- and postsynaptic parts of the excitable membrane (Tomas *et al.*, 1975).

In the light of all these advances it is difficult to understand that some investigators still use "the absence of AcCh-esterase in many conducting fibers" as a pertinent argument against the theory of the essential role of the enzyme in conduction. Even more surprising is the use of staining techniques in combination with light microscopy for claiming "quantitative" differences between various types of fibers.

B. AcCh-Receptor

1. Notion of an AcCh-Receptor

Since the beginning of this century it was recognized that strong effects on cells produced by any chemical compound, drug, toxin, must be due to the reaction with specific molecules. Whatever their nature may be, these molecules must have specific sites capable of binding the chemical compound or drug before it produces the characteristic response in the cell. Some of the early formulations of this notion were expressed by Ehrlich (1900, 1909), the founder of modern chemotherapy, and much of his work was based on the idea that drugs must react with specific receptors in the effector cell. With the advances in the knowledge of cellular mechanisms, it became possible to explain some pharmacological effects by specific chemical reactions. The notion of receptors has been frequently revised and has been formulated in increasingly specific terms. An early attempt at a quantitative approach to the interaction of drugs and receptors was presented in the review of Clark (1937). He applied the formalism of the Langmuir isotherm to the reactions of drugs with receptor molecules and tried to relate drug concentration and the response of the cell to the number of receptors occupied. Clark himself questioned some of his own assumption, and owing to the developments of the last few decades about the interaction of drugs and receptors, many of his ideas had to be revised. Some more recent notions of receptor mechanisms are discussed in the review of Furchgott (1964) and in the book "Molecular Pharmacology" edited by Ariëns (1964). A still more recent evaluation of this rapidly advancing field is the lucid review of Mautner (1967) in which classical views of receptors are analyzed and integrated with modern concepts and notions of proteins and biopolymers in general. Several of the facts and problems discussed in Mautner's review are pertinent to the specific receptor discussed in the following.

In the early phase of the hypothesis postulating AcCh to be a neurohumoral transmitter, the target of this transmitter was not yet discussed. It was just suggested that the transmitter stimulates the second cell, nerve or muscle, in a way which was not specified; the inactivation of AcCh was specifically attributed to the hydrolysis by an esterase (see, e.g., Dale, 1937). Only in the last two decades has the notion of a receptor of AcCh gradually emerged, particularly when, owing to the widespread interest in organophosphates, the effects of a great variety of inhibitors of AcCh-esterase was studied. It became apparent to several investigators that it was difficult to explain some of the pharmacological actions and electrophysiological observations simply in terms of enzyme inhibition. Wescoe and Riker (1951), e.g., clearly recognized that the effects of 3-hydroxyphenylammonium, although a strong inhibitor of AcCh-esterase, could not be readily attributed to enzyme inhibition (for a more elaborate discussion, see Riker, 1953). In the following decade, an increasing number of observations with inhibitors of AcCh-esterase were reported by many investigators, supporting the assumption of an action of some of these inhibitors on a receptor as distinct from and preceding that on the enzyme (for a review see, e.g., Werner and Kuperman, 1963). The notion of the receptor was essentially an operational term.

The notion of a specific AcCh-receptor protein was first postulated by the author (Nachmansohn, 1952, see also Fig. 2). There was no experimental evidence for this assumption, it was based on the notion that only a protein has the ability to recognize with a high specificity a ligand such as AcCh.

The assumptions of a receptor were based essentially on pharmacological and electrophysiological studies. The complexity of the preparations used (various types of neuromuscular junctions, frog rectus muscle, guinea pig ileus, etc.) offered serious difficulties for quantitative evaluations and even greater ones for the analysis of the chemical nature and properties of the AcCh-receptor. If the AcCh-receptor protein has the function of initiating the permeability changes in excitable membranes during electrical activity, it must be localized there. No appropriate experimental conditions were available for an analysis of the protein nature of the receptor. It must be mentioned that studies with isolated electroplax of electric organs were performed in the late 1930's (see, e.g., Auger and Fessard, 1939; for summaries, see Fessard, 1946, 1958). However, the preparation did not yet offer appropriate conditions for the analysis of the chemical nature of the receptor.

2. Monocellular Electroplax Preparation

A new chapter in the studies on the nature and properties of the AcCh-receptor was opened in the 1950's by the development of the monocellular electroplax preparation by Schoffeniels (1957, 1959; Schoffeniels and Nachmansohn, 1957). Details of this preparation and some early results have been described on pp. 153–162. Although this preparation is still the fundamental basis of all the later work, many refinements were subsequently introduced, such as, e.g., the use of intracellular electrodes, a switching device permitting measurements of membrane potentials across the innervated (excitable) and noninnervated (nonexcitable) membranes and across the whole cell introduced by Higman, Bartels and Podleski (see Higman and Bartels, 1961, 1962; Higman *et al.*, 1963, 1964). These refinements have greatly increased the sensitivity of the preparation for the analysis of the reaction between the receptor protein and specific ligands; they permit precise and reproducible titrations of the dose-response curves and the evaluation of dissociation constants between ligands and the protein.

The primary molecular event controlling the permeability changes was proposed to be a conformational change of the AcCh-receptor protein induced by AcCh. AcCh and several related compounds act and simultaneously depolarize the membrane. This type of compound is therefore referred to as receptor activators ("agonists"). Another type of compound which reacts with the active site is apparently unable to induce a conformational change: it blocks the response of the membrane, but does not depolarize. In analogy with enzyme chemistry it is referred to as a receptor inhibitor ("antagonists," "antimetabolites"). Receptor activators have usually one structural feature in common: most of them are methylated quaternary ammonium derivatives. Their tertiary analogs are either poor activators or, most frequently, receptor inhibitors. Evidence for these two types of compounds reacting with the receptor was offered with the use of isolated rows of electroplax (see pp. 148–151). However, the use of the monocellular electroplax preparation permitted a much more conclusive distinction and a quantitative evaluation. (For a summary describing the use of the monocellular preparation in more detail, see Nachmansohn, 1966).

3. Evidence for the Protein Nature of the Receptor

The use of the monocellular electroplax preparation offered the possibility to bring experimental evidence for the postulated protein nature of the AcCh-receptor. A great variety of ligands was used for this demonstration; only some of them will be described as illustrations. Webb (1965) determined the apparent dissociation constants of a series of *N,N′*-bis(diethylaminopropyl) quinone (benzoquinonium) (I) and *N,N′*-bis(diethylaminoethyl)oxamide-bisbenzylhalide quaternary salts (ambenonium) (II) derivatives with the AcCh-receptor and compared them with those obtained with an active AcCh-esterase preparation (structures I and II). As seen in Table I, when R in the benzoquinonium is an

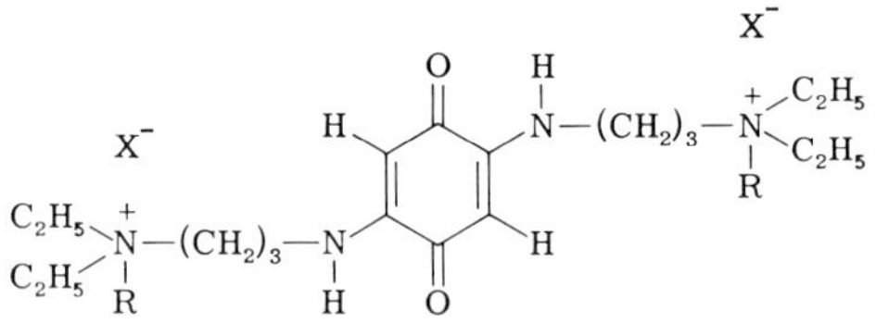

(I)

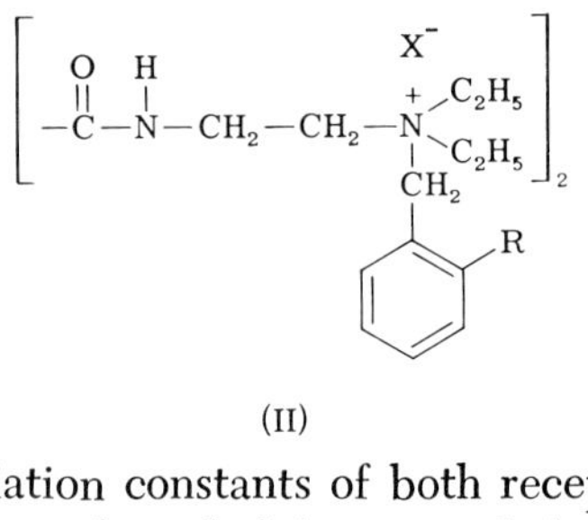

(II)

ethyl group, the dissociation constants of both receptor and esterase are about equal. Substituting the ethyl by a methyl group hardly affects the affinity to the esterase, but decreases significantly that to the receptor. Substitution of the ethyl by a phenyl group increases affinity to both, but an affinity that is stronger to the receptor than to the esterase. Most pronounced is the effect of the addition of a Cl atom to the phenyl group; in the ortho position the affinity to the receptor is hardly affected, that to the enzyme is sixfold increased, whereas in the para position

TABLE I
DISSOCIATION CONSTANTS OF BENZOQUINONIUM AND AMBENONIUM DERIVATIVES

R	$K_{\text{ACh-R}}$[a]	$K_{\text{ACh-E}}$[a]	$K_{\text{ACh-R}}/K_{\text{ACh-E}}$[b]
Benzoquinonium derivatives			
C_2H_5	3.8×10^{-7}	2.9×10^{-7}	1
CH_3	1.2×10^{-6}	2.6×10^{-7}	5
C_6H_5—CH_2— (benzyl)	1.4×10^{-8}	8.1×10^{-8}	0.2
o-Cl-C_6H_4—CH_2—	1.6×10^{-8}	5×10^{-9}	3
Cl—C_6H_4—CH_2— (p-)	3.1×10^{-8}	3.2×10^{-7}	0.1
Ambenonium derivatives			
Br	1.2×10^{-6}	2.1×10^{-10}	6000
Cl	1.6×10^{-6}	5.1×10^{-10}	3000
H	3.4×10^{-6}	2.1×10^{-7}	16

[a] Constants with ACh-receptor (ACh-R), tested on the monocellular electroplax preparation, and with ACh-esterase (ACh-E), tested in purified enzyme solution prepared from electric tissue of *Electrophorus*.

[b] Indicates the strong differences in affinity to the two proteins expressed by the ratio of $K_{\text{ACh-R}}/K_{\text{ACh-E}}$.

the affinity to both is decreased, but that to the esterase much more strongly than that to the receptor.

Still more striking are the differences observed with the ambenonium derivatives. With either Br or Cl on the benzene ring, the affinity to the enzyme is three orders of magnitude greater than to the receptor. Substitution by a proton brings the two affinities very close to each other. The analysis of these two groups of compounds offers an excellent illustration how similar some of the affinities are and how they may be modified by small substitutions of one group or even one atom, sometimes similarly for the two proteins, sometimes differently, but sometimes in the same and sometimes even in the opposite direction. Since the compounds are receptor and enzyme inhibitors, they are favorable for comparing only the binding forces between them and the two components. The results make it difficult to assume another type of cell component for the receptor than a protein, since only proteins would be able to distinguish between the subtle modifications of the two types of molecules tested and to react to them with affinities sometimes close and sometimes different from those measured for AcCh-esterase.

The depolarizing strength of a series of *n*-alkyltrimethylammonium ions was measured on the electroplax by Podleski (1966, 1969). The potency increased from tetramethylammonium to trimethylbutylammonium, which was found to be the strongest activator. The ability of this compound to depolarize the membrane is as strong as that of AcCh. Further increase of the length of the side chain—up to the trimethylhexylammonium—led to a decrease in potency. Bergmann and Segal (1954) had measured the inhibition of AcCh-esterase by this series of *n*-alkyltrimethylammonium ions and had found that increase of the chain length from one carbon through seven resulted in an increase of inhibitory strength. It is true that in the experiments with the enzyme only binding strength of the compounds is compared, whereas in those on the receptor biological activity is recorded. Nevertheless, the marked differences of increased chain length on the reactions with the two components support the assumption that the molecular groups in the active sites are different. In particular, the equal strength of AcCh and trimethylbutylammonium suggests that the active site of the receptor does not have an esteratic site. Of special interest is the strong difference of potency between the action of tetramethylammonium and trimethylammonium ions on the receptor found in these studies. The quaternary ion is a thousand times stronger activator than is its tertiary analog. This remarkable difference in biological activity which is due to the presence or absence of one methyl group strongly supports the assumption that an anionic group is present and functionally important in both proteins, as is borne out by many other observations.

Tests with the series of aryltrimethylammonium ions also offer pertinent data to compare the reactions with receptor and enzyme (Podleski, 1966, 1969). Wilson and Quan (1958) had found 3-hydroxyphenyltrimethylammonium to be 120 times more effective than phenyltrimethylammonium (PTA) as an inhibitor of AcCh-esterase; 3-methoxyphenyltrimethylammonium was only 5 times more effective, whereas benzyltrimethylammonium was about half as effective. The addition of the methyl group in the 3- or 4-position of PTA did not markedly alter the inhibitory strength. The large increase in the inhibitory strength of 3-hydroxyphenyltrimethylammonium is attributed to a hydrogen bond which is oriented so that the bond is probably formed with the esteratic site of the enzyme.

When tested on the isolated electroplax, the compounds are receptor activators, i.e., they depolarize the membrane. 3-hydroxyphenyltrimethylammonium ion was found, as previously observed by Bartels, to be one-fifth as active an activator as PTA; 3-methoxyphenyltrimethylammonium is a much weaker activator, approximately 65 times less effective than PTA. Also, the maximum depolarization effected is markedly reduced by the presence of the 3-methoxy group, but not by the addition of the hydroxyl group. Again, as with the *n*-alkyltrimethylammonium series, it is difficult to relate directly the measurements of the activation of the receptor to the inhibition of the esterase, since in the latter case only binding strength is measured. However, one would have expected some similarity in the order of decreasing or increasing strength if the two sides of the receptor and enzyme were identical, whereas there is no similarity at all and no correlation exists in any respect between the inhibitory strength on the enzyme and the potency as activator of the receptor. Another compound may be mentioned which has structural similarities, although it is not a derivative of aryltrimethylammonium. This is *N*-methylpyridinium; it possesses a quaternary nitrogen, but only one methyl is free; the other carbons on the nitrogen form part of the ring structure. As an inhibitor of AcCh-esterase, *N*-methylpyridinium is about one-half as strong as PTA. When tested on the electroplax, the compound in 10^{-3} M was found to be without effect either as an activator or as an inhibitor.

The 120-fold increase in inhibitory action of AcCh-esterase by the presence of the 3-hydroxyl group in PTA is equivalent to a decrease of about 3 kcal/mole in the free energy of binding. Such a large effect and the much poorer binding of the 3-methyoxy derivative and other data strongly support the assumption that a hydrogen-bond formation is involved. The phenolic hydroxyl group apparently forms a hydrogen bond with the nucleophilic oxygen of the serine in the active site of the esterase.

In the 3-hydroxyphenyltrimethylammonium, the (N^+---OH) distance is about 5 Å. As we have seen, tested on the electroplax, the presence of the hydroxyl group in the 3-position in PTA decreased the potency as an activator. It appeared desirable, since inhibition and biological activity are not comparable, to test with a receptor inhibitor having a proper (N^+---OH) distance whether the compound would form a hydrogen bond with an atom in the active site.

The hydrolysis of a series of isomeric 1-methyl-7-acetoxyquinolinium iodides catalyzed by AcCh-esterase has been recently analyzed by Prince (1966) using a newly developed and highly sensitive spectrophotometric method. When isomers were compared with an acetoxy group in the

5-, 6-, 7-, and 8-position, it was found that the K_m of 1-methyl-7-acetoxyquinolinium was the lowest. Estimating the distances of the nitrogen to carbonyl-carbon atoms (N^+---C=O) in the 4 isomers with the aid of Fisher–Hirschfelder and Dreiding models, Prince found that the distance in the 7-position was about 4.8 to 5.9 Å, which is, among the isomers tested, closest to the distance of about 5 Å separating, according to previous estimates, the anionic binding site and the atom to which the carbonyl-carbon is bound to the enzyme.

The hydrolysis products of 1-methyl-7-acetoxyquinolinium iodides are receptor inhibitors. They thus offer the possibility of testing the question of whether the (N^+---OH) distance influences the inhibitory strength and whether there is an indication for the formation of a hydrogen bond between the receptor and the inhibitor. A series of isomeric 1-methyl-7-hydroxyquinolinium were tested as inhibitors of the receptor on the electroplax in the usual way, determining the repolarizing strength after the cell had been depolarized by carbamylcholine (Podleski and Nachmansohn, 1966). Surprisingly, evidence was obtained for a hydrogen bond formation between the receptor and 1-methyl-7-hydroxyquinolinium. At pH 6.9, the K_I of the 1-methyl-7-hydroxyquinolinium is 11 times lower than the K_I of 1-methylquinolinium; the 6-hydroxyquinolinium has the lowest K_I of the isomers tested. In order to determine whether the ionized or unionized form was the more active inhibitor, the pH of the solution was altered. Changing the pH of the solution from 5.9 to 8.8 reduced the inhibitory strength of 1-methyl-7-hydroxyquinolinium, whereas the pH changes had no effect on the action of 1-methylquinolinium. Since the pK of the 7-hydroxyl compound is 5.9 (Prince, 1966), the most active form is the un-ionized hydroxyl group. The increase in inhibitory strength by the presence of the hydroxyl group in the 7-position over the 1-methylquinolinium is 110-fold, comparable to the effect previously observed with the corresponding enzyme inhibitors and equivalent to a decrease in the free energy of binding of about 3 kcal/mole. The distance of the hydroxyl group from the quaternary nitrogen is about 5 Å, i.e., about the same (N^+---OH) distance as in 3-hydroxyphenyltrimethylammonium. The evidence thus supports the indication that a hydrogen bond is formed in the reaction with both active sites.

In view of the evidence that the active sites of enzyme and receptor are different, except for the presence of an anionic group, the similarity of the hydrogen bond formation in both components was unexpected. It appeared desirable to elucidate this problem and to test whether the hydrogen bond formation between the quinolinium derivative and the receptor was due to a reaction with the esteratic site of the enzyme or with some group in the active site of the receptor. This was tested by Podleski (1967) by the following procedure: the electroplax was exposed to methanesulfonyl fluoride (MSF) which forms an irreversible complex with the enzyme, a sulfonyl enzyme comparable to the phosphoryl enzyme formed with organophosphates (Myers and Kemp, 1954). The advantage of blocking the esteratic site with this compound for the analysis of the problem is the stability of the sulfonyl enzyme complex and the smallness of the sulfonyl group, which reduces the possibility of steric interactions with the anionic group. The action of MSF was tested by the potentiation of the AcCh effect in the electroplax. The specificity of the effect was, moreover, tested by addition of tetraethylammonium (TEA), which had been shown by Kitz and Wilson (1963) to accelerate the sulfonylation of the enzyme by MSF approximately 30 times. The presence of TEA plus MSF greatly potentiated the effect of AcCh on the electroplax. The strength of the effect remained unchanged 30 and 80 min after removal of TEA.

After the irreversible sulfonylation of the active sites of AcCh-esterase, the electroplax was exposed to either tetramethylammonium (TMA) or 3-hydroxyphenyltrimethylammonium ion. The depolarizing strength of both compounds was not affected. These ions had activity identical to that observed on control cells not exposed to MSF plus TEA. When the repolarizing strength of 1-methyl-7-hydroxyquinolinium was now tested, it was found to be exactly the same as in control cells. The K_I of receptor inhibitors had previously been shown to be the same, independent of the depolarizing ions used (Podleski, 1966, 1969). Since the data show that the pretreatment with MSA plus TEA had no influence whatsoever on the repolarizing activity of 1-methyl-7-hydroxyquinolinium, they appear incompatible with the view that the hydrogen bond is formed with the esteratic site of the enzyme.

Another test for the protein nature of the AcCh-receptor and to the analysis of its active site was applied by Karlin anl Bartels (1966). If the receptor were a protein, it might react with reagents known to act on side chains of proteins, such as sulfhydryl groups or disulfide bridges. In several observations, it had been described that the inhibition of sulfhydryl groups blocked action potentials in frog and lobster nerves (Smith, 1958) and in the perfused squid axon (Hunneus-Cox and Smith, 1965), and inhibited the action of AcCh on the frog heart (Nistratova and Turpaev, 1959; Pohle and Matthies, 1959). Karlin and Bartels (1966) tested the effect of the block of sulfhydryl groups by *p*-chloromercuribenzoate (PCMB) and of the reduction of disulfide bridges by 1,4-dithiothreitol (DTT) on the response of the electroplax to AcCh, carbamylcholine, and trimethylbutylammonium (TMB). DTT is a reagent designed to reduce disulfides with a minimum formation of the mixed disulfide (Cleland, 1964).

Both PCMB and DTT were found to inhibit the depolarization of the electroplax by the three receptor activators used. Exposure to 0.5 mM PCMB for 5 min reduced the response by about 40%. The inhibition due to PCMB is not affected by extensive washing with either Ringer's solution (pH 7.0) or Tris-Ringer's solution (pH 8.0). The inhibition is reversed, however, by reducing agents such as 5 mM β-mercaptoethanol and by 5 mM L-cysteine in Tris-Ringer's solution at pH 8.0. The inhibition of the response of the cell to carbamylcholine due to PCMB is characterized by a shift of the dose response curve toward higher concentrations of activator. The maximum response is not significantly affected. Thus, as described for reactions of PCMB with other proteins, it increases the dissociation constant. Reduction of the S—S bridges by DTT proved to be most effective in inhibiting the response of the electroplax. Exposure to 1 mM DTT (pH 8.0) for 10 min reduces the subsequent response to 50 μM carbamylcholine to more than 30%. The response to AcCh or TMB is similarly reduced. The inhibition due to DTT is not reversed by 40 min washing with either Ringer's solution (pH 7.0) or with Tris-Ringer's solution (pH 8.0). It is completely reversed by oxidizing agents such as, for instance, 1 mM potassium ferricyanide (pH 8.0) applied for 10 min. The disulfide compound DTNB [5.5′-dithiobis(2-nitrobenzoic acid)], a compound designed for the assay of sulfhydryl groups (Ellman, 1959), also completely reverses the inhibition by DTT. As with PCMB, the inhibition due to DTT is characterized by a shift to the right of the dose response curve of carbamylcholine with little change in the maximum response.

4. Effects of Quaternary Groups, Conformational Changes

In spite of all the dramatic advances of protein and enzyme chemistry during the last two decades, the extraordinary catalytic power of enzymes remains one of the most fascinating, but still not fully understood activities in living cells. The lock-key concept of Emil Fisher assumed a rigid protein which absorbs the substrate to specialized catalytic groups. The resulting formation of an enzyme substrate complex seemed, at least to some extent, to account for the specificity of enzymes. However, the progress in the analysis of enzyme mechanisms has made it apparent that the simple lock-key theory did not provide a satisfactory explanation. A notion referred to as induced-fit theory was introduced by Koshland and his associates (Koshland, 1969; Koshland *et al.*, 1962; Koshland and Neet, 1968). He postulated on the basis of a large amount of experimental data, using a variety of enzymes and substrates, that ligands may induce conformational changes in the protein, producing a precise orientation of catalytic groups and thereby permitting the reaction. Binding alone is not sufficient without the favorable change in conformation. During the last decade, this notion has found strong support by a variety of developments, such as the exploration of the tridimensional structure of proteins and enzymes in association with kinetic and chemical studies and the introduction of the notion of allosteric effects of the role of subunits and enzymes. Conformation and conformational changes of proteins during activity form an integral part of biochemical thinking today.

The possibility that a molecule such as a protein may not be a rigid structure and that local changes of conformation may take place during activity occurred to the writer some 25 years ago, long before any experimental evidence was available (Nachmansohn, 1955a,b). It appeared striking that quaternary ammonium derivatives have so much more powerful pharmacological actions than their tertiary analogs, sometimes as much as a hundredfold, as has been known for nearly a century. How could one extra methyl group increase the biological response so strongly? As long as this phenomenon was limited to observations of pharmacological effects, the complexity of the system excluded interpretations on a molecular level. However, when the enzymes associated with the hydrolysis and the formation of AcCh became available in purified solutions, it became possible to explore experimentally the problem of the effect of the extra methyl group in quaternary compounds.

The tetrahedral structure of a methylated quaternary ammonium group is more or less spherical. A direct contact between all four methyl groups of such a structure and the protein surface would not be readily possible, since one of the groups would be located away from the protein. Thus, it appeared possible that such a quaternary compound may induce a conformational change in the protein, possibly a localized one, permitting the macromolecule to have contact with all four groups of the tetrahedral structure and that this conformational change increased the catalytic efficiency.

As described on p. 110, the effect of the extra methyl group was first analyzed with a highly active, partially purified AcCh-esterase solution. When the inhibitory strength of ammonium and hydroxyethylammonium ions, with an increasing number of methyl groups substituting for the protons, were tested, the introduction of two or three methyl groups, respectively, markedly increased the binding strength. The extra methyl group of the two quaternary compounds hardly increased the binding strength (Wilson, 1952) or, as was found later, did so to a much lesser degree than that observed with other methyl groups. Although the extra methyl group did not contribute significantly to the binding, strong differences of enzyme activity were observed between AcCh and its tertiary analog: the rate of formation of acetyl enzyme was about 10 times higher in the presence of the extra methyl group (Wilson and Cabib, 1956). This finding seemed to support the possibility of a change of conformation of the protein, maybe a local one, whereby the extra methyl group of the quaternary group would be enveloped and thus would contribute to the catalytic efficiency.

An equally strong difference of catalytic efficiency between quaternary and tertiary analogs as substrates has been observed with choline *O*-acetyltransferase. The rate of acetylation of choline was found to be more than 12 times as high as that of dimethylethanolamine (Berman *et al.*, 1953; Berman-Reisberg, 1957) (see p. 141).

However, by far the strongest difference between AcCh and its tertiary analog was observed in tests on the AcCh-receptor with the use of the monocellular electroplax preparation. By the removal of the extra methyl group from AcCh, the depolarizing action decreased two-hundred-fold (Bartels, 1962).

Another support for the possibility of conformational changes, also still very indirect, came from studies of the enthalpies and entropies of activation of AcCh-esterase activity (Wilson and Cabib, 1956) described on p. 144. The entropy of activation, ΔS_a, was found to be extremely favorable for the quaternary compound as compared to that of the tertiary compound. While ΔS_a is negative for the tertiary analog, it is positive for the quaternary group; the difference amounts to about 30–40 entropy units. This favorable entropy of activation and the higher rate of hydrolysis of quaternary esters by enzyme catalysis became particularly significant when Chu and Mautner (1966) found that the nonenzymatic hydrolysis of the tertiary ester is very much faster than that of the quaternary analog. Thus, the actual difference of efficiency between the enzyme-catalyzed hydrolysis of the tertiary and quaternary analog is much greater than was apparent on the basis of comparing the two enzyme-catalyzed reactions.

The suggestion of a conformational change of a protein as an essential factor in its function, made for the first time in 1953 by the author in respect to the postulated AcCh-receptor protein, was in fact a speculation; it was firmly established experimentally for proteins in general only in the following decade by the many spectacular advances in enzyme chemistry mentioned above, and it is today a generally accepted notion. It might be, nevertheless, of interest to the reader to learn about the reasoning on which this first speculation was based since, as happened with other speculations in the history of science, it turned out to be correct. In the last few years several observations by other investigators on the proteins processing AcCh have offered new support for conformational changes. The latest developments of the properties of these proteins will be described in the new monograph.

5. Cooperativity and Allosteric Action

Regulatory enzymes act at critical metabolic steps and have specific functions in regulation and coordination. They are activated or inhibited by metabolic effectors that are not substrates catalyzed by the enzymatically active site; they act on different sites. Monod *et al.* (1963) introduced the notion of allosteric effectors and referred to their binding sites as allosteric sites. The effectors do not act directly on the catalytic reaction, but produce an alteration of the molecular structure of the protein referred to as an allosteric transition which modifies the properties of the active site and changes kinetic parameters. When the reaction velocity of such enzymes is plotted against substrate concentration, the curve is not the usual hyperbola corresponding to a Langmuir isotherm, as is to be expected from a simple first-order reaction, but has a sigmoid form indicating a second-order reaction. There is a cooperative effect in the binding of more than one molecule to the enzyme. Cooperativity and allosteric systems have been extensively studied in recent years in many laboratories. The symmetry model of Monod *et al.* (1963, 1965) has found support in studies of a number of enzymes. Koshland and his associates analyzed cooperative properties on the basis of the induced-fit theory and suggested a modified and more flexible interaction between ligands and protein which may induce a new conformation of the subunit (Kirtley and Koshland, 1967; Koshland and Neet, 1968). The prediction of their "sequential" model was borne out in experiments with several enzymes which do not fit the symmetry model.

Changeux and his associates were the first to recognize certain common properties of regulatory enzyme systems and excitable membranes. The biological activities of both systems depend upon threshold concentration of the regulatory ligand, and both systems exhibit cooperative phenomena and other similarities of behavior (Changeux *et al.*, 1967). The dose-response curve of the electroplax membrane to receptor activators has a sigmoid shape (Higman *et al.*, 1963). On the basis of the notion that biological membranes are an ordered collection of repeating globular lipoprotein units organized into a two-dimensional crystalline lattice, Changeux *et al.* (1967) interpreted the sigmoid shape of the response of the membrane in terms of allosteric systems and cooperativity (see also Changeux and Thiéry, 1968). Similar ideas were discussed by Karlin (1967). A detailed analysis of the response of the electroplax membrane to AcCh and its congeners was performed by Changeux and Podleski (1968). Dose-response curves were obtained with several different activators within a large range of concentration. In all cases, the characteristic sigmoid shape was obtained. The Hill coefficient was 1.7; this is widely accepted as an indication of the cooperative character of the response. The maximal responses of three activators tested—decamethonium, carbamylcholine, and phenyltrimethylammonium—were different. The maximal response seems to be directly related to the structure of the ligand. Changes of the ionic environment, either low Na^+ or high K^+ medium, as compared to the usual medium of the Ringer's solution, did affect the absolute values of the maximum depolarization, but the relative differences between the different activators remained the same. Thus, the differential amplitude of the maximal response is determined by the elementary interaction between the receptor activator and the receptor protein in the membrane.

6. Affinity Labeling

Compounds forming a covalent bond with a molecular group in the active site of an enzyme have long been important tools in the analyses of these sites. Such type of compounds provided some information long before the exploration of the tridimensional structure of proteins permitted a precise analysis of the composition and function of the active site. They are still useful tools for the active site of the many proteins, including enzymes, where the tridimensional structure is not yet known. Organophosphates, for example, were used in the studies of AcCh-esterase and other ester-splitting enzymes. A different type of compound forming covalent bonds, referred to as affinity labeling, has been introduced by Wofsy *et al.* (1962). The reagent, because of its steric complementarity to the active site, first combines specifically and reversibly with the site with which it forms a reversible complex; then a small and reactive group reacts with one or more amino acid residues to form irreversible covalent bonds. Affinity labeling of the AcCh-receptor was first attempted by Changeux *et al.* (1968b) with *p*-(trimethylammonium) benzenediazonium fluoroborate (TDF) applied to the electroplax membrane. TDF is a structural analog of phenyltrimethylammonium (PTA), a potent activator of the receptor. It was, therefore, reasonable to assume that it would form a reversible complex with the anionic site of the receptor; the diazonium would form a covalent bond with an amino acid residue in or near the active site. Exposure of the electroplax produced an irreversible block of the response to receptor activators, although the concentration required appeared high, 10^{-4} M. A competitive action for the receptor was demonstrated with reversible receptor inhibitors, such as *d*-tubocurarine and others.

While these observations indeed supported the assumption of a typical affinity labeling, more recent observations on the reaction of TDF and related compounds with the electroplax membrane suggest a modification of the reaction mechanism proposed on the basis of earlier experiments. When phenyltrimethylammonium was replaced by an uncharged *p*-nitro group, the effect on the electroplax response was just as strong as that of TDF (Mautner and Bartels, 1970). When the *p*-nitro group is replaced by an acetoxy group, which attracts electrons less strong than either the *p*-nitro or the trimethylammonium groups and thereby decreases the positive charge of the diazonium, the blocking strength decreases by an order of magnitude. The experiments suggest that it is the positively charged diazonium group which is attracted to the anionic subsite of the receptor.

Another more potent affinity labeling was obtained in a two-step procedure (Karlin and Winnik, 1968). As mentioned before, the response of the electroplax to AcCh and its congeners is blocked when disulfide bridges are reduced by DTT; the block is reversed by oxidizing compounds. When *N*-ethylmaleimide (NEM) is applied after DTT, the reversal by oxidizing compounds is prevented. Presumably NEM forms a covalent bond with the exposed sulfhydryl groups. Since the block of action on the receptors suggested a location of the disulfide bridges near the active site, it appeared possible to increase the strength of the reaction by introducing a group attracted by the anionic site, as was done in various earlier studies with AcCh-esterase, and as was also the reasoning for the use of TDF. The ethyl group of the maleimide was substituted by phenyltrimethylammonium:4-(*N*-maleimido)phenyltrimethylammonium iodide (MPTA). This compound appeared to be a potent affinity label when applied to the receptor reduced by DTT; the response of the electroplax to carbamylcholine is abolished by MPTA at 10^{-7} M. The tertiary analog of MPTA is not more potent than NEM. Adding the extra methyl group enhances the alkylation of the receptor about three-hundredfold, which is about the same difference which was found between tertiary and quaternary analogs in other derivatives of receptor activators, as mentioned before.

Two other affinity labeling compounds for the AcCh-receptor were applied to the electroplax on the suggestion of Silman: bromoacetylcholine bromide and (*p*-nitrophenyl)-*p*-carboxyphenyl trimethylammonium iodide. Both compounds are receptor activators; they depolarize the electroplax, but they form covalent bonds with the receptor only after exposure to DTT, i.e., presumably reacting with the nucleophilic sulfhydryl groups (Silman and Karlin, 1969; Silman, 1970).

Drastic changes of biological effects may result from the reduction of the disulfide bridges in the active site of the receptor by exposure to DTT. Hexamethonium, a bisquaternary reversible receptor inhibitor, becomes a receptor activator after reduction (Karlin and Winnick, 1968). The compound TDF, an irreversible inhibitor of the receptor at 10^{-4} M, becomes after reduction, a potent reversible activator at 10^{-6} M (Podleski *et al.*, 1969). *Thus apparently opposite biological actions, leading to excitation or inhibition, may be due to relatively small changes in the state of the receptor and to other factors in the membrane, and not, as it is widely assumed, to different excitatory or inhibitory transmitters.*

7. Special Ligands

a. Oxygen, Sulfur, and Selenium Isologs of AcCh and Its Analogs. Small changes of structure and configuration of ligands reacting with proteins and enzymes may strongly modify the reaction and greatly affect biological activity. For a systematic study of these factors, Mautner and his associates prepared a great variety of sulfur and selenium isologs of AcCh, choline, benzoylcholine, and many of their congeners (Mautner and Günther, 1961; Chu *et al.*, 1972). The oxygen, sulfur, and selenium isologs are similar in size, but may greatly differ in electron distribution and configuration. The role of these two aspects of the ligands in their reaction with AcCh-esterase and -receptor have been analyzed in the last few years by Mautner and his associates; for quantitative tests of the effects on the receptor, the electroplax preparation was generally used in collaboration with the writer's laboratory.

Kinetic, spectroscopic, and dipole measurements of isologous esters performed by Mautner and his associates offer evidence that the electron distribution in the oxygen, sulfur, and selenium isologs is different (for a summary, see Mautner, 1967). Therefore, their ability to bind to the active sites of the two proteins and of inducing conformational changes could be different. Remarkably, strong differences have been found in the biological activity of the isologs. For instance, choline, even in 10^{-1} M quantities, is virtually unable to produce a depolarization. This is about 40,000-fold the concentration at which AcCh has a strong depolarizing action. The sulfur isolog cholinethiol, has, on the other hand, only a slightly less depolarizing effect than AcCh itself; the difference is less than tenfold (Mautner *et al.*, 1966). Although the choline isologs increase in potency by the substitution of the oxygen atom by sulfur and selenium, the opposite is true for the esters. AcCh is more effective than acetylthiocholine and very much more than acetylselenocholine. It is interesting that the biological ester is by far the most active of the three isologs, whereas its hydrolysis product is the most inactive. On the other hand, the rates of hydrolysis by the enzyme are similar. The great contract in potencies between the reactions of the isologs with receptor and those with esterase is another support for the difference between the active sites of the enzyme and the receptor.

The differences of conformation of the isologs were studied by X-ray analysis in order to distinguish between the effect of this factor on biological activities and the effect of electron distribution (Shefter and Mautner, 1967, 1969). The results indicate that the structures of crystals of oxygen and sulfur isologs tend to differ, whereas those of sulfur and selenium are so similar as to make such molecules isosteric. AcCh, choline, and a series of related molecules show in the solid state that, in general, the N^{+}—C—C—O grouping is in the gauche formation (Canepa *et al.*, 1966; Shefter and Mautner, 1969); this is also the case in solution (Culvenor and Ham, 1966). In contrast, in acetylselenocholine crystals the nitrogen and selenium are trans to one another (Günther and Mautner, 1965; Shefter and Kennard, 1966). The trans conformation of the S—C—C—N^{+} group in acetylthiocholine, and the corresponding group in acetylselenocholine are rather stable. Both compounds retain their trans conformation in solution, as was shown with nuclear magnetic resonance (Cushley and Mautner, 1970). In view of the almost identical conformation of sulfur and selenium isologs, the marked differences in biological activities, both with the receptor and with the enzyme, may be ascribed to differences in electron distribution. These and other observations support the assumption that both factors, conformation and electron distribution, play an essential role in biological activity.

The novel idea of Mautner of using isologs as a tool for the analysis of the properties and function of biopolymers offers new perspectives for many fields of importance to biology and medicine. It may be used, for instance, in pharmacology for obtaining subtle and specific changes of toxicity of drugs, making them more efficient and useful for the action desired. One such possibility, of interest in context with the problems discussed in this monograph and to be elaborated in the upcoming monograph, would be the modification of the properties of organophosphates by changing one atom and inducing thereby a modification of the toxicity which may greatly improve their usefulness as insecticides.

b. Photoregulation of Membrane Potential by Photochromic Substances. The strong effects of sunlight on growth and development of plants has long been known. One of the compounds known to function by photoregulation in many plant processes is the chromoprotein phytochrome that seems to act as an absorber and transducer of light energy (Hendricks and Borthwick, 1967). The activity of phytochrome has been suspected for some time to result in changes of membrane permeability. Recent observations of Jaffe (1970) suggest that AcCh regulates the phytochrome-mediated phenomena by effecting changes in ion fluxes across cell membranes of the plant which he tested.

The classic example of photoregulation by photochromic substances in animals is vision. It is based on the cis–trans isomerization of retinal (Wald, 1968). *cis*-Rectinal reacts in the dark with the protein opsin to from rhodopsin. Light-induced isomerization of the *cis*-retinal to the all-trans configuration leads to nerve excitation. Absorption of only a few quanta of light leads to a neural response (Hecht *et al.*, 1941). Indications begin to accumulate that photoregulation by photochromic substances may be a more widespread mechanism than has been suspected, not only in plants but also in animals.

Recently, Erlanger and his associates prepared a number of diazo compounds which exist as cis and trans isomers and are intraconvertible by means of light. The compounds share a *p*-phenylazophenyl group. Light causes a reversible shift in the cis–trans equilibrium about the nitrogen–nitrogen double bond of the compounds. The trans isomer predominates at 420-nm wavelength in the light of a photoflood lamp. The cis predominates at 300-nm wavelength in ultraviolet light. When these light-sensitive ligands were applied to chymotrypsin (Kaufman *et al.*, 1968) or to AcCh-esterase (Bieth *et al.*, 1969), the potency of the two isomers as enzyme inhibitors was different. A photoregulation of the activity of the two enzymes was obtained on exposure of the enzyme system in presence of the azo compounds to different wavelengths.

In view of the effect of cis–trans isomerism in vision leading to neural stimulation and the effects obtained on AcCh-esterase, it appeared of interest to test whether photoregulation of the potential across the excitable membrane of the electroplax could be obtained with ligands which are structural analogs of AcCh-receptor activators and which isomerize by exposure to different wavelengths. Two diazo compounds were tested on the electroplax membrane: *N*-*p*-phenylazophenyl-*N*-phenylcarbamylcholine chloride (azo-CarCh) and *p*-phenylazophenyltrimethylammonium chloride (azo-PTA) (Deal *et al.*, 1969). The compounds, derivatives of strong receptor activators, are strong receptor inhibitors. Both trans isomers, predominating at 420-nm, are markedly stronger inhibitors than the two cis forms, predominating at 320 nm. When the electroplax is depolarized by 20 μM carbamylcholine, the two compounds depolarize

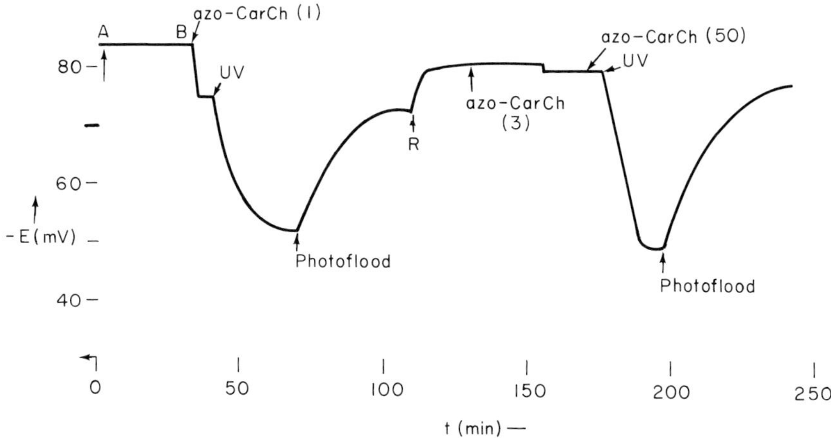

FIG. 14. Photoregulation of the potential across the innervated electroplax membrane of *Electrophorus*. Ultraviolet (Spectroline Model B-100) and photoflood lamps were turned on at the times indicated. (R) Ringers. (A) Carbamylcholine (CarCh), 20 μM; (B) azo-CarCh, 1 μM; azo-CarCh, 3 μM; CarCh, 50 μM. (Deal *et al.*, 1969.)

the membrane in low concentrations. Azo-CarCh in its trans form has a strong repolarizing action in 1 μM concentration, whereas the cis form has a weak effect. The two isomers of azo-PTA have comparable effects, but at ten times higher concentrations. When the electroplax is exposed to carbamylcholine in the presence of 1 μM azo-CarCh under ultraviolet irradiation, a marked depolarization takes place. When a steady state is reached and the preparation is then exposed to a photoflood lamp, the membrane is repolarized by 20–30 mV (Fig. 14). Similar results were obtained with 20 μM azo-PTA.

The system may be considered as a model illustrating how one may link a cis–trans isomerization, the first step in the initiation of a visual impulse, with substantial changes in the potential difference across an excitable membrane. The changes of 20–30 mV in membrane potential are comparable to those occurring in the visual process, but it takes minutes to obtain the effects on the electroplax membrane, whereas in the eye the biological response to light occurs in a millisecond. This difference in efficiency is not surprising. The active compounds in the retina are highly specialized; the light-sensitive molecules form part of the membrane and are almost certainly structurally organized; the potency of intracellularly active biological compounds is usually small when applied from the outside (see Section V). The interesting feature of these experiments is the demonstration of a neural stimulation induced by a change in the conformation of a ligand of a kind similar to that occurring on the biological process of vision.

Among a number of new phenylazophenyl derivatives recently synthesized by Erlanger and Wassermann, two of the new compounds proved to be of particular interest in several respects. The one is 3,3′-bis[α-(trimethylammonium)methyl]azobenzene dibromide (bis-Q) and the closely related 3-(α-bromoethyl)-3′-[α-(trimethylammonium)methyl]azobenzene bromide (QBr):

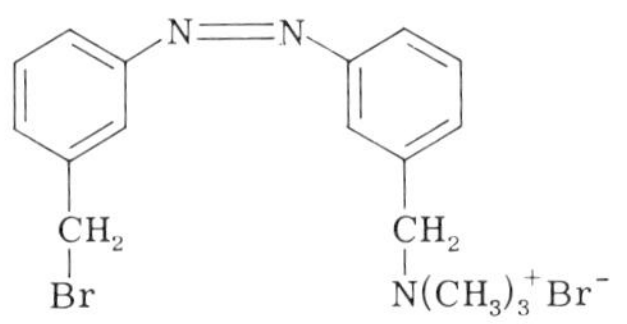

N=N
CH_2 — Br CH_2 — $N(CH_3)_3^+\,Br^-$

QBr

N=N
$Br^-(CH_3)_3{}^+N$ — CH_2 CH_2 — $N(CH_3)_3^+\,Br^-$

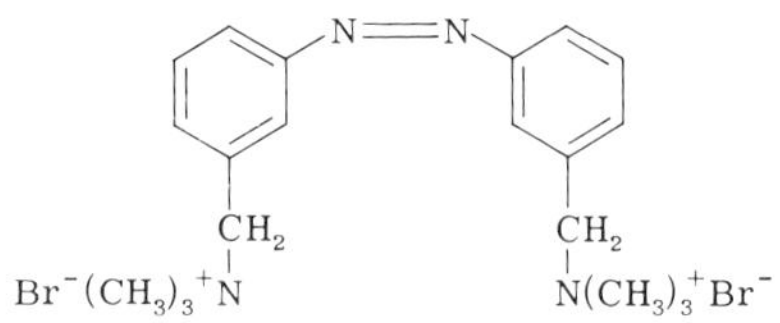

Bis-Q

trans-bis-Q tested on the electroplax was found to be an extremely potent AcCh-receptor activator, one of the most potent ever found (Bartels *et al.*, 1971). When the concentration at half-maximal response ($6\text{–}8 \times 10^{-8}\ M$) was used for comparison, this isomer was found to be 500 times more potent than carbamylcholine. No pure cis isomer was available, but the calculated activity for the cis isomer from the mixture indicates very low potency; it is possible that pure cis isomer may even lack activity. Increase of the concentration of *trans*-bis-Q to 2×10^{-6} M produces repolarization. *d*-Tubocurarine blocks the depolarization by the low concentration of *trans*-bis-Q. The repolarization induced by $2 \times 10^{-6}\ M$ *trans*-bis-Q takes place both in the presence and in the absence of *d*-tubocurarine, suggesting the possible existence of two binding sites of which only one competes with *d*-tubocurarine. The response to *trans*-bis-Q is inhibited by the reduction of the receptor sites with dithiothreitol.

Exposure of the electroplax to *trans*-QBr at a concentration of $2 \times 10^{-7}\ M$ causes a slight depolarization of 5–10 mV. This action is readily reversed. *trans*-QBr inhibits the response to carbamylcholine. However, prior treatment of the cell with dithiothreitol results in irreversible inhibition of the effect of carbamylcholine: thus it is evident that *trans*-QBr is irreversibly attached to the reduced receptor, forming a covalent bond with a sulfhydryl group of the reduced receptor comparable to that with covalently attached *p*-(carboxyphenyl)trimethylammonium iodide reported by Silman (1970) and Silman and Karlin (1969).

The high potency of *trans*-bis-Q, depolarizing the membranes at exceedingly low concentrations (about $10^{-7}\ M$) indicates a remarkably high degree of specificity, emphasized by the finding that the cis isomer has very little or no activity. Receptor activators, just as enzyme substrates, require greater specifications of molecular structure than receptor (or enzyme) inhibitors: factors such as electron distribution, configuration, complementarity, etc., apparently play a stronger role. Inhibitors may act by less specific interactions. It may be estimated from the now known approximate number of molecules of enzyme and receptor in the electroplax membrane, that only about 10^4 molecules of *trans*-bis-Q are required in the solution for the reaction with the receptor (see Bartels *et al.*, 1971). This remarkable degree of specificity has been very useful for the isolation and characterization of the receptor (Chang, 1974).

8. Isolation, Purification, and Characterization

The evidence that the AcCh-receptor is a protein, as postulated in 1953, appeared conclusive. It was also obvious that there is a marked difference between the active site of the receptor and that of the enzyme. However, there was no conclusive evidence that receptor and enzyme are two different proteins, although the author had always considered such an assumption as the most reasonable one in view of the different functions postulated in the AcCh cycle. It appeared more likely that two active sites are not part of the same protein, but belong to two different proteins. A decisive answer to this question required, however, the separation of the two proteins. An early report of Ehrenpreis (1960) to have isolated the receptor protein by using *d*-tubocurarine for binding, mentioned on p. 162, turned out to be due to experimental mistakes clarified by Beychok (1965). A separation was achieved by Changeux and his associates by using the α-toxin of *Naja nigricollis* coupled to Sepharose: they were able in this way to separate the receptor protein from AcCh-esterase in soluble extracts of electroplax membranes (Meunier *et al.*, 1971). The result opened the way to the isolation, purification, and characterization of the receptor protein. Such studies were carried out in the following years in many laboratories and by a great and ever increasing number of investigators.

Before discussing the results of the studies devoted to the purification of the receptor, the use of another snake venom toxin must be mentioned since it turned out to be an important tool in these studies: α-bungarotoxin, a polypeptide of molecular weight of 8000 purified from the venom of the snake *Bungarus multicinctus*, a toxin which had been studied by Lee and his associates (Chang and Lee, 1963; Lee and Chang, 1966; Lee *et al.*, 1967). They found that the toxin produces irreversible neuromuscular block; *d*-tubocurarine protects against the action of α-bungarotoxin. On the basis of their findings Lee and Chang (1966) concluded that the toxin combines irreversibly with the AcCh-receptor at the motor endplate. This toxin was introduced by Changeux *et al.* (1970) for the identification and characterization of the receptor. It has become an extremely valuable tool in all the efforts of receptor purification.

Significant progress has been achieved in the isolation, purification, and characterization of the AcCh-receptor protein with the use of electric tissue of *Torpedo* and *Electrophorus*. In most procedures the binding of radioactively labeled AcCh or α-bungarotoxin was used for identification (see, e.g., Franklin and Potter, 1972; Fulpius *et al.*, 1972; Karlsson *et al.*, 1972; Meunier *et al.*, 1972, 1973; Olsen *et al.*, 1972; Biesecker, 1973; Bourgeois *et al.*, 1973; Klett *et al.*, 1973; Eldefrawi and Eldefrawi, 1973; Hucho and Changeux, 1973; Karlin and Cowburn, 1973, 1974; Schmidt and Raftery, 1973; Lindstrom and Patrick, 1973; and many others). A critical evaluation of work on the AcCh-receptor as well as many other membrane receptors may be found in the review of Cuatrecasas (1974). In spite of the many impressive advances, the establishment of sufficient criteria for the molecular properties has not yet been achieved, nor do the reports mentioned offer conclusive evidence that the receptor has been obtained in its active form. The binding of various radioisotope labeled snake neurotoxins has ben widely accepted as a specific marker for the receptor (see, e.g., Franklin and Potter, 1972; Fulpius *et al.*, 1972). These toxins may, however, be a necessary but not a sufficient criterion by itself. Evidence by H. W. Chang (unpublished work) from this laboratory and by other investigators (see, e.g., Karlsson *et al.*, 1972; Raftery *et al.*, 1973; Chiu *et al.*, 1973) indicates that there is more than one type of α-bungarotoxin-binding component present in the excitable membrane. Moreover, when the purity of the isolated protein is judged by the extent of AcCh binding alone (Eldefrawi and Eldefrawi, 1973), this criterion may be misleading without the knowledge of the AcCh binding to other proteins of the AcCh cycle, which may easily contaminate the purified preparation.

A detailed evaluation of the procedures and the large amount of information obtained about the properties of the receptor protein will be presented in the new monograph in preparation. However, the results of Chang (1974) may be briefly discussed. Her procedure is one among many other convenient methods for obtaining highly purified receptor protein. However, the more compelling reason for describing her results in the context of this monograph is the observation of the presence of two types of AcCh-receptor proteins, tentatively referred to as AcChR-I and AcChR-II, with two distinctly different properties. The use of this preparation has recently led to experiments by Neumann and Chang (unpublished) that are of paramount importance for the further development of the role of the AcCh cycle and offer the first experimental support for the role of the receptor in the control of the release of Ca^{2+} ions, postulated on purely hypothetical basis by the author for many years. The experiments also may indicate that the protein obtained is indeed in an active form. Moreover, it is possible that AcChR-II is the long postulated storage protein.

The principle of the procedures used by Chang may be outlined. Two affinity resins were used: [*N*-(6-aminocaproyl)-*p*-aminobenzyl]trimethylammonium bromide (III), or methyl[*N*-(6-aminocaproyl-6′-aminocaproyl)-3-amino]pyridinium bromide hydrobromide (IV). The

$(CH_3)_3N^+H_2C$—C_6H_4—$NHC(=O)(CH_2)_5NH$—gel

(III)

H_3C—$N^+C_5H_4$—$[NHC(=O)(CH_2)_5]_2$—NH—gel

(IV)

electric organ of *Electrophorus* was homogenized in buffer, centrifuged and the pellet was extracted twice with 1 *M* NaCl in buffer, thereby removing most AcCh-esterase in addition to other soluble proteins. The pellet was then washed twice with 1% Triton X-100 in buffer. The turbid supernatant was applied, after centrifugation, to the affinity columns. The column was washed with at least three column volumes of 1% Triton X-100 in buffer and subsequently with 1% Brij in buffer. Applying linear NaCl gradients removed loosely bound proteins but not the protein referred to as AcChR-I. The AcChR-I was selectively eluted from either resin with the use of bis-Q, the photochromic substance described before (p. 278) with the highest known affinity to the receptor. Bis-Q was then removed by dialysis from the protein solution.

AcChR-I binds with much lower affinity to the dicaproyl column (I) than AcCh-esterase. Therefore most of the enzyme is retained tightly during the elution of AcChR-I, which with the use of bis-Q may be selectively eluted in a relatively small volume. An approximate 200-fold purification of AcChR with a specific activity ranging from 7 to 10 nmoles of α-bungarotoxin binding sites per mg protein was obtained. The recovery corresponds to about 35% (average) of Triton X-100 crude extract. Homogeneity of the protein was tested with polyacrylamide gel electrophoresis in the absence of sodium dodecyl sulfate. Some aggregation had occurred which increased after several weeks of storage. Such aggregations of homogeneous receptor protein has also been reported by other investigators. Free SH groups present in AcChR-I rapidly decreased with storage, suggesting that the aggregation phenomenon may be due to intermolecular disulfide bond formation caused by air oxidation.

Electrophoretic analysis of purified receptor in 1% sodium dodecyl sulfate in the absence of reducing agent showed several protein-staining bands. The molecular weights of the lower three distinctive bands correspond to 49,000, 42,500, and 37,000, respectively. In the presence of a disulfide reducing agent there was a sharp increase only in the 41,500 dalton band. The results resemble those obtained by Biesecker (1973) with the densitometric scan. Several other laboratories reported a major polypeptide species of 40,000–43,400 daltons. The 41,500-dalton subunit may correspond to the subunit polypeptide of 40,000 daltons which is affinity labeled in intact cells or purified AcCh-receptor (Karlin and Cowburn, 1973). The present data suggest that the 41,500-dalton species is the major subunit of AcChR-I.

The binding of [^{14}C]bis-Q to Triton X-100 crude extracts of electroplax shows two distinct dissociation constants, one of low affinity (K_D = 2.9 μM) and one of high affinity (K_D = 6.4 nM), on a Scatchard plot. The high-affinity sites are blocked by preincubation with α-bungarotoxin. After purification only the high-affinity sites are associated with AcChR-I, but the Scatchard plot is not linear. The possibility of negative cooperativity (Levitzki and Koshland, 1969) is indicated by a double-reciprocal plot and a Hill coefficient of 0.7. Negative cooperativity in the binding of AcCh to purified receptor protein has been reported by Eldefrawi and Eldefrawi (1973).

Many purification experiments of Chang have shown that another component meets the present criteria for AcCh-receptor. Large variations in the [^{3}H]AcCh binding capacity but not in that for α-[^{125}I]bungarotoxin were observed among different preparations. Finally a reasonable fractionation of two components was achieved both of which bind AcCh and α-bungarotoxin competitively. The major component is AcChR-I. While this component has a low dissociation constant, the second one referred to as AcChR-II has high binding capacity, but a markedly higher dissociation constant. This second component can easily contaminate the preparation of AcChR-I; even if the contamination is small, its presence is important because of its high binding capacity. Some preliminary experiments indicated that the AcCh sites on AcChR-II, unlike those on AcChR-I are completely blocked by sulfhydryl reagents.

C. *Choline O-Acetyltransferase* (*Choline Acetylase*)

The discovery of the enzymatic acetylation of choline in a soluble system, in 1942, by an enzyme first referred to as choline acetylase, now as choline *O*-acetyltransferase (EC 2.3.1.6) has been described on pp. 83–95. On the basis of bioenergetic considerations ATP was used as source of the required energy for the synthesis and found to be effective. It was the first acetylation achieved in a soluble system. Since the finding was so unexpected and the exact mechanism of the use of the energy of ATP hydrolysis was not yet understood, the first report of this discovery was rejected by several journals as described in the Prefatory Chapter (Nachmansohn, 1972a). The mechanism of acetylation was finally explained in the following decade when the role of CoA was analyzed in the laboratories of Lipmann, Ochoa, Lynen, and many others. The intermediary steps with ATP as energy source were described by Berg (1956). The distribution and concentration of the enzyme was also immediately explored and its presence in a large variety of different types of excitable cells was demonstrated (see p. 90). Again no exception was found.

In general, the concentration of this enzyme is one to two orders of magnitude lower than that of AcCh-esterase. In the frame of the chemical hypothesis of excitability this is not surprising. The hydrolyzing enzyme must be directly associated with electrical activity, and requires not only a fast reaction but possibly also a larger margin of excess than an enzyme acting in the recovery period which may be of long duration. In some fibers choline *O*-acetyltransferase appears to be low. It was found that in some sensory fibers only about 30 μg of AcCh are formed per gram fresh tissue per hour. The low concentration in this fiber was considered by Hebb (1957) as a contradiction to the theory that acetylcholine is associated with conduction. Referring to activity of the enzyme to gram fresh tissue, in general, does not give a satisfactory indication of the actual concentration of AcCh formed at or near the membranes. Membranes may form only a small fraction, say 10^{-4}, of the total mass of the tissue. Therefore, the statement that the concentration is too low to be compatible with the assumption of the role of AcCh in conduction does not appear to be justified (see Nachmansohn, 1963); in other sensory fibers choline *O*-acetyltransferase is present in high concentrations (Davis and Nachmansohn, 1964). Of particular importance is, moreover, the recognition that the enzyme is very unstable. Although the AcCh-esterase is relatively stable, it is nevertheless extremely difficult to obtain quantitative data for the AcCh-esterase concentration in a tissue (see below pp. 295–297); the problems of a quantitative evaluation of an unstable enzyme are even greater. Some properties were tested with a partially purified enzyme extracted from squid head ganglia (see pp. 137–141). A significant finding was the evidence of SH groups in the active site (Berman-Reisberg, 1954). However, the specific activity of the preparation was still rather low, 40–80 μg of AcCh formed per mg protein per hr.

For several years relatively little attention was paid to the enzyme, in spite of its importance for the formation of a compound that plays a crucial role in nerve function. Recently, however, Mautner and his associates have initiated new studies and have achieved pertinent results in several directions. Husain and Mautner (1973) applied affinity chromatography to the purification of the enzyme extracted from squid head ganglia. Two types of Sepharose columns were used: mercurial-Sepharose (V) and columns to which the enzyme inhibitor *p*-(*m*-bromophenyl)vinyl pyridinium (VI) was attached.

Sepharose-NH—$(CH_2)_3$—NH—$(CH_2)_3$—NH—CO—$(CH_2)_3$—CO
NH
$HgOCOCH_3$

Mercurial-Sepharose
(V)

Sepharose-NH—$(CH_2)_3$—NH—$(CH_2)_3$—NHCO$(CH_2)_2$—CO—NH
$(CH_2)_3$
NH
$(CH_2)_3$
Br
CH—CH—N^+—$(CH_2)_5$—CO—NH

Styryl pyridinium-Sepharose
(VI)

Purification by affinity chromatography was applied in combination with several other procedures. These experiments led to the isolation of three fractions of choline *O*-acetyltransferase, ranging in specific activity with the different procedures used from 1000 to 4000 μmole/mg of protein per hr. This is a remarkable progress if one compares these values to the specific activities obtained previously with the enzyme extracted from various sources: 110 μmole/hr for the squid enzyme (Prince, 1967), 7 μmole/hr for the ox brain (Glover and Potter, 1971), 37 μmole/hr for the rat brain (Potter *et al.*, 1968), 140 μmole/hr for the human placenta (Morris, 1966), and 44 μmole/hr for the fly brain (Mehrotra, 1961).

The molecular weights obtained varied in fractions obtained with different procedures: in the fraction with a specific activity of 4000 it was 125,000, in that of 1000 it was 120,000, whereas in the fraction of 2800 it was 200,000 ± 125,000. These variations are not surprising in view of the different procedures applied. It is now well known for many enzymes that the molecular weights may vary depending on the methods used and may appear in multiple forms. An excellent illustration are the multiple forms described for AcCh-esterase as discussed before. A critical evaluation, clarifying these differences, may be found in the review of Rosenberry (1975).

Some efforts were recently devoted to the elucidation of the mechanism of the catalytic action (see, e.g., Morris *et al.*, 1971; Rama Sastry and Henderson, 1972). Some evidence was presented by the investigators quoted that the enzyme catalyzed transacetylation from acetyl-CoA to choline involves an ordered or a Theorell–Chance mechanism with acetyl-CoA as the leading substrate. It has also been proposed that imidazol may play a catalytic role in the function of the enzyme (White and Cavallito, 1970; Burt and Silver, 1973).

In recent studies on the mechanism of the action of the enzyme by Currier and Mautner (1974) with highly purified preparations, it was shown that photooxidation in the presence of methylene blue or rose bengal rapidly inactivates the enzyme. These observations strongly support the assumption that a histidine residue is involved at the catalytic site. The authors suggest that general base catalysis by imidazole enhances the ability of enzyme-bound choline to react with a thiolester group. The mechanism of catalysis may thus be pictured as follows: acetly-CoA is first bound to a suitable specific binding site. Choline is bound to an anionic site and reacts with the N of the imidazole in close neighborhood of the anionic site attracting the H of the OH group of choline, making the oxygen atom of choline a better nucleophile for attacking the carbonyl group of actyl-CoA; the transacetylation thus yields acetylcholine. The enzyme is restored to its original condition. The mechanism of the catalytic action proposed seems well supported by the data.

The observations described in these studies also show that considerable specificity must be involved in the catalytic action of the enzyme. Homocholine, e.g., with just one more methylene group than choline, is not acetylated. Neither cholinethiol nor homocholinethiol was acetylated. This is not surprising, since it is known that thiolesters, while relatively resistant to hydrolysis, are highly suceptible to thiolysis. Therefore, acetyl-CoA will undergo rapid thiolysis in the presence of either cholinethiol or homocholinethiol even without the enzyme. Aminocholine is acetylated, but at a much lower rate, about one-half of that of choline. Thiolesters are also susceptible to aminolysis. Thus aminocholine may also produce spontaneous aminolysis of acetyl-CoA at pH's at which the amino group is not protonated. The major role of the enzyme apparently resides in increasing the reactivity of the hydroxy group of choline with regard to the thiolester group. This may explain why aminocholine may be acetylated, although at a lower rate, even at pH's at which the amino group is protonated.

D. *Storage Site*

The storage site for AcCh in the excitable membrane is most likely a protein. Only proteins would account for the specificity of binding. A relatively high binding constant is suggested by the great difficulty of removing AcCh from the membrane. It appears likely that the storage protein (S) forms part of the protein assembly located close to the receptor (R) and the enzyme (E). The SRE complexes surround the gateway as pictured in Fig. 18 which shows a basic excitation unit proposed by Neumann (1975b). The release of AcCh from the protein may be effected by electric fields and will be described in Supplement II.

IV. Experimental Basis for the Direct Link of the AcCh-Cycle with Electrical Activity

A. *Effects of Specific Inhibitors of AcCh-Esterase on Electrical Activity of Axons*

Neither the universal distribution of AcCh-esterase in all excitable tissues, well documented when this monograph was first written, nor the localization of this enzyme in excitable membranes as established during the last decade, are an indication for the direct role of the enzyme in the electrical activity of axons. Even the remarkably high concentration of AcCh-esterase in the excitable membrane of such a functionally specialized cell as the electroplax is no evidence for this assumption. At best all these factors may be considered as suggestive. Both are, however, necessary prerequisites. Another prerequisite for the postulate of a control of electrical activity by the AcCh cycle is a high rate of the enzyme activity, since the speed of the two activities must be similar. The high turnover number of the enzyme and its small turnover time, first tested by the author in the early 1940's (p. 37) (see Rothenberg and Nachmansohn, 1947) have been confirmed by experiments of Lawler (1961) and Wilson and Harrison (1961). According to the best estimates, available at present, the turnover number of the enzyme is 8×10^5 min^{-1}, its turnover time is about 60 μsec (see Rosenberry, 1975). This astonishingly high speed is adequate for the function postulated.

However, as has been repeatedly emphasized, the chemical theory of excitability is based on an entirely different type of experimental fact. One important approach by biochemists for the demonstration of a direct link between an enzyme and a specific cell mechanism has been the use of potent and specific enzyme inhibitors in order to test whether or not they affect the particular function investigated. This approach has been tried by the author since 1945 (see pp. 52–67). It has been possible to demonstrate in an unequivocal way that specific inhibitors of either AcCh-esterase or AcCh-receptor affect and eventually block electrical activity.

The author was, however, aware from the beginning of the many difficulties and pitfalls of this approach. In view of the complexity of living cells, structural barriers may drastically reduce or even completely prevent the action of many compounds applied externally. The many additional factors which do not permit simple extrapolation from reactions *in vitro* to a complex structure such as cell membrane have become apparent only in the last decade and were discussed in Section I,A; thus the difficulties and pitfalls are even much greater than could possibly have been foreseen by the author some 30 years ago, and they are even today not realized by many investigators. Applying, e.g., AcCh or neostigmine to the squid giant axon, a so-called "unmyelinated" axon, should have been anticipated to have no effect, since the excitable membrane is surrounded by a Schwann cell of 4000 Å thickness which is rich in lipid. AcCh and related quaternary ammonium derivatives are lipid insoluble. They could not be expected to penetrate this structural barrier. The preparation readily permits such tests, since the interior of the axon, the axoplasm, may be extruded; therefore, the entrance of externally applied compounds may be determined. When these experiments were performed (see, e.g., Bullock *et al.*, 1946; Rothenberg *et al.*, 1947) the prediction was fully borne out (see p. 178): neostigmine and ^{15}N-labeled AcCh, applied in highest concentration for prolonged periods of time, failed to penetrate and obviously had no effect on electrical activity. In contrast, physostigmine (eserine), a potent inhibitor of AcCh-esterase, with a K_I of 10^{-7} M, did enter the axoplasm and did affect and eventually abolish electrical activity. When the tertiary analog of neostigmine is used, it also decreased and finally blocked conduction. While neostigmine is an equally strong inhibitor as physostigmine, the tertiary analog is about 100-fold less potent, but, nevertheless, it had an effect on electrical activity comparable to that of physostigmine. The concentrations of physostigmine required for block of conduction appeared high, about 10^{-3} M. This is not surprising in view of the differences discussed between reactions in solution and in intact structure.

The fraction of assayable enzyme in reference to the total amount present is extremely difficult if not impossible to evaluate with presently available methods, even in normal tissue, as will be discussed later. Experiments with exposure of squid axons to the strong, lipid soluble, and irreversible inhibitor DFP, in 10^{-4} *M*, using a new assay method show that there was still 30% of the assayable enzyme present when electrical activity was blocked (see below). This result may explain the necessity of using 10^{-3} *M* of the poorly lipid soluble physostigmine to inhibit a sufficiently high fraction of the enzyme to become incompatible with its function in electrical activity. Physostigmine was applied to a great variety of different types of axons and was demonstrated to affect and eventually block conduction in all of them without any exception (see pp. 51–60). The effects were all easily reversible as would be expected from an easily reversible competitive inhibitor.

Organophosphates, such as DFP and many others, block the enzyme "irreversibly." In contrast to the carbamoylated enzyme, the phosphorylated enzyme regains its activity either very slowly or not at all, and thus blocks conduction irreversibly. Only by chemical reactivation with nucleophilic agents such as 2-PAM may a reactivation be achieved in time to restore electrical activity. Here again the external concentrations required appeared high, but the same reasoning applies as to the "high" concentrations of physostigmine required, although with these compounds many more complications are encountered as will be discussed later. As an example, the experiments on the squid giand axon (see p. 65) may be mentioned. The outside concentration of DFP required for irreversible block was 5×10^{-3} *M*, but the amount which had penetrated at the time of irreversible block was found to be extremely low, of the order of 10^{-6} *M*.

About 20 years later, in the 1960's, these experiments were repeated with the use of radioactively labeled compounds and fully confirmed (Rosenberg and Hoskin, 1963). Experiments with radioactively labeled DFP provided also an unexpected factor for the explanation of the high outside concentration required for block of conduction by a compound lipid and water soluble (Hoskin *et al.*, 1966). These axons are rich in DFP-ase first described by Mazur (1946) and later more properly referred to as phosphorylphosphatase by Augustinsson and Heimburger (1955). The enzyme occurs in many tissues and has been extensively studied (for a summary, see Mounter, 1963). The low concentration of DFP in the axoplasm (about 10^{-6} *M*) at the time of conduction block found previously was confirmed with the highly sensitive magnetic diver technique. It is interesting that recently it was found, with a radiometric assay which permits estimates of intact axons, that the DFP in

10^{-4} *M* concentrations inhibited only about 30% of the assayable enzyme concentration in the squid axon (Kremzner and Rosenberg, 1971), while electrical activity was already irreversibly blocked. Further strong support for the importance of the structural barriers in the evaluation of the effects of AcCh, curare, and related compounds was provided by Rosenberg and his associates with the use of phospholipase in snake venoms. The most potent venom seems to be that of the cottonmouth moccasin (for summaries, see Rosenberg, 1965, 1966). It was found that after exposure of the squid giant axons to a few micrograms of phospholipase A for brief periods, about 15–25 min, AcCh and curare (*d*-tubocurarine) blocked reversibly electrical activity. Radioactively labeled compounds were found in the interior of the axon (Hoskin and Rosenberg, 1964). The treatment itself did not affect electrical activity. Physostigmine blocked conduction in markedly lower concentrations than it did without treatment. Examination with the electron microscope revealed structural alterations in the outer layer of the Schwann cell; the plasma membrane seemed not to be affected (Martin and Rosenberg, 1968). Interestingly, neostigmine and carbamylcholine, much more closely related to AcCh in structure than *d*-tubocurarine, did not affect electrical activity even after exposure of the axons to phospholipase A and was not found in the axoplasm (Brzin *et al.*, 1965). One may even speculate that within the membrane there exist barriers which while permitting some compounds to react with the active site of a protein, prevent this action of even closely related compounds. In view of our ignorance of the complex molecular organization of membranes, such questions may be raised but not answered at present. This is another illustration of the complexity of the factors determining the effects in intact structures. The effect of phospholipase A is fully accounted for by the formation of lysolecithin which by itself produces the same result as the treatment by the venom (Condrea and Rosenberg, 1968; Rosenberg and Condrea, 1968).

In this context experiments may be mentioned in which squid giant axons were perfused and compounds were added to test whether they affect electrical activity. Rapid and reversible block of electrical activity was observed with physostigmine and tetracaine; *d*-tubocurarine also blocked electrical activity in 5 m*M* concentration (Tasaki *et al.*, 1965). Recently, Yeh and Narahashi (1974) repeated the perfusion experiments of squid axons, applying a few quaternary compounds to the perfusion fluid, using a technique different from that of Tasaki *et al.* (voltage clamp). The actions of applying AcCh, carbamylcholine, hexa- and decamethonium, and *d*-tubocurarine were tested as to their effect on electrical activity. The first three compounds had only insignificant effects on electrical activity, decamethonium none; with *d*-tubocurarine in 10 m*M* concentration, the action potential was decreased by about 50% of the controls. The failure of these compounds (added to the inside of the axon) to prevent impulse conduction is considered by Yeh and Narahashi as incompatible with the author's theory and with the integral model of nerve excitability (Neumann *et al.*, 1973). These investigators still do not realize a very fundamental phenomenon of nerve fibers: many quaternary compounds do not enter intact structures, cells or membranes, as has been repeatedly discussed. Even after treatment with phospholipase A, which partially disintegrates protective layers covering axons, only AcCh and *d*-tubocurarine affected electrical activity, i.e., acted on the AcCh cycle; but neostigmine and carbamylcholine did not act. Neither this difference nor the difference in the efficiency of *d*-tubocurarine in the earlier and recent observations can presently be explained, because we do not know how these compounds interact within the (still unknown) molecular organization of membranes in general (see Section I,A). The difficulties to interpret the results of externally applied compounds, especially poorly lipid soluble compounds, are not surprising in view of the extraordinary complexity of intact structures. Moreover, within the membrane, proteins are tightly bound to lipids. Yeh and Naharashi neglected to test lipid soluble compounds such as physostigmine, AcCh analogs such as local anesthetics, DFP, etc. Their data are in contrast to those obtained by Tasaki *et al.* in their perfusion experiments. There are many additional examples in which it has clearly been documented that the AcCh cycle plays a key role in electrical activity. The necessity to use lipid soluble compounds to perturb the AcCh cycle is all the more indicated because in the last few years it was found that the AcCh-receptor proteins—in contrast to the major portion of AcCh-esterase—are tightly bound to the membrane. The extraction of those proteins requires the use of Triton X-100 or other detergents; this is clearly an indication of their firm association with lipids. It is, therefore, understandable that the active sites of the basic excitation units (Neumann *et al.*, 1973 and Supplement II) may be reached only with great difficulty, or not at all, by poorly lipid soluble compounds applied, e.g., to the inside of the axon. Thus, the conclusion of Yeh and Naharashi seems to be unjustified because the authors have no evidence that the compounds they applied did actually reach the target protein within the excitable membrane. Moreover, once it has become evident that the assumption of a simple diffusion process is unable to explain electrical activity, the use of pharmacological tests only aimed at analyzing the molecular events within the membrane becomes clearly inadequate.

Another support for the assumption that the irreversible block of conduction by organophosphates must be attributed to the irreversible block of AcCh-esterase has been offered by experiments in which 2-PAM was applied to restore irreversibly blocked electrical conduction. If the block of conduction is due to the strong P—O bond of the phosphorylated enzyme, potent reactivators capable of breaking the P—O bond and specific for phosphorylated AcCh-esterase such as 2-PAM, may be able to restore the irreversible block. Such a reactivation of electrical activity would be a strong evidence for the interdependence of electrical and enzyme activity, formation of the P—O bond blocking it, breaking of the P—O bond restoring it.

Obviously this is not an easy type of experiment to carry out for several reasons. It is difficult to determine the exact period at which the block of electrical activity becomes irreversible. This usually varies in certain limits from one preparation to the other, even if the same type of fiber is used, in view of many biological variables. Too long exposure, in some cases, one minute or two, after the irreversible block may, and almost certainly will, result in secondary reactions which may make the return of the electrical activity impossible even if a large fraction of the enzyme is reactivated. If the action of one chemical compound on the enzyme in an intact structure is a complex phenomenon, the complexity is potentiated if two different types of compounds are applied competing for the same molecule, one of them poorly lipid soluble (2-PAM).

In spite of all these difficulties, restoration of irreversibly blocked electrical activity was obtained with 2-PAM in a great number of cases and with a variety of preparations. Two strips of frog sartorius muscle were exposed to Paraoxon until electrical activity was abolished. One strip was used as control, the other exposed to benzoyl-2 and 4-PAM. These two compounds are only slightly less effective reactivators of phosphorylated AcCh-esterase in solution than 2-PAM, but both are more lipid soluble; these two structures proved to be effective, whereas 2-PAM was not. Out of 11 experiments performed, in 9 irreversible block of electrical activity was restored by PAM (Hinterbuchner and Nachmansohn, 1960). Later this type of experiment was successfully performed on the axons of the walking leg of lobster by Dettbarn *et al.* (1964). The possibility of restoring irreversible block of electrical activity by 2-PAM was further tested on squid giant axons by Rosenberg and Dettbarn (1967). Although the results were less regular than those observed with other preparations, a significant difference was found between the reversibility in the presence and absence of 2-PAM. However, the differences became much more impressive and the data became quite convincing when the axons were first exposed to phospholipase A, apparently facilitating the access of the poorly lipid-soluble 2-PAM. In view of the many complex factors involved, it appears pertinent that restoration of electrical activity, after its irreversible block by an organophosphate, has been achieved in a significant number of experiments with various preparations.

In view of the structural barriers, protecting all conducting parts of excitable membranes, the high concentration of AcCh-esterase inhibitors required for affecting electrical activity is not surprising and explained by the data in the preceding pages. There are, however, preparations in which at least a small part of the axonal membrane is only poorly protected and therefore more comparable to the conditions prevailing at junctional parts of the membrane. This is the case at the nodes of Ranvier of myelinated axons. In contrast to the 30–50,000-Å thick myelin sheath surrounding most of the axons in a frog sciatic nerve fiber, at the node there is no myelin, although there is still a base membrane of the Schwann cell and a poorly defined "porous" structure. As discussed on p. 177, the powerful action of the AcCh on the junction (Brown *et al.*, 1936) was considered to be one of the strongest supports for its special role there, since AcCh completely fails to affect axonal conduction. This limitation of the action of AcCh paralleled that of the effect of curare, the striking observation of Claude Bernard a century before. In both cases frog sciatic nerve fibers were used. The frog sciatic nerve fiber is formed by several thousand heavily myelinated axons surrounded by a connective tissue sheath impervious to many compounds, as has been demonstrated by the strong effects of compounds after the removal of the sheath, compounds completely inactive on the intact fiber. However, in the 1950's, Staempfli (1956) developed a method which permitted one to measure the electrical activity of a single node of Ranvier of an isolated axon (Fig. 15).* When W.-D. Dettbarn, trained

* The author has avoided discussion of statements which ignore elementary and well-established facts described in textbooks. One example that may be quoted is the denial of structural barriers surrounding axons in contrast to junctions (see Fig. 15). The author's only modest contribution in this respect was the demonstration that even relatively much less compact structural barriers, for instance the Schwann cell surrounding "unmyelinated" squid axons, prevents compounds such as AcCh and curare from reaching the excitable membrane. This monograph has been written for readers familiar with a certain amount of current knowledge of ultrastructure, biochemistry, and physicochemistry. It is not at all intended to convince opponents. Such attempts are always doomed to failure. A fruitful scientific discussion requires a common basis of a minimum of knowledge. The only aim of this book is to provide information to those investigators interested in the chemical and molecular events in excitable membranes.

in Staempfli's laboratory, joined the author's laboratory, the author suggested the use of this preparation for tests on the role of AcCh and curare in the conducting membrane to him. Dettbarn (1960a, 1961) exposed single fibers to physostigmine in low concentrations, 5×10^{-5} to 5×10^{-6} *M*. Effects on the electrical activity appeared rapidly, in seconds. Moreover, the electrical parameters are modified in a way that has been postulated by electrophysiologists if the proposed theory were proven to be correct. First, the spike height was increased and the descending phase was prolonged. On longer exposure, the amplitude increased, but the prolongation remained or became even greater. At a concentration of 2×10^{-4} *M* (60 μgm/ml) the spike height was first reduced and the spike duration was increased; after exposure of 15–20 min, block of conduction occurred. With 10^{-3} *M*, conduction was blocked in 25 sec. The effects of AcCh-esterase inhibition on the nodes are thus comparable to those observed on junctions. Physostigmine has a p*K* of 8.2. At neutral Ringer's solution only a fraction is in the uncharged form and capable of penetrating the structure. When Dettbarn tested the effectiveness of physostigmine as a function of pH, he found it to increase with increasing pH. A roughly linear relationship exists between the concentration of the conjugate base in solution and the percentage attenuation of the action potential, which is consistent with the assumption that only the neutral molecule may penetrate; no such relationship was observed with the cationic form. It would be difficult to attribute the effects observed on the electrical activity of the axon to some unknown or some unspecific effect in view of the high affinity of physostigmine to AcCh-esterase and the rapidity, effectiveness, and type of action. The blocking effect of curare on the nodes of Ranvier will be described in the discussion of the effects of inhibitors of AcCh-receptor.

In summary, the data presented demonstrate the essential role of AcCh-esterase in axonal conduction and its direct link to electrical activity in the way postulated by the AcCh cycle.

B. Problem of Minimum Requirements of AcCh-Esterase Activity for Electrical Activity

1. Determination of AcCh-Esterase Activity in Tissues

There exists today a great variety of good and widely used methods for determining AcCh-esterase activity in solution. For almost three decades the author considered a thoroughly homogenized suspension of a tissue as an adequate procedure for the estimation of the total amount

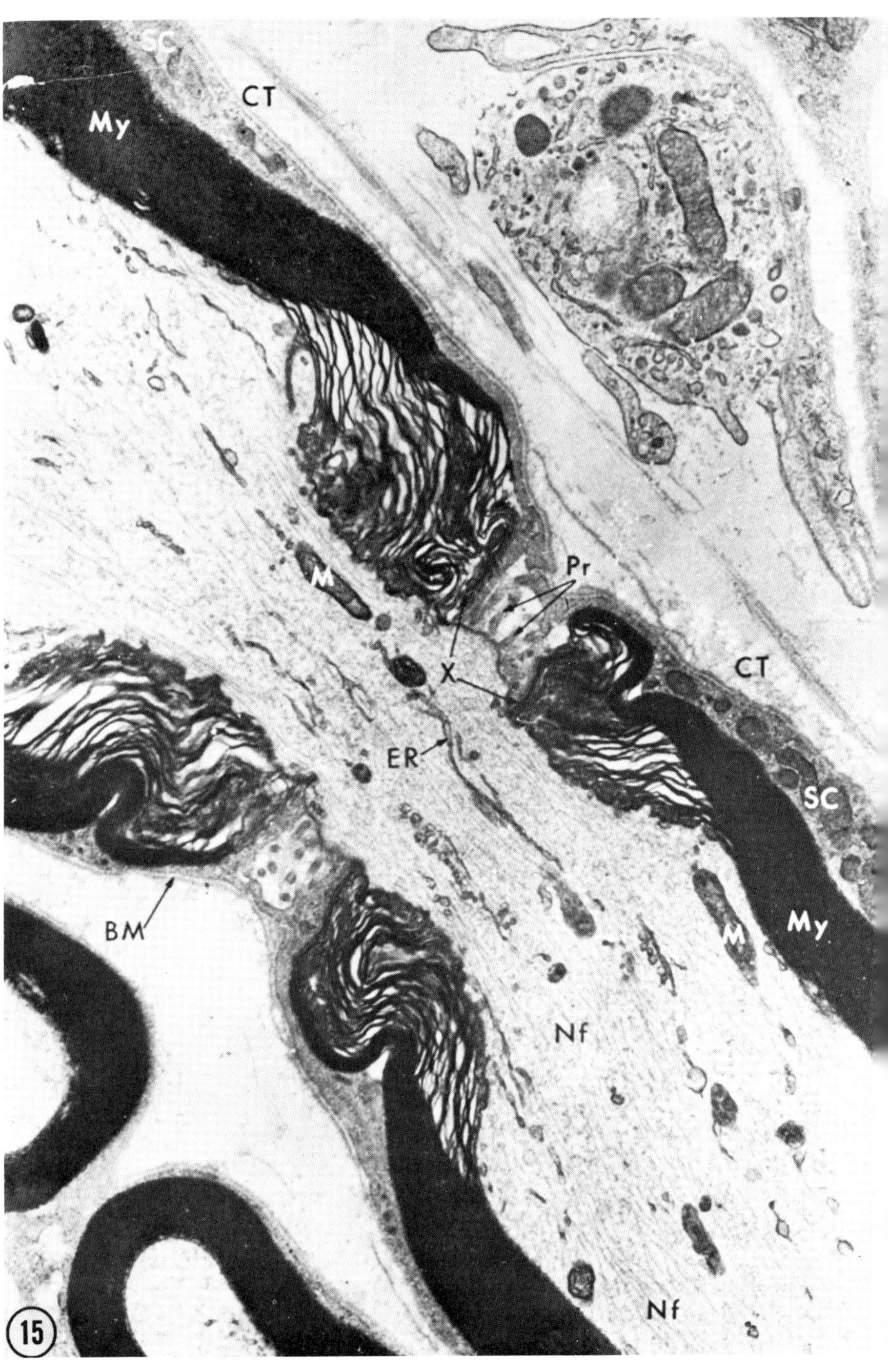
SC
CT
My
M
Pr
X
ER
CT
SC
BM
M
My
Nf
Nf
15

present in a tissue, nerve, muscle, or brain. Most of the values obtained with this procedure are presented on pp. 26–30.

About a decade ago observations began to accumulate that seriously questioned the adequacy of the use of homogenized suspensions for obtaining precise values of the actual total amount of the enzyme present in a tissue. A few examples may be given. When unmyelinated nerve fibers are homogenized the increase of the value compared with that of the intact fiber may range from 100 to 400%, even when the same fiber is used. These great variations may be in the biological range, but the question could be raised whether, in addition, the enzyme in the particles of the suspension is not fully and not equally available to the substrate, and whether tissue components attached to the enzyme slow down the rate to a varying degree. However, when about a decade ago a myelinated fiber (frog sciatic nerve) was used by Brzin and Dettbarn (1967), the difference between intact fiber and homogenized suspension was surprisingly small, only about 25% higher in the suspension. In view of the very high lipid content of these fibers, this small increase suggested to the author that even in the small particles the lipid content is high and actually prevents or slows down the reaction of the substrate with the enzyme, hence, the almost negligible increase in activity in the homogenized suspension. This observation, in addition to the discrepancy between the chemical data and the electron microscope examination, suggested the possibility of structural barriers in myelinated axons even in slices of 500 Å thickness. When Brzin (1966) added Triton X-100 in his electron microscope studies he indeed found the presence of the enzyme in the excitable membrane as has been shown in Fig. 9. Another pertinent observation was the use of small pieces of the envelope (100 μg weight) of the "unmyelinated" squid axon for determining enzyme activity with magnetic diver technique. When these pieces, containing the excitable membrane, were exposed to sonic oscillation, the activity increased by 250% (Brzin *et al.*, 1965). However, by far the strongest and most conclusive evidence was provided by experi-

---

FIG. 15. Electron micrograph showing a node of Ranvier of a single fiber from the sciatic nerve of mouse. The sheath of myelin forms a compact tube (My) over most of the internodal area. In the region of the node, fingerlike processes (Pr) of neighboring Schwann cells (SC) interdigitate and cover the nodal area. A basement membrane (BM) and connective tissue fibers (CT) of the endoneurium complete the wrappings of the fiber. At the node the membrane of the axon is free of myelin and is exposed to the interstitial fluids which diffuse through the basement membrane and between the Schwann cell processes. Axoplasm is rich in neurofilaments (Nf) and contains slender elements of the endoplasmic reticulum (ER) and small number of mitochondria (M). (Porter and Bonneville, 1964.)

ments of several investigators with isolated membrane fragments of the electroplax of *Electrophorus* (see, e.g., Silman and Karlin, 1967; Silman, 1969). When the preparation was suspended in 1 *M* NaCl, the activity measured increased fourfold, indicating a greatly enhanced solubilization. Today tissues are sometimes even suspended in 2 *M* NaCl, increasing at least in some preparations the activity still further. A further increase may sometimes be observed by addition of Triton X-100. While in the electroplax under these conditions virtually the total concentration may be measured, there are most likely many preparations with much more complex structure with which even this treatment may not be adequate to solubilize all the enzyme and/or expose it to the substrate; hence the values obtained still may not represent the actual total activity of the enzyme present in the intact tissue (see also Nachmansohn, 1971, pp. 37–48).

A systematic study of the effects of various solubilization procedures on the assayable activity and on the release of AcCh-esterase from electric organ tissue of *Electrophorus* was recently made by Dudai and Silman (1974a). The procedures employed include homogenization in different ionic media, in the presence of detergents, and treatment by enzymes. Studies were performed with intact single electroplax as well as with homogenized suspensions of pieces of tissue.

AcCh-esterase activity of whole intact isolated electroplax was measured by the titrimetric method at pH 6.8. After addition of 3 m*M* phosphate buffer at the same pH the activity is enhanced two to four times. Addition of trypsin to the reaction mixture gradually leads to an approximately ten-fold increase in the observed AcCh-esterase activity. All the activity revealed by the tryptic treatment appeared to be due to solubilized enzyme, since almost no decrease in the rate of enzymic hydrolysis was observed when the cell was removed from the reaction mixture after the activity had reached plateau. When the electroplax was kept in an assay medium containing 1 *M* NaCl instead of 0.1 *M* NaCl, the enzyme activity was again observed to increase gradually to about the same level as after trypsin treatment. The level seemed to be even slightly higher, although the significance of this apparent increase is still open to question. Again, almost all the increase was due to the solubilization of the enzyme.

AcCh-esterase in homogenized suspensions of electric eel tissue was studied in a variety of media. The results of the total homogenized suspension were compared to those obtained with the supernatant fluid after centrifugation at 30,000 *g* for 20 min at 4°C. The data of a few representative experiments may be given as an illustration. The figures represent the AcCh-esterase activity in units per milliliter, one unit being defined as the amount of enzyme hydrolyzing 1 μmole of AcCh/min. The figures in parentheses are the corresponding data obtained in the supernatant fluid.

Homogenization media	Units
H_2O	48 (28)
5 m*M* EDTA	75 (47)
1 *M* NaCl, 0.05 *M* Tris buffer, pH 7.5	97 (75)
1 *M* NaCl, 0.05 *M* Tris buffer, pH 7.5, Triton X-100	158 (107)
0.05 *M* Tris buffer, pH 7.5, 0.5% sodium deoxycholate	181 (114)
0.05 *M* Tris buffer, pH 7.5, followed by incubation with 0.6 mgm/ml trypsin for 2 hours at 25°C	93 (83)

A special study was made of the effects of various Triton X-100–NaCl combinations on the solubilization of AcCh-esterase activity. By far the highest values were obtained when the tissue was homogenized in 1 *M* NaCl + 0.5% Triton X-100, centrifuged and the activity then determined in the supernatant fluid. The activity of solubilized enzyme was about 50% higher than with the use of 1 *M* NaCl without Triton X-100. A marked increase of solubilized enzyme was also observed when the homogenized suspension of electric tissue was incubated with 300 μgm/ml (final concentration) of collagenase, prior to centrifugation. Although Ca^{2+} ions were reported to enhance collagenase activity, the solubilization of AcCh-esterase was promoted when EDTA was added with collagenase. Ca^{2+} ions actually inhibit AcCh-esterase release by collagenase from electric tissue homogenates. The solubilizing effect of collagenase is of interest in view of the reports that this enzyme detaches AcCh-esterase from synapses and muscles (Hall and Kelly, 1971; Betz and Sakmann, 1971).

The effects of the various procedures on the molecular forms of AcCh-esterase have also been investigated and several interpretations are proposed. These aspects need not be discussed here, since they have no direct bearing on the topic of this section. The data illustrate the drastic effects of the various homogenization media, of enzymes and of the use of detergents on the activity of AcCh-esterase measured. They support the view of the author, emphasized for several years, that homogenized suspension in the commonly used buffer solutions do not provide information about the actual value of the enzyme activity present in a tissue and that a reevaluation of the figures obtained several decades ago has become a necessity. It must be emphasized that the experiments described above are limited to the electric tissue of *Electrophorus*. The effects of the procedures applied may greatly vary with the different types of tissues investigated (nerve cells and muscle fibers, ganglia, different parts of brains, neuroblastoma, etc.) and with different species in view of the wide range of differences of chemical composition and probably molecular organization. The realization that homogenized suspension does by no means reveal the total amount of enzyme activity present in a tissue is obviously of crucial importance for those problems for which quantitative evaluations are essential.

2. Evaluation of Enzyme Activity in Tissues after Exposure to Inhibitors

In view of the difficulties for a reliable quantitative evaluation of the total enzyme activity even in normal tissue, the question at what level of enzyme activity inhibitors of the enzyme block electrical activity cannot be answered at present, since exposure to inhibitors adds many additional factors and greatly complicates the problem. Some of them were recognized from the beginning, but the number of complicating factors revealed over the years has greatly increased and unexpected new ones are continuously emerging.

A correlation between the effects of inhibitors on electrical and on enzyme activity, one of the main problems in the present context and discussed on pp. 51–68, cannot be achieved with a reversible type such as physostigmine. In order to overcome the barriers preventing the inhibitor to reach the excitable membrane, as discussed before, it is necessary to apply the inhibitor in large excess. If the exposed tissue is homogenized for enzyme determination without removal of the excess inhibitor, i.e., without washing, all enzyme activity will be inhibited in the homogenized suspension. Some inhibitors will reach the enzyme much more readily even in the suspension than the substrate: physostigmine, e.g., is a tertiary amine whose conjugate acid has a dissociation constant of 8×10^{-9}; at pH 6 the molecule is almost exclusively a cation, whereas at pH 10 it is predominantly neutral. In contrast, AcCh is a cation charged at every pH. Moreover, many of the inhibitors have a dissociation constant several orders of magnitude lower than the substrate. Low inhibitor concentration applied to the tissue externally will have, under these conditions, strong effects in homogenized suspensions. Removal of the excess by first washing the tissue must, on the other hand, result of necessity in total restoration of the initial activity in view of the reversible nature of this inhibition (see p. 52).

When organophosphates became available it seemed that their use might permit one to determine the minimum level of enzyme activity required for electrical activity: since the enzyme is irreversibly blocked, removal of the excess at the time of the block may provide this information. In the beginning this hope was shared by the author. At that time, i.e., almost three decades ago, very little was known about the complexity of the problem and the serious obstacles to overcome in order to provide a satisfactory answer. But the knowledge accumulated in this field since then, particularly in the last few years, necessitates a complete revision of the assumptions made at that time and makes it obvious that many data reported in the literature, including those by the author, have become obsolete.

It would be difficult, and it appears unnecessary for the particular problem of the interdependence between electrical and enzyme activity, to discuss in some detail the many complex problems of the chemistry, the toxicity, and the biological and physicochemical properties of organophosphates. Hundreds of organophosphates have been prepared and investigated from a variety of aspects. They differ greatly in their properties and in their reactions with different types of esterases. A bewildering amount of literature has appeared in this field during the last 20 years in view of the theoretical interest as well as the practical implication of their use as insecticides. In the early period of investigation of their pharmacological actions, little was known about the many factors involved, and many effects were poorly or not at all understood. It was difficult to avoid pitfalls and errors. Even today, there exist many differences of opinion and interpretation. A huge amount of data on inhibitors of AcCh-esterase including organophosphates may be found in the "Handbook of Pharmacology," edited by Koelle (1963a). Although many additional data have been obtained since then, and even if one disagrees with many views presented there—as the writer does—the book is still today a most important and valuable source of information. Critical discussions about the many attempts of correlating the action of organophosphates with toxic effects may be found in the monograph of O'Brien (1960). Some more details specifically limited to the reactions of organophosphates with AcCh-esterase may be found in a recent summary (Nachmansohn, 1971, pp. 42–48) and will be more extensively treated in the new monograph (D. Nachmansohn, in preparation).

For almost three decades attempts were made to test the effect of organophosphates on AcCh-esterase in axons in relation to electrical activity. After exposure to the various organophosphates the fibers were washed in saline and the enzyme activity was then tested in a homogenized suspension. As discussed before, this procedure turned out to be unsatisfactory for a quantitative test even in a normal tissue without any exposure. Among the many additional complications resulting from the exposure to organophosphates is the difficulty, or even impossibility, of complete removal of the organophosphates accumulated in the tissue exposed either by prolonged washing, by extraction with organic solvents, or by various other procedures tried so far. This fraction may have been located in the interstitial tissue, Schwann cell, blood vessels, capillaries, fat deposits, or even the axon interior and may still never have acted *in vivo* on the enzyme located in some parts of the excitable membrane. We know that even in the excitable membrane of the electroplax, so exceptionally rich in enzyme, this protein forms only a very small fraction of the total membrane, only a few percent, and some of its active sites may not be readily reached by external application, even if some compounds enter the interior. For instance, using some particularly potent organophosphates, with a K_I of about 10^{-10} *M*, Hoskin *et al.* (1969) had found that they had penetrated into the axoplasm of squid giant axons. No AcCh-esterase activity was detected in the homogenized tissue prepared after prolonged washing. when experiments were performed to test the preparation for residual inhibitor, it was found that about 99% had been removed by the washing. However, the fraction of 1% or less of the organophosphate used still remaining in the tissue was in one case 5 and in the other 50 times as high as the concentration required to inhibit all the enzyme activity in view of the high potency of the inhibitor. The compounds applied blocked electrical activity irreversibly, but only after treatment with phospholipase A, clearly an indication that *in vivo* the enzyme was protected against the action. It took many years before the implications of this "homogenization artifact," as it was referred to by Van Asperen (1958), were fully and widely recognized. Even when procedures should be developed which will permit a reliable test of the total activity in a homogenized tissue, the complication of removing the residue once the tissue is exposed to organophosphates will remain. This additional obstacle must be overcome before the problem of the minimum level of AcCh-esterase required for electrical activity can be resolved. In the light of this information, the many reports which have appeared over decades, claiming that *all* the enzyme activity was destroyed in fibers exposed to organophosphates, whereas electrical activity was still normal, have lost their validity and do not require any discussion. But also the determinations, in which AcCh-esterase activity was found in the axons after exposure to organophosphates (see, e.g., p. 66) cannot be considered anymore as a quantitative indication of the total activity actually remaining or of the minimum requirement of the enzyme for electrical activity.

Another factor interfering with the establishment of a correlation between electrical and enzyme activity and not recognized in the early phase by this type of investigation, is the complication introduced by the use of multifiber preparations. In a preparation of several thousand axons such as, e.g., the frog sciatic nerve, the large amounts of interstitial tissue, Schwann cells, myelin, and blood vessels may retain quite considerable amounts of extracellularly applied organophosphates which react with the membrane-bound enzyme only after homogenization. This source of error may be smaller in a preparation formed by a few axons or even a single one, because they are less protected by extracellular structures. The sources of error will be even greater when the inhibitor is injected into the whole animal and the intact tissue is taken out subsequently, since in such a case there is virtually no control of the concentrations of inhibitor to which the tissue was exposed in the intact animal. The effects on electrical activity are also less ambiguous with a single axon than with a multifiber preparation. The inhibitor may reach the excitable membranes of the many fibers of such a preparation at different rates so the effects may not proceed simultaneously. Under such conditions the relationship between electrical parameters and enzyme activity cannot be properly evaluated even when procedures do become available for reliable quantitative tests of the enzyme activity before and after exposure. A more detailed discussion of this particular aspect may be found in the paper by Dettbarn and Rosenberg (1962).

A few remarks may be appropriate about a recent report of Kremzner and Rosenberg (1971). Realizing some of the limitations of determining AcCh-esterase activity in homogenized tissue for correlating enzyme and electrical activity, the authors used a sensitive radiometric assay and acetyl-β-methylcholine as substrate to measure AcCh-esterase activity in intact squid giant axons. However, their assay gives values comparable to those obtained with homogenized tissue, not treated with the now available techniques which greatly increase the amount of solubilized enzyme in the homogenized suspension and markedly increase the activity measured. As we know today, high ionic strength (1 *M* NaCl) solubilizes the enzyme of a homogenized suspension both in a homogenized suspension and in membrane suspensions, and a greatly higher activity is measured (see p. 297). When, in addition, Triton X-100 is applied, the activity obtained increases still further to a considerable degree. For experiments aimed at a really quantitative data of the total amount present, such treatment of the homogenized suspension should be tried and only then the values compared with those obtained on the intact fiber. Another useful control may be the use of much more lipid-soluble substrates. The authors used instead of AcCh, acetyl-β-methylcholine. This is a rather poor substrate of AcCh-esterase even in solution, although it seems to be more lipid soluble than AcCh. A poor substrate is usually a drawback for quantitative evaluation with activities on the borderline and a method with a high standard error. In recent experiments of Dudai and Silman (1974a) it was found that indophenol acetate is a very good substrate of AcCh-esterase and its use increases the assayable enzyme activity of the membrane-bound enzyme by 50%. Thus, it appears questionable whether the new assay really indicates the total amount of enzyme present or, what appears more likely, indicates only a fraction. Moreover, the high standard errors, about 10–20%, are much too high for a reliable quantitative estimate of the minimum requirement of enzyme activity for electrical activity. Without going into the details of the results reported, it seems to the author that their data form, in general, one of the strongest supports for the view that organophosphates irreversibly block electrical activity by irreversibly blocking AcCh-esterase activity. For instance, after exposure to DFP in 5×10^{-4} M for 1–5 min in four experiments irreversible block of electrical activity was obtained, but the assayable enzyme activity was still 30% of the control. The actual total amount may have been much higher. In four experiments in which DFP was used in 1×10^{-2} M concentration followed by exposure to PAM, 45% of the assayable enzyme activity was found and electrical activity was restored; in other experiments PAM failed to restore electrical activity. The difficulties involved in this type of experiment were discussed before, but it is remarkable that the authors found such a high activity of the assayable enzyme and restored electrical activity after exposure to high DFP concentrations for periods of time which without PAM regularly blocked electrical activity irreversibly. With the tertiary analog of phospholine $(C_2H_5O)_2P(O)SCH_2CH_2N(CH_3)_2$ almost all the assayable enzyme was inactivated without block of electrical activity, but when the axon was first exposed to phospholipase A (cottonmouth moccasin venom) in two experiments electrical activity was irreversibly blocked, but the activity of the assayable enzyme, most likely only a part of the total, was still 7% of the control; when phospholine after the treatment failed to block electrical activity, the assayable enzyme activity was still 23%. In view of the high standard error of the technique and the uncertainty what fraction of the total enzyme activity has been assayed with this method, it is obvious that this method is unable to provide a quantitative evaluation of the precise level of enzyme activity at which electrical activity is blocked.

While it was essential to establish a link between electrical and enzyme activity by the use of either reversible or irreversible inhibitors of AcCh-esterase, blocking conduction either reversibly or irreversibly, the precise minimum level of enzyme activity appears to be not really relevant to the central question of whether or not the AcCh cycle controls the ion permeability in the gateway. It is well known that in living cells most enzymes are present in excess, usually about 3–6 times. Once the essential role of AcCh-esterase has been established, it appears of secondary interest to know the precise excess of enzyme present. The amount of excess enzyme may, and probably does, vary in different types of membranes, or even in the same neuron at its different membrane sites (dendrites, cell body, axon, terminal).

C. *Effects of Specific Inhibitors of the AcCh-Receptor in Electrical Activity*

1. CURARE

For more than a century Claude Bernard's famous experiments with curare blocking transmission in the frog from nerve to muscle, but leaving conduction in nerve and muscle unaffected, seemed to support the assumption that special conditions may prevail at the junction. As is generally accepted today, curare (*d*-tubocurarine) is a potent and specific inhibitor of the AcCh-receptor. The failure of curare to affect conduction was, as in the case of AcCh, considered as one of the pillars of the hypothesis that AcCh is a neurohumoral transmitter and has no role in conduction.

As discussed before, almost all conducting membranes are strongly protected by structural barriers, particularly against poorly lipid-soluble compounds such as AcCh and curare, which are quaternary ammonium derivatives. But when Dettbarn (1960b) applied curare to single unmyelinated axons of frog sciatic nerve fibers and measured the electrical activity at the nodes of Ranvier, as described before for his tests with physostigmine, he obtained rapid and reversible block. Even at the nodes of Ranvier the covering structure varies considerably in different fibers (see Fig. 16). This structure is again less pervious to quaternary than to tertiary derivatives. Neostigmine, with a K_I similar to that of physostigmine, must be applied in tenfold-higher concentrations than the latter compound in order to obtain comparable results. It is, therefore, not surprising that *d*-tubocurarine, a rather large molecule (MW about 800) with two positive charges, must be applied in 10^{-5} *M* concentrations, and sometimes in even still higher concentrations being dependent apparently on the thickness and surface area of the covering layer. Moreover, as already mentioned when the Schwann cell covering the squid giant axon is exposed for short periods of time to small amounts of phospholipase A, AcCh and curare penetrate into the interior of the axon and block electrical activity, indicating that the AcCh-receptor, as well as the esterase, are present and essential for conduction.

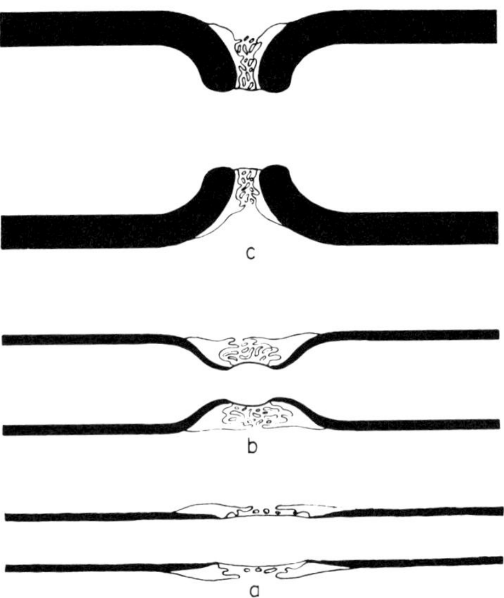

FIG. 16. Diagram illustrating the arrangement of the collar of nodal processes of Schwann cells about nodes of Ranvier in fibers of different sizes. In the smallest fibers the collar is relatively thin as in (a) and (b), but it is spread over a relatively greater length. In larger fibers, the thickness of the collar is much greater, but its extent along the axon is reduced (c). (Robertson, 1960b).

2. "LOCAL ANESTHETICS"

A series of compounds that are close analogs of AcCh in structure have long been known as "local anesthetics" (e.g., procaine, tetracaine). They differ from AcCh in two respects: (1) they are receptor inhibitors, i.e., they block electrical activity without deplarizing the membrane; (2) they are able to reach the receptor not only in the membranes of the junctions, but also in the conducting parts of the membrane, owing apparently to their greater lipid solubility which permits them to penetrate structural barriers protecting conducting membranes (see p. 287). The electroplax preparation permits a quantitative evaluation of the effects of the molecule substitutions that transform AcCh step by step into receptor inhibitors that act on junctional as well as on conducting parts of the membrane.

A systematic analysis of a great number of compounds in which a group of either the quaternary ammonium or the acyl part of AcCh has been substituted, has provided information about the molecular groups that transform one type of compounds into the other type (Bartels, 1965; Bartels and Nachmansohn, 1965). Table II gives some of

TABLE II

TRANSFORMATION OF ACETYLCHOLINE EFFECTS BY SUBSTITUTIONS[a]

Compound	Synaptic junctions: Activator	Synaptic junctions: Inhibitor	Conducting membrane
	Concentration (M)		
$CH_3-\overset{\oplus}{N}(CH_3)_2-CH_2-CH_2-O-\overset{\oplus}{C}(CH_3)-O^{\ominus}$ Acetylcholine	2.5×10^{-6}	0	0
$CH_3-\overset{\oplus}{N}(CH_3)_2-CH_2-CH_2-O-\overset{\oplus}{C}(C_6H_{11})-O^{\ominus}$ (cyclohexyl ring) Hexahydrobenzoylcholine	5×10^{-4}	0	0
$CH_3-\overset{\oplus}{N}(CH_3)_2-CH_2-CH_2-O-\overset{\oplus}{C}(C_6H_5)-O^{\ominus}$ (benzene ring) Benzoylcholine	5×10^{-4}	5×10^{-4}	1×10^{-3}
$CH_3-\overset{\oplus}{N}(CH_3)_2-CH_2-CH_2-O-\overset{\oplus}{C}(C_6H_4NH_2)-O^{\ominus}$ (benzene ring with para NH_2) *p*-Aminobenzoylcholine	0	1×10^{-3}	2.5×10^{-3}

[a] Substitutions of AcCh gradually transform the molecule from a receptor activator acting on the synaptic junctions only into a receptor inhibitor blocking electrical activity both of junctions and of the conducting membrane ("local anesthetic"). Benzoylcholine is a transitory form, both in structure and in biological activity. For details, see text.

TABLE II (*Continued*)

Compound	Synaptic junctions: Activator	Synaptic junctions: Inhibitor	Conducting membrane
	Concentration (M)		
$C_2H_5\overset{\oplus}{-}N(C_2H_5)(H)-CH_2-CH_2-O-\overset{\oplus}{C}(-O^{\ominus})-C_6H_4-NH_2$ Procaine	0	2.5×10^{-4}	5×10^{-4}
$CH_3\overset{\oplus}{-}N(CH_3)(CH_3)-CH_2-CH_2-O-\overset{\oplus}{C}(-O^{\ominus})-C_6H_4-NH-C_4H_9$ Tetracainemethiodide	0	2×10^{-5}	1×10^{-5}

the most characteristic features of the transformation of the biological action in relation to chemical structure. When the methyl group on the carbon of the carbonyl groups, for example, is substituted by a saturated benzyl ring, the potency of the compound as an activator is decreased about two-hundredfold as compared to that of AcCh; but the compound is still an activator, and the action is limited to the junction. Even in high concentration it does not act on the receptor of the conducting membrane, apparently because it is unable to reach it. However, such a small modification as a substitution of a saturated by an unsaturated benzyl ring (i.e., benzoylcholine) changes the properties of the molecule. It may act as a receptor activator or as an inhibitor according to the experimental condition used, and in concentrations of 10^{-3} M, benzoylcholine acts also on the conducting membrane. Thus, the compound is a typical transitory form in chemical structure and in biological action between AcCh and local anesthetics. Addition of a single amino group to the phenyl ring in para position transforms the analog into a receptor inhibitor and increases the penetrating ability; the molecule has acquired the two features characteristic of a local anesthetic. Both inhibitory potency and ability of penetration may be increased by further substitutions in either the quaternary or in the acyl part of the molecule.

These experiments lend further support to the presence of the receptor in the conducting parts of the membrane, since these analogs of AcCh act similarly as receptor inhibitors in both junctional and conducting parts of the electroplax membrane. The usual type of competitive action between local anesthetics and carbamylcholine or curare can only be demonstrated at the junction, because these quaternary compounds, when applied externally, do not reach the receptor in the conducting part; there the action is limited to the competition with internally released AcCh. Moreover, since it is known that local anesthetics block all electrical activity in all types of conducting fibers, the observations with these receptor inhibitors, typical "antimetabolites," offer evidence that the receptor is not only present but essential for electrical activity.

It has been suggested in several reports that local anesthetics affect electrical activity by reacting with phospholipids and that their effects may be potentiated or antagonized by Ca^{2+} ions. While reactions with other compounds are certainly possible, the question arises as to which particular reaction the block of electrical activity must be specifically attributed. Almost all "specific" inhibitors are not 100% specific; they react with a number of other components present in the cell. The effect of *d*-tubocurarine, e.g., is generally contributed to the competition of AcCh for the receptor; however, it is well known that *d*-tubocurarine reacts with a number of negatively charged molecules by coulombic and van der Waals' forces, but that the binding is usually weaker than in the case of the receptor. Once the typical competitive nature between the action of curare and carbamylcholine and that of tetracaine has been unequivocally demonstrated, at the junction of the electroplax, the only part of the membrane where such experiments can be performed (Podleski and Bartels, 1963), the blocking effect must obviously be attributed to the action on the same cell constituent. This conclusion is further supported by the observation that tetracaine acts on both synaptic and conducting parts of the membrane in similar concentrations. The slightly lower concentration of tetracaine required for block of the conducting part of the membrane as compared to that at the junction, 3.3×10^{-5} *M* and 5×10^{-5} *M*, respectively, may be attributed to a lower density of basic excitation units in the conducting part, or to the fact that the internally released AcCh in the conducting part requires slightly lower concentrations than the competition with externally added *d*-tubocurarine or carbamylcholine. In the context of all the other data it appears unacceptable to attribute the effect of the "local anesthetics" on electrical activity to the reactions with other cell constituents, and it

TABLE III
PERMEABILITY BARRIERS PROTECTING CONDUCTING (AXONAL) PARTS OF EXCITABLE MEMBRANES AGAINST THE ACTION OF QUATERNARY (BUT NOT OF TERTIARY) ANALOGS OF AMMONIUM DERIVATIVES[a]

Compound	Axon (squid)		Synapse (electroplax)
	Untreated	After exposure to phospholipase A	
	Tertiary Compounds		
Atropine	2×10^{-3}	3×10^{-4}	3×10^{-4}
Physostigmine	7×10^{-3}	1×10^{-3}	0.7×10^{-3}
Procaine[b]	3×10^{-3}		1×10^{-3}
Dibucaine[b]	3×10^{-5}		3×10^{-5}
	Quaternary Compounds		
Acetylcholine	$(\geq 10^{-1})$	2×10^{-4}	3×10^{-6}
d-Tubocurarine	$(\geq 10^{-2})$	3×10^{-5}	3×10^{-6}
Benzoylcholine	2×10^{-2}		1×10^{-3}

[a] Data taken from Rosenberg (1966).
[b] Local anesthetics (analogs of acetylcholine).

is not surprising that the reports claiming this mode of action are full of contradictions when data of various laboratories are compared (see, e.g., Nachmansohn, 1973, pp. 32–117). Table III illustrates with a few representative compounds, the differences and similarities between the actions of tertiary and quaternary nitrogen derivatives on the axonal membranes of squid axon and the junctional parts of the monocellular electroplax preparation (Rosenberg, 1966). [Exposure of axons to phospholipase A (in a few micrograms for a few minutes) strongly reduces the protection by the barriers. Apparently no such barriers exist at synaptic parts of many excitable membranes. The squid giant axon has been used to test the barrier-reducing effect of phospholipase A in conducting (axonal) parts. This axon is unmyelinated; however, it is covered, as are all axons, by a layer of Schwann cells, which in this particular axon is 4000 Å thick. Schwann cells are rich in phospholipids. Reliable values for testing the effects of compounds on the synaptic parts can be obtained from the monocellular electroplax preparation. The molar concentrations of a few representative tertiary compounds are given, at which impulse conduction is blocked (without depolarization) in the axons, before and after exposure to phospholipase A. Many quaternary ammonium derivatives, such as AcCh and *d*-tubocurarine (in the concentrations given in brackets), did not affect conduction at all. However, after exposure of the axon to phospholipase A the concentrations of AcCh and *d*-tubocurarine are close to the concentration values required to block impulse transmission across the synaptic junction.] It would be difficult to construe these figures as support of the assumption of an entirely different role of the AcCh cycle in conducting and synaptic parts of the membrane.

Recently it has been found that tetracaine and [^{3}H]AcCh act in a typical competitive way when tested on membrane fragments of *Torpedo* (Weber and Changeux, 1973). When membrane fragments of *Electrophorus* were used for testing the competitive action of several local anesthetics, the effects were interpreted not to be as strictly competitive as with *Torpedo* fragments. As just discussed, Podleski and Bartels (1963), testing tetracaine and other local anesthetics, found also a *competitive* type of action between tetracaine (or curare) *exclusively* at the synaptic junction. In *Torpedo* electroplax a very large fraction of the excitable membrane is synaptic, in contrast to that of *Electrophorus*. It is, therefore, not surprising that membrane fragments of *Torpedo* are a much more favorable material than those of *Electrophorus* (formed by conducting membrane parts) for testing competitive action between receptor activators and competitive inhibitors. The double reciprocal plot applies strictly only for tests of competitive action *in solution*. Extrapolation to tests on complex structures is always difficult and risky if not impossible, as is now well known. This is particularly true when the structure used contains lipid and when lipid-insoluble versus lipid-soluble compounds are tested. The data of Weber and Changeux seem rather to support than to contradict the view that local anesthetics are receptor inhibitors. Their main effect on electrical activity can be attributed to their reaction with the AcCh-receptor, even when in tests on lipid-containing structures the effects show deviations from a strictly competitive action. It appears premature at present to advocate definite statements about a different mode of action until more is known about the chemical reactions involved and the modifications caused by the molecular organization of membranes.

There are some preparations in which even a direct action of AcCh on the electrical activity of the axonal membrane may be found. One such preparation is, e.g., the axons of the walking leg of lobster. An extended series of careful experiments on the action of AcCh and related compounds was performed by Dettbarn and Davis (1963). AcCh applied externally to those areas first markedly prolongs the descending phase of the action potential and then blocks electrical activity and depolarizes the membrane. Under proper conditions this effect is antagonized by atropine. Dettbarn (1963) was even able to demonstrate an effect on curare on these axons. This preparation is usually tested in seawater, which has a high concentration of Ca^{2+} and Mg^{2+} ions. Since these ions are known to antagonize the action of curare, Dettbarn reduced the concentration of Ca^{2+} and omitted Mg^{2+} ions in the test solution. Under these conditions he obtained marked and rapid effects on the electrical activity with curare (*d*-tubocurarine); the effects were readily reversible. Apparently the Schwann cell in these axons does not offer the complete protection which it does in other preparations. This assumption is supported by the findings with electron microscopy (De Lorenzo *et al.*, 1968). When the axons were exposed to cottonmouth moccasin venom prior to the application of AcCh and related structures, the potency of the compounds increased 20 to 50 times, indicating the existence of a barrier that is reduced by phospholipase A.

Another preparation which reacts to AcCh is the vagus nerve of rabbit (Armett and Ritchie, 1960). After the removal of the sheath surrounding these axons, AcCh affects electrical activity. This effect was considered to be a pharmacological curiosity (Ritchie, 1963).

Recently, it was found that fibers of certain neuroblastoma cells produce action potentials both upon electrical stimulation and upon electrophoretic application of AcCh (Harris and Dennis, 1970; Nelson *et al.*, 1971; Hamprecht, 1974).

3. Oxygen, Sulfur, Selenium Isologs

The introduction of these isologs by Mautner, discussed before, has proved to be a powerful tool for the analysis of the properties of AcCh-esterase and -receptor. These isologs offer also a possibility for testing similarities or differences of their reaction with the AcCh-receptor in junctional and conducting parts of the membrane. If, as appears likely, the active sites of the receptor protein are similar in both parts of the membrane, the potency of oxygen, sulfur, and selenium isologs applied to the parts should exhibit comparable differences, provided the compound is able to penetrate through the barriers surrounding the conducting part and there reach the receptor. When 2-dimethylaminoethylbenzoate and its sulfur and selenium isologs (the ether oxygen was replaced by sulfur or selenium) were tested on the electrical activity of the electroplax membrane and on the squid giant axon, the differences in potency obtained with these three isologs on the two preparations were found to be remarkably similar (Rosenberg *et al.*, 1966). Similarly marked differences of potency were obtained in the reactions with the receptors of both preparations when the carbamyl oxygen was replaced by sulfur and selenium (Rosenberg and Mautner, 1967). The data offer a new kind of support for the presence and functional similarity of the AcCh-receptor in conducting and junctional parts of the excitable membrane.

D. Involvement of Other Proteins in Excitability

The effects of reagents reacting with SH and S—S groups of the AcCh-receptor has been discussed before. In view of the important role of these groups in many proteins it appears *a priori* highly unlikely that such reagents acting on excitable membranes would be limited to the reaction with the AcCh-receptor protein or the other proteins processing AcCh. Even when this fraction is high, let us assume the certainly exaggerated figure of 20%, many more proteins will almost certainly be involved in the process of conduction, e.g., proteins in the gateway, proteins in the close vicinity of the proteins of the AcCh cycle or even more removed proteins which, as is well known today, may have far-reaching effects by allosteric action, cooperativity, and regulatory actions. As mentioned before, effects on axonal conduction by agents reacting with SH and S—S groups have been described long ago, before their reaction with the AcCh-receptor was studied.

Recently an interesting effect of this type of compounds on squid giant axons was described by Marquis and Mautner (1974). A strong acceleration of the block of conduction by repeated, brief stimulation of the nerve fiber was observed after exposure to the following compounds: mercuric chloride, *p*-chloromercuribenzoate, mercurochrome, and fluorescein mercuric acetate, which affect the action potential, but whose effect is greatly enhanced by stimulation. For instance, after exposure to 5×10^{-6} M *p*-mercuribenzoate for 10 min the axons become completely inexcitable after stimulation at 6 pulses/sec for about 15 sec, while without stimulation the decrease in action potential amplitude amounts only to 10–20% and no further inhibition is observed after exposure of one hour without stimulation. The effects are reversible with β-mercaptoethanol. *N*-Ethylmaleimide, 2-dimethylaminoethylselenolbenzoate, and 5,5′-dithiobis(2-nitrobenzoic acid), a thiol oxidizing agent (Ellmans reagent) also exhibit the stimulation effect, but these blocks are irreversible. Iodoacetate and iodoacetamide block nerve conduction, but stimulation has no effect on their action.

Three possible interpretations of these effects of stimulation are suggested: altered permeability, unmasking of buried SH groups in the membrane, and electrolytic reduction of disulfides. While permeability increase by stimulation is still an open question in view of the structural complexity of the preparation, the two other possibilities discussed seem to the author most attractive. It has been known for some time that electrolytic reduction of disulfides to thiols may occur in peptides (Dohan and Woodward, 1939) and proteins (Cecil, 1963; Leach *et al.*, 1965). A rearrangement of *intra*chain disulfides to *inter*chain disulfides has been recently postulated to take place in the association of trypsin and pancreatic trypsin inhibitor (Vincent and Lazdunski, 1972). Pontremoli and Horecker (1970) have suggested that disulfide exchange may be an important regulatory factor in the regulatory mechanism of fructose diphosphatase; a reversible inactivation of glycogen synthetase D by sulfhydryl–disulfide exchange has been recently described by Ernest and Kim (1973).

Another series of experiments may be mentioned in this context, namely the microelectrophoretic application of compounds reacting with SH and S—S groups to the synaptic part of the excitable membrane of the electroplax, using the monocellular preparation, by Del Castillo and Bartels (Del Castillo *et al.*, 1972). The evidence of disulfide bonds in the vicinity of the anionic site of the receptor has been discussed before. Experiments applying reducing and oxidizing, as well as alkylating, agents were performed on the motor endplates of frog muscle with the aid of the microelectrophoretic technique (Del Castillo *et al.*, 1971). The results obtained in both preparations were, in general, in good agreement in both preparations, as one would expect in view of the fundamentally similar properties of excitable membranes when tested at junctions. There was, however, one notable and interesting difference between the results obtained by the two different techniques. The experiments on the electroplax using the standard technique had indicated that in the AcCh-receptor the SH groups are in the disulfide form when the protein responds to AcCh-receptor activators, such as AcCh, carbamylcholine. The experiments on the frog motor endplates using the electrophoretic technique indicated that the activation of the AcCh-receptor is apparently accompanied by a partial oxidation of the protein, suggesting that the presence of reduced SH groups may be required for activation: receptor activation had been achieved by oxidizing SH groups when oxidizing agents were applied, such as *o*-iodosobenzoate, organic mercurials and divalent metal ions, particularly Hg^{2+} ions. All these compounds produce a rapid transitory change of membrane potential (depolarization) quite similar to that obtained with AcCh. In contrast, when these compounds are added to the bath, even in relatively low concentrations, they act essentially as inhibitors. This discrepancy was discussed in terms of the difference of the techniques used in the application, rather than to a basic difference of a membrane property.

In the evaluation of the differences of effects observed between standard and electrophoretic techniques, it is stressed that the high potency of standard techniques with some compounds, which are active at micromolar concentrations, is due to their high chemical affinity to the receptor, resulting in a high number of molecular interactions in a short period of time and in their high efficacy, i.e., their special ability to induce activation of the receptor. Microelectrophoretic application, on the other hand, assures that high concentrations of the compound tested can be rapidly built up in the environment of a relatively small number of the membrane constituents reacting with the test compounds. When used with potent agents, the electrophoretic technique has the advantage of allowing a great spatial and temporal resolution. Thus, this technique is helpful to demonstrate interactions between different compounds. Del Castillo *et al.* (1972) emphasize most appropriately that it *does not lend itself well to quantitative studies,* an aspect frequently ignored in the literature.

The electrophoretic technique is moreover of great value in the study of the reactions of the membrane to compounds that have either a very low affinity for chemical groups and functions found not only in the receptor but also in many other proteins, including enzymes. Neither of these two types of compounds can be readily studied by adding them to the bath. The first group will act only at high concentrations which may exert a variety of unspecific effects unrelated to the receptor, the second group may produce widespread effects difficult to interpret in terms of specific receptor interactions. The electrophoretic application may be used to bring these compounds directly to the receptor environment while keeping the concentration in the bulk of the solution transiently at a virtually zero level. As an illustration, the effect of *p*-chloromercuribenzoate may be mentioned, adding it to the bath in 0.5 m*M*. The most prominent effect of this compound (0.5 m*M*) is an inhibition of the receptors to receptor activators. This inhibition can only be reversed chemically. However, electrophoretically applied to a locally limited membrane area, it produces a rapid and reversible depolarizing action similar to that of AcCh.

Similarly, adding *o*-iodosobenzoate, in 50 μM, to the bathing solution restores the sensitivity of the electroplax to reactions with AcCh-receptor reduced by dithiothreitol, while at much higher concentrations it has no effect on the resting potential (E. Bartels, unpublished observations). This is again in contrast to the electrophoretically produced action resembling that of AcCh.

The method of electrophoretic application of any ionic test compound is certainly of great value, but it has also limitations. In this context it may be appropriate to comment on the "failure" of microelectrophoretically applied K^+ ions to affect the frequency of miniature endplate potentials (mepps). Such a "failure" is considered by Rosenberry (1975) as a difficulty for the integral model of bioelectricity. In this model it is suggested that mepps may be reflections of the continuous subthreshold activity of the cholinergic control system in excitable membranes.

It should be realized that at present there is nothing known about a correlation between the frequency of mepps and microelectrophoretically (i.e., locally) applied K^+ ions. It appears likely, however, that small amounts of locally injected K salt will rapidly diffuse from the site of injection before the membrane potential is changed by electrodiffusion across the membrane. Thus, the effect of locally applied AcCh, which *interacts* with membrane components, is certainly not comparable to those which may be caused by transient local changes in the K^+ ion concentration.

The ability of K^+ ions, which were injected close to the spatially limited neuromuscular junction, to change the membrane potential of the postsynaptic membrane just as by nerve stimulation, has been known for decades. The rapid and transient effect of K^+ injection is widely considered as evidence of a release of AcCh by the injected K^+. The relatively short-lived potential changes indicate a rapid inactivation of AcCh by an esterase, since K^+ ions alone induce a rather prolonged change in membrane potential.

It is recalled that the integral model does not claim that the mepps *are* due to the cholinergic system. The model suggests a possibility—gives a proposal which is specific enough to be explored experimentally. On the other hand, an alternative hypothesis is the correlation of mepps with presynaptic vesicles. This interpretation does, however, not account for the experimental observation that denervated muscles (i.e., in the absence of presynaptic vesicles) also show mepps.

In this context it is appropriate to recall that the integral model does *not* attribute to K^+ ions a transmitter role (as stated, e.g., by Rosenberry, 1975) as may be the case for epinephrine, or GABA, or others. In the description of the integral model it is mentioned that any working hypothesis for the investigation of the information transfer across synaptic gaps should be based on an experimental fact and not on an assumption. The neurohumoral transmitter hypothesis for AcCh *assumes* that AcCh crosses the synapse, whereas the integral model points at the measurable accumulation of K^+ ions close to the gap during nerve activity. The transient increase in the K^+ concentration in the spatially limited gap volume may well lead to changes in the potential of the postsynaptic membrane which then, in turn, may affect the cholinergic control system.

The discussion of the differences owing to the use of two different types of techniques is very lucid and illuminating. It is in no way a contradiction to the views presented when the author would like to raise the question of whether the SH oxidizing agent may act in a way which has not been contemplated. The protein referred to by Chang (1974) as AcChR-II may be the storage protein. In contrast to the AcCh-receptor, this protein appears to react with SH oxidizing agents. Thus it appears possible that compounds, such as *o*-iodosobenzoate and *p*-chloromercuribenzoate, may lead to a release of AcCh from the storage protein and permit the translocation of AcCh to the receptor protein, as suggested in the AcCh cycle (see also Supplement II). The effect on the electrical parameters would be quite similar. This idea obviously requires further investigations.

Some compounds tested previously only with the standard techniques on the electroplax by adding them to the bath, were applied electrophoretically. One of these compounds is choline disulfide, which exerts a reversible curarelike action on the electroplax (Bartels *et al.*, 1970). The same result was obtained with electrophoretic application. The other compound is the photochromic substance bis-Q, described by Bartels *et al.* (1971). It may have stimulating and blocking effects. Again the same results were obtained with both techniques.

E. Parallelism between Chemical Action on the AcCh-Receptor of Isolated Membrane Fragments of the Electroplax of Electrophorus and the Electrical Stimulation of their Intact Membrane

1. Isolated Fragments of Excitable Membranes (Microsacs)

Changeux *et al.* (1969) have prepared isolated fragments of excitable membranes which permit *in vitro* studies of a variety of properties of these membranes and of the proteins located there. The membranes are prepared in the following way. Homogenized suspensions of electric organ of *Electrophorus electricus* are sonicated and centrifuged in sucrose gradients. A particulate fraction rich in AcCh-esterase contains membrane fragments free of cytoplasm. The membranes form closed microsacs or vesicles. Most of them are derived from excitable membranes and are rich in AcCh-esterase. Some of the microsacs, although only a minor fraction, are parts of the nonexcitable membranes; they contain ATPase. The two types differ not only by their chemical but also by their structural features as found by electron microscopy.

The microsacs are incubated over night in a ^{22}Na-containing medium. The suspension is then diluted into a nonradioactive medium. The ^{22}Na content of the microsacs as a functon of time is then followed by rapid filtration on Millipore filters. The rate of ^{22}Na efflux is increased in the presence of AcCh-receptor activators (cholinergic "agonists") and blocked by receptor inhibitors such as *d*-tubocurarine or tetracaine. Thus the preparation permits to analyze the action of these compounds *in vitro* in a cell-free medium in a well-defined environment. The use of these isolated membranes on the effects of excitation by cholinergic agents on ion fluxes has provided pertinent information and may lead to further advances, especially when the techniques will be further improved.

2. Chemical Stimulation of Microsacs and Electrical Stimulation of Axons

An important result of these observations is the remarkable parallelism found with receptor activators and inhibitors when the effects on the membranes *in vitro* were compared with those obtained on steady-state potentials of the intact electroplax (Fig. 17) (Kasai and Changeux, 1970). The electrophysiological data are taken from Changeux and Podleski (1968). The same affinities are observed with the two activators used, carbamylcholine and decamethonium, for controlling the ion effluxes from the microsacs and the electric response in the intact cell. Even the sigmoid shape of the dose-response curve is similar in both systems, indicating cooperativity and allosteric effects. The Hill coefficients for AcCh (in the presence of eserine), carbamylcholine, decamethonium, and phenyltrimethylammonium were found to be the same *in vivo* (intact cells) and *in vitro* (microsacs).

As was discussed before, exposure of the electroplax to dithiothreitol (DTT) reduces S—S bridges in the neighborhood of the active site of the AcCh-receptor and modifies the response to chemical excitation, sometimes transforming receptor inhibitors into activators or vice versa. Similar effects were obtained with the isolated membrane fragments. After *in vitro* exposure of the microsacs to 10^{-3} *M* DTT for 15 min, the response to carbamylcholine is decreased, whereas that to hexamethonium before exposure to DTT and inhibitor, is increased.

Among the other observations with the microsacs the effects of pH and temperature appear of particular interest and may be briefly mentioned. The rate of Na efflux, $1/\tau_0$ min^{-1}, increased fourfold between pH 5.5 and 8.5, while the excitability, $(\tau_0/\tau) - 1$, remained constant; τ and τ_0 are the times for half-equilibration in the presence and in

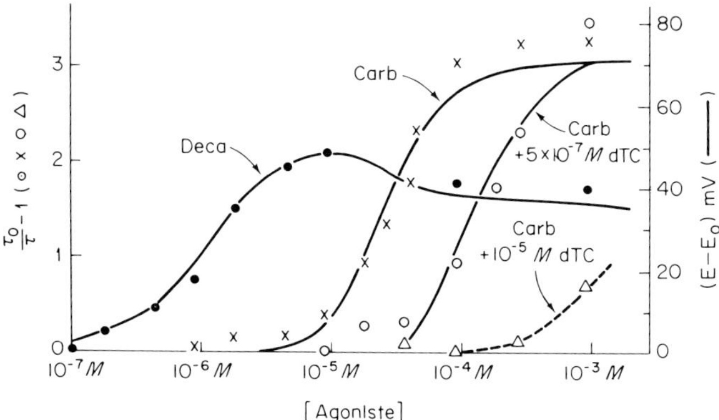

FIG. 17. Parallelism between chemical stimulation of isolated membranes ("microsacs") and electrical stimulation of the intact isolated electroplax of *Electrophorus*. The dose-response curves of the chemical stimulation of the microsacs were evaluated by measuring the rate of ^{22}Na efflux from the microsacs. τ is the period of time required for the 50% loss of ^{22}Na of the vesicles in the presence of carbamylcholine, τ_0 in its absence. The dose-response curves on the electroplax of activators [carbamylcholine (carb), decamethonium (deca)] or inhibitors [*d*-tubocurarine (dTC)] have been taken from Changeux and Podleski (1968); they were evaluated as usual by electrical parameters (voltage changes are given on the right-hand side). The data show the striking similarity of the affinities to the AcCh-receptor tested on cellular level (isolated electroplax) and of those tested on subcellular level (isolated membranes, "microsacs"). (Kasai and Changeux, 1970.)

the absence of cholinergic agonists respectively. These data support the assumption that the sites for the recognition of the signal (the receptor protomer) differ from the permeability gateways that are responsible for the accelerated ion movements. Similarly, the rate of Na efflux increased with temperature in the presence, as well as in the absence, of decamethonium, but the relative response to 3×10^{-6} *M* of the compound did not change. The Q_{10} for the change of ion flux was found to be about 1.8.

Another pertinent result of these studies with isolated membranes are the selective permeability changes effected by the action of carbamylcholine. The permeability to Na^+, K^+, and Ca^{2+} ions is increased, but not that to large ions, such as choline or tetraethylammonium, or uncharged permeants. The permeability to negatively charged ions is either slightly (Cl^-) or not at all (SO_4^{2-}) affected.

Kasai and Changeux (1971) estimated also the Na^+ and K^+ fluxes across the isolated excitable membranes of the microsacs stimulated chemically by carbamylcholine and compared them with the figures of Hodgkin and Huxley (1952) for the fluxes across the membranes of the squid axons stimulated electrically. The values estimated were quite comparable. This similarity supports the assumption that the AcCh cycle controls the permeability changes in both preparations. The presence of the proteins processing AcCh and the evidence for their role in the electrical activity hardly justify the conclusion of two different molecular mechanisms involved. Estimates of the permeabilities (cm/sec) and conductances (mho/cm^2) for Na^+ and K^+ ions also indicate that the values are similar. However, in view of the uncertainties of the present technique used for the microsacs one has to wait for more precise data. A final evaluation requires further improvement of the method.

Recently, a reconstitution of the microsacs has been described (Hazelbauer and Changeux, 1974). Membrane fragments rich in AcCh-receptor sites, purified from homogenates of electric organ of *Torpedo,* were dissolved in Na cholate in Tris buffer containing Na ethylenediaminetetraacetate. After extensive dialysis to remove cholate, the solution is supplemented with lipid extracts from native membrane fragments, and $MgCl_2$ and $CaCl_2$ are added. The reconstituted macrosacs obtained retain $^{22}Na^+$. Release of Na from these microsacs increase in the presence of carbamylcholine. The α-toxin from *Naja nigricollis* blocks this effect. This new preparation may be useful for various tests; it will be interesting to compare a number of properties of the reconstituted microsacs with those of the native ones.

F. Basic Excitation Units

A concept of fundamental importance for the understanding of nerve excitability and the integral model has been suggested by Eberhard Neumann: that of the basic excitation unit (BEU). In this concept the ion permeation zone (gateway) through which the Na^+ ions pass during the action potential, is not surrounded by one, but by several complexes of S (storage), R (receptor), and E (enzyme) proteins (6, for instance, in Fig. 18). The idea that a complex process such as the increased permeability of a gateway is not controlled by the action of a single molecule, but requires the simultaneous action of a number of molecules, is a basic tenet of biology (see, e.g., Schrödinger, 1944). In fact, the model suggests that the action potential results from the activation of several clusters of BEU's (see Supplement II). This concept is of decisive value for the interpretation of many electrophysiological observations, such as threshold, graded or all-or-none response, accommodation, and many others. The concept is only briefly mentioned here, since

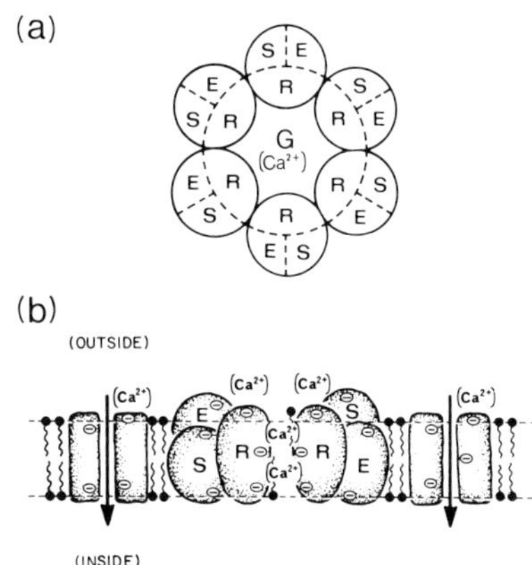

FIG. 18. Scheme of the AcCh-controlled gateway, G. (a) Basic excitation unit (BEU) containing in this example 6 SRE-assemblies, viewed perpendicular to the membrane surface. S, AcCh-storage site; R, AcCh-receptor protein; E, AcCh-esterase. (b) Cross section through a BEU flanked by two units which model ion passages for K^+ ions; the arrows represent the local electrical field vectors due to partial permselectivity to K^+ ions in the resting stationary state. The minus signs, ⊖, symbolize negatively charged groups of membrane components. (Neumann and Nachmansohn, 1975.)

it is also essential for the understanding of the following section discussing the role of AcCh in synaptic junctions.

V. Role of the AcCh Cycle at Junctions

The neurohumoral transmitter theory attributes to AcCh a function at synaptic junctions only: after its release from the nerve terminal, and after having crossed the nonconducting gap it stimulates the postsynaptic membrane. Thus, the view explicitly postulates two fundamentally different mechanisms in excitable membranes: a chemical one at junctions, a physical one in axons. As discussed in the description of the early history of the hypothesis of neurohumoral transmission by AcCh (pp. 14–18), it was this assumption of a basic difference, the postulate of two mechanisms, which appeared unacceptable to many prominent neurobiologists. The electrical characteristics of excitable membranes exhibit too many similarities in their properties to justify such a far-reaching assumption and the interpretation offered for an explanation of the experiments on the role of AcCh performed.

In 1959, when this monograph was first printed, many biochemical data already contradicted the special role at junctions attributed to AcCh. However, at that time the information about many structural features of all membranes and of junctions and about many biophysical and biochemical aspects was very incomplete. Many pertinent questions were controversial and open to discussion and three chapters of the monograph were devoted to the problem of synaptic transmission. As in so many other fields of biological sciences, the various advances of the last decade profoundly changed our views of the role of AcCh in excitable membranes in their conducting and synaptic parts. Although many questions concerning the molecular events in excitability still remain unresolved, the particular problem of the role of AcCh at junctions appears to be clarified by a great variety of new data, which have removed many difficulties. Strong support is now available for the assumption that the AcCh cycle plays the same role of controlling ion permeabilities in both axonal and synaptic parts of the excitable membrane. This aspect will be discussed in the following in the light of the new data.

A. *Observations that led to the Assumption of AcCh as a Neurohumoral Transmitter*

1. External Application of AcCh

One of the most fundamental facts supporting a transmitter role of AcCh appeared to be the powerful action of externally applied AcCh on junctions, in contrast to the total failure of AcCh to affect conduction of fibers even in high concentration. As Dale explained to the author in a discussion, during a visit at Columbia in 1946, it was this particular fact which had impressed him for decades. This pharmacological action has been fully discussed and explained in Section IV. This aspect requires, therefore, hardly any additional comments. It is the presence of strong structural barriers surrounding the axon, such as myelin or Schwann cells, impervious to AcCh and related compounds, which fully explains the limitation of the action by preventing AcCh to reach the excitable membrane. Where these protective barriers are poor or incomplete (nodes of Ranvier, lobster walking leg, rabbit vagus) or when they are reduced chemically or by enzyme action (phospholipase A on the squid axon), AcCh and related compounds may reach the membrane and act on the conducting part in a way similar to that observed at junctions.

2. The Artificially Induced Appearance of AcCh outside Nerve Fibers

Under physiological conditions there is no release of AcCh from the nerve terminals. The appearance of AcCh in the perfusion fluid of junctions has only been demonstrated under nonphysiological conditions. The observations of Otto Loewi (1921) on the *Vagusstoff* described in the textbooks as the classical foundation of the neurohumoral transmitter role of AcCh, are only reproducible in severely deteriorated heart preparations, as was shown by Ascher (1925). AcCh is found outside nerve fibers in many other nerve preparations, including for instance the ganglion nodosum, which has no synapses, when they are severely damaged (Lorente de Nó, 1938). It is known that Otto Loewi tried for more than ten years in the United States to repeat his experiments originally performed in Europe and finally abandoned his efforts. He attributed his failure to a species difference (*Rana pipiens* in the United States).* But even if we disregard Loewi's experiments, Dale and his associates emphasized that they were unable to find any trace of AcCh in the junctional perfusion fluid *unless eserine,* a potent inhibitor of AcCh-estrase, was present. Obviously, in the presence of an inhibitor (of the rapid and powerful removal mechanism for AcCh), AcCh will escape from the membrane and appear in the surrounding gap. It should be recognized that the experiments of Dale and his associates are in direct contradiction to Loewi's interpretation of an actual *release* of the *Vagusstoff* (AcCh) in the *absence of an inhibitor* (Loewi, 1921). Nevertheless, the *release of AcCh from damaged nerve is today still considered as one of the key evidences for a neurohumoral transmitter role.*

In view of the insulating barriers surrounding all conducting membrane parts, AcCh appears even in the presence of eserine only in the perfusion fluid of junctions. However, preparations in which AcCh may reach the conducting membrane because of incomplete protection (De Lorenzo *et al.*, 1968), such as the axons of the walking leg of lobster, AcCh is released even in resting condition when they are kept in saline solution containing physostigmine (Dettbarn and Rosenberg, 1966). This is in full agreement with the observations of Lorente de Nó mentioned before, who demonstrated the appearance of AcCh on several sites outside synaptic junctions. Moreover, it was shown, in the 1930's, by several investigators, as discussed on p. 180, that AcCh is released from the cut surfaces of axons into the surrounding fluid when they are stimulated because at the cut axon surface there is no barrier. In his review Rosenberry (1975) expresses the view that an *inter*cellular role of AcCh at the junction still may be possible. This view appears to the author difficult to reconcile with two considerations. First, in spite of many efforts covering a period of half a century, nobody has ever found a

* Bruecke in Vienna and his associates unsuccessfully tried to reproduce Loewi's experiments using *Rana esculenta* (personal communication to D. Nachmansohn in 1958).

trace of AcCh outside the cell except in the presence of eserine as discussed before. The second reasoning is based on a more fundamental biochemical notion. Once the existence of a complex AcCh cycle has been demonstrated to exist within the membrane to control ion permeability to Na^+ and K^+ ions, which is now recognized to be similar in all parts of the excitable membrane, it is difficult to assume that such a complex cycle, acting with high speed and precision, is divided at the junction into two parts: one part which acts in one membrane forming and releasing AcCh, the second part which acts in the membrane of the effector cell reacting with the target protein and being removed there. Highly developed and complex cycles associated with specific cellular mechanisms, such as motility, vision, are amazingly similar in all living cells, as will be discussed later. To assume that the AcCh cycle functions differently along the axonal membrane and its terminal would only be acceptable if very compelling evidence were presented. In view of the complete failure of all such attempts, it seems more reasonable to accept the similarity of the AcCh cycle and its control of permeability in all membrane parts.

3. Origin of AcCh Release after Nerve Terminal Degeneration

This problem has been fully discussed on p. 185. The data of McIntyre (1959) and his associates have convincingly demonstrated that the AcCh found in the perfusion fluid is not released exclusively from the nerve terminal, as originally proposed (Dale *et al.*, 1936) and considered as evidence of its transmitter function, but appears also after complete denervation of the nerve terminals when the muscle is stimulated directly. This is also in agreement with the just mentioned experiments in which appearance of AcCh from the cut surface of stimulated axons was observed.

In summary, the interpretation of the original observations on which the role of AcCh as a neurohumoral transmitter were based, has become untenable in the light of the facts discussed.

B. Evidence Supporting a Similar Role of AcCh Cycle in Pre- and Postsynaptic Junctional Membranes

In contrast to the contradictions facing the assumption that AcCh is a neurohumoral transmitter across junctions, the evidence for a similar role of the compound in both conducting and synaptic parts of the membrane has found increasingly strong experimental support. In the

1930's knowledge of the structural aspects of synaptic junctions were extremely limited; the only important aspect which was widely accepted was the existence of a nonconducting gap between nerve terminal and effector cells, although the extent and form of this gap were quite obscure. This situation changed only in the 1960's mainly owing to the rise of electron microscopy. No information existed about the biochemical aspects of cell membranes in general, including obviously excitable membranes in conducting as well as in synaptic parts. The hypothesis of neurohumoral transmission *simply assumed that AcCh is released from the nerve ending, and after crossing the gap, is hydrolyzed by an esterase located in the effector cell.*

Since that time, information about structural aspects and the chemical composition of membranes in general and of excitable membranes in particular, both in conducting and synaptic parts, has made tremendous progress. Barrnett (1962) was the first to demonstrate by electron microscopy combined with histochemical staining techniques that AcCh-esterase is located in both junctional membranes, i.e., the nerve terminal and the postsynaptic membrane. This result was consistent with the conclusion based on a series of indirect biochemical data (see pp. 188–189). In the last decade both the methods of electron microscopy and of staining techniques (see, e.g., Koelle *et al.*, 1974) were remarkably improved. A particularly elegant picture of the neuromuscular junction and the localization of AcCh-esterase in pre- and postsynaptic membranes is shown in Fig. 12 of Koelle (1971). It strikingly demonstrates the presence of AcCh-esterase in both junctional membranes and illustrates the complexity of shape and structure of a junction. The presence of the enzyme in synaptic and conducting membrane parts of the electroplax of *Electrophorus* has been discussed before.

The presence of an AcCh-receptor in its function in the nerve terminals were first indicated by the results of Masland and Wigton (1940): they found that AcCh, neostigmine, and curare act in a similar way on pre- and postsynaptic membranes of the neuromuscular junction, and that AcCh produces antidromic impulses. Since these experiments did not fit the current notions of neurohumoral transmission, they were completely ignored for more than a decade. But in the 1950's these observations were confirmed and greatly extended by many investigators. Today it is well recognized that nerve terminals are just as sensitive as the postsynaptic membrane to AcCh and its structural analogs (for details, see, e.g., Riker *et al.*, 1959; Werner and Kuperman, 1963). Antidromic impulses in motor nerve produced by AcCh and analogs injected into the neuromuscular junctions were recorded at the ventral roots. Thus, both AcCh-esterase and -receptor are not only present at both junctional membranes, but they function in the same way in both membranes. Quite recently also biochemical evidence has been obtained for the presence of the receptor in the nerve terminal membrane. B. W. Festoff, presenting data of A. Bender *et al.*, to be reported at the *Amer. Acad. of Neurol.*, showed in a seminar an electron micrograph of a rat neuromuscular junction to which immunochemical procedures were applied in combination with α-bungarotoxin: the receptor was present in both pre- and postsynaptic membranes.

All these data strongly support the assumption of the similar role of the AcCh cycle and of the basic excitation units in both the synaptic and other parts of the excitable membrane as proposed in the integral model of bioelectricity (see Supplement II).

K^+ Ion Efflux from Nerve Endings. Cowan (1934) demonstrated a strong efflux of K^+ ions from stimulated axons. Similarly, Feldberg and Vartiainen (1934) found an efflux at synaptic junctions following stimulation. Eccles (1935) considered at that time these findings as evidence that K^+ ions are the transmitters carrying the impulse from the nerve terminal to the postsynaptic membrane. The question of the mechanism by which the K^+ ions are released from the nerve terminal was and could not be raised since no information was available permitting even speculations, neither as to the efflux from axons nor to that from the nerve terminal. It may be useful to mention in this connection that Nastuk (1954) found that there was no synaptic transmission in absence of Na^+ ions.

It was difficult to understand that for many years no current flow from nerve terminals could be detected. This failure was considered as a strong evidence for the assumption that, in contrast to axons, transmission of impulse across junctions requires chemical transmitters. But in 1963, action potentials along nerve terminals were demonstrated by Hubbard and Schmidt (1963); these findings were confirmed by Katz and Miledi (1965). Thus, another objection which had been used as support for two fundamentally different mechanisms of "conduction" and "transmission," had been removed.

C. Alternative Interpretation of the Function of AcCh at Junctions

1. Similarity of the AcCh Function in Conductional and Synaptic Membrane Parts

In view of the remarkable advances of the information about cell membranes in general and the information about the presence and function of the proteins processing AcCh in excitable membranes both in their conducting and synaptic parts, a reevaluation of the role of AcCh at synaptic junctions appears necessary and unavoidable. The observations on which the original hypothesis of AcCh as neurohumoral transmitter were based have found an entirely different explanation and have made the early interpretation untenable. On the other hand, the data in the preceding section strongly support a similar role of AcCh in synaptic as well as in conducting membrane parts.

Thus, the alternative interpretation, postulated for many years, appears more satisfactory than the original hypothesis proposed: AcCh released on the arrival of the nerve impulse within the nerve terminal from a storage protein acts on the receptor and controls the ion movements across the membrane in a way similar to that proposed for the conducting membrane. The signal given by the release of AcCh is greatly amplified: for 1000 molecules of AcCh released in a nerve terminal, many millions of Na^+ ions will enter the axon and an equivalent amount of K^+ ions will flow into the nonconducting gap. The amount of K^+ ions in the nonconducting gap of about 200 to 400 Å may easily reach the concentration required to produce the change of potential across the postsynaptic membrane (Neumann *et al.*, 1973). The K^+ concentration in the synaptic cleft will become high enough to finally cause the conformational change of the storage protein in the postsynaptic membrane leading to a release of AcCh and thus initiate the same sequence of events as in the conducting and terminal membranes. The BEU's proposed for the axonal membrane are assumed to be present in the nerve terminal as well as in the postsynaptic membrane (Fig. 19).

In summary, the view presented postulates that the AcCh cycle, the basic elementary process controlling the permeability changes required for ion fluxes, is the same in the entire excitable membrane. This concept applies, on the basis of experimental evidence, to excitable membranes throughout the animal kingdom and apparently even to membranes of plants; it is a universal characteristic of excitability. In fact, it appears possible that the AcCh cycle even controls ion permeability in other, nonexcitable tissue, such as placenta, red blood cell membranes, and other structures which require rapid transient change in permeability. The high concentrations of the two enzymes, AcCh-esterase and choline *O*-acetyltransferase, found in nonexcitable tissue, are in favor of such a possibility. In the course of evolution, the AcCh system may have been applied to the propagation of the nerve impulse. On the other hand there is no direct evidence that AcCh at junctions becomes a neurohumoral transmitter across the synaptic cleft in contrast to its suggested function in conducting membranes controlling ion permeability.

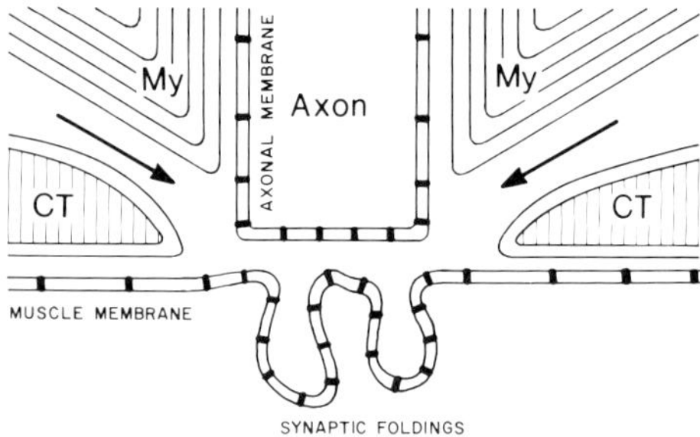

FIG. 19. Scheme of a cross section through a synaptic junction between a neuron and a muscle cell. The bars across the excitable membranes represent cross sections of the suggested basic excitation units (BEU); the density of BEU is assumed to be higher in the synaptic region than in the axonal parts. My, membrane layers of myelin protecting the axonal membrane; CT, protective layers (e.g., connective tissue) of the muscle membrane. The arrows indicate the sites of relatively easy access for external application of chemicals to the excitable membranes of nerve and muscle. (Neumann and Nachmansohn, 1975.)

2. DIFFERENCES BETWEEN CONDUCTION AND TRANSMISSION

However, it must be emphasized that the view of the universal role of AcCh in nerve excitability in no way contradicts the existence of marked and important differences between conduction along axons and transmission across junctions: there are striking differences of structure, shape, organization, and environment. Many of the differences of electrical parameters such as, e.g., the duration of the action potential may be attributed to such structural factors. Inspecting the complex shape of the neuromuscular junction shown in Fig. 12 we may readily imagine that geometrical factors alone must influence the time interval in which the different parts of the postsynaptic membrane are affected (e.g., by presynaptically released K^+ ions).

Owing to the lack of insulating barriers at most junctions the pharmacological action of externally applied AcCh and curare is usually restricted to junctions. This apparent limitation of action may apply to most drugs.

Finally, it must be recalled that we do not know the molecular organization of cell membranes. This factor, however, may well play an important role in the differences between conduction and transmission. In the framework of our model, the BEU may not be evenly distributed; the BEU density may be greater at junctions than at axons, but at present no unequivocal quantitative data on esterase and receptor densities are available. In addition, neuroeffectors, such as for instance catecholamines, developed at a later stage of evolution, GABA (γ-amino-butyric acid) and other amino acids, cyclic AMP, prostaglandins, and other compounds may influence the elementary processes taking place at junctions. These compounds may act, for instance, as modulators or regulators of the cholinergic control or on the processes in the gateways, directly or indirectly. The control of the permeability changes in axonal and synaptic parts of the membrane by the AcCh cycle has been occasionally referred to as the "unified concept of transmission and conduction." This interpretation is an oversimplification and may cause serious misinterpretation. The notion of a unified role concerns exclusively the specific intramembraneous control action of AcCh in junctional and axonal membrane parts. This concept is the main difference between the integral model and the hypothesis of neurohumoral transmission. Attributing to ATP, myosin, and actin a central role in the molecular events during muscular contraction does not exclude marked differences between many parameters of the contractile processes of rabbit striated and smooth muscle, or the muscles of holothuria and worms; a great variety of additional factors are involved which will strongly influence and modify the parameters observed, even if the underlying elementary chemical process is the same.

3. Excitatory and Inhibitory Effects of Nerve Stimulation

Nerve stimulation may have excitatory and inhibitory effects. This difference is widely attributed to excitatory and inhibitory transmitters. AcCh usually initiates a series of reactions leading to increased membrane permeability causing depolarization and excitation. But in some preparations such as, for instance, in the heart, vagus stimulation or artificial external application of AcCh have inhibitory effects: the membrane is hyperpolarized, presumably by further decreasing the Na^+ permeability. It has been speculated that these opposite results of stimulation, i.e., depolarizations and hyperpolarizations, may be attributable to different types of neurohumoral transmitters, in spite of the evidence that a compound such as AcCh may have both excitatory and inhibitory effects.

However, an understanding of these seemingly opposite effects requires an understanding of the precise molecular events. We are at present very far from having this information. We do not know why AcCh usually increases, but in some cases apparently, it decreases Na^+ permeability or increases the permeability to Cl^- ions. Various factors, such as structural organization, small changes in environment, reactions involving S—S or S—H groups of membrane proteins surrounding the gateways, may readily account for the apparently opposite effects of the same control mechanism. It should be recalled that ATP is required both for muscular contraction and for relaxation. For many years this phenomenon escaped a satisfactory interpretation until it was established by many investigators that the effect is controlled by Ca^{2+} ions. In the case of the AcCh-receptor there are S—S groups near the anionic group of the active site (Karlin and Winnik, 1968). Reducing these S—S groups by dithiothreitol (DTT) drastically changes or modifies the effect of compounds acting on the receptor. As mentioned before, it was found, for instance, that trimethylbenzene diazonium fluoroborate (TDF) applied to an electroplax, forms a covalent bond with the AcCh-receptor at or near the active site and blocks irreversibly the response to receptor activators. After exposure of the electroplax to DTT, which reduces tht S—S groups in the neighborhood of the active site, TDF in 10^{-6} *M* becomes a potent reversible receptor activator depolarizing the membrane (Podleski *et al.*, 1969). Similarly, hexamethonium, a curarelike reversible inhibitor of the receptor of the electroplax, becomes an activator of the receptor after exposure to DTT (Karlin, 1969). Thus, such a minor change as the reduction of S—S groups in the receptor protein may transform an inhibitory into an excitatory action. Before we know more about the specific molecular processes taking place in the membrane and in its proteins, the assumption of excitatory and inhibitory transmitters based simply on changes of electrical parameters or on release or external application of metabolites such as amino acids, hormones, etc. seems premature. On the basis of the present knowledge there appears to be little justification for the proposal of various *transmitters acting independently of the specific AcCh control mechanism* (in the pre- and postsynaptic membranes).

It may be briefly mentioned in this connection that the action of some compounds, such as, e.g., veratridine, and of certain neurotoxins, such as tetrodotoxin or batrachotoxin, used in recent years to affect ion permeabilities is also still not understood in molecular terms. These toxins act apparently on the gateway directly, but do not affect the control mechanism effected by the AcCh cycle.

4. Synaptic Vesicles

One problem should be briefly discussed, namely the role of synaptic vesicles in the function of AcCh, because the author is frequently asked about his views on this subject. When it was discovered by electron microscopy, that there are vesicles near the axon terminals (De Robertis and Bennett, 1955), it was immediately suggested by the supporters of the hypothesis of AcCh being a neurohumoral transmitter that these vesicles are like little glands containing AcCh and releasing it during activity (see p. 209). Just on the basis of the structural findings, it was proposed that these glands have "doors" and move during nerve activity towards "doors" located in the terminal membrane; the doors open simultaneously by some unknown mechanism and release packages of AcCh which then act on the postsynaptic membrane.

The difficulties of interpreting structural patterns seen with the electron microscope in terms of function has been severely criticized by Sjöstrand (1962). He vigorously objects to speculations and uncritical interpretations based *exclusively* on structural patterns. Some investigators stimulated by these patterns are carried by their imagination, he writes, to describe rather nice stories of functional sequences and events, without even any serious effort of proving or disproving their ideas. While it seems obvious that the vesicles, like any other cell structure, must have a functional significance, their role in synaptic transmission was first based exclusively on their presence, i.e., on the interpretation of a picture without any evidence of biochemical or biophysical data in support of such a speculation. It is remarkable that vesicles have been found in other parts of the neuron; sometimes they are present at a high concentration on the postsynaptic site (Edwards *et al.*, 1958).

Only several years later, in the early 1960's, did efforts start to identify the chemical contents of the vesicles. One well-established and documented fact is the presence of catecholamines and ATP in many of these vesicles. Catecholamines, which appear to be not directly associated with the elementary process of conduction, may act as modulators or regulators, accelerating or slowing down the primary events according to the needs of the organism. The action may take place in milliseconds or even seconds, and their availability in functionally distinct units appears a reasonable way of storing them and releasing them whenever needed. The presence of ATP suggests that it provides the energy for the release.

The efforts to find evidence for the presence of AcCh in these vesicles and their role in transmission were started by Gray and Whittacker (1962) and De Robertis *et al.* (1962). But so far these efforts have failed to provide a satisfactory answer. The results of the biochemical data offer more questions and difficulties rather than an elucidation of the problem. One serious difficulty already raised by Whittacker (1968), a strong supporter of the synaptic transmitter role of AcCh, is that of the mechanism by which AcCh gets into the vesicles. There is no choline *O*-acetyltransferase in the vesicles, isolated from synaptosomes, complex particles from synaptic membranes, and containing vesicles. Radioactively labeled AcCh and choline penetrate into the cytoplasm of the synaptosomes, but only negligible amounts enter into the vesicles. According to Marchbanks (1968, 1970) these facts suggest a permeability barrier between cytoplasm and vesicles. Another serious difficulty is the question of how AcCh is released from the vesicles, a mechanism still completely mysterious. Also once released, how would the ionic ester be able to reach the outside of the cell, i.e., the synaptic cleft, where it is supposed to act as the transmitter to the second cell? The ester released from the vesicles should pass, within microseconds, through the cytoplasm and penetrate a membrane in which AcCh-esterase, as is well established, is present in high concentrations and which is rich in lipids poorly pervious to quaternary ammonium ions. Having passed the membrane, they must reach the target protein in the membrane of the second cell, also rich in AcCh-esterase and lipids. Whittacker makes the *ad hoc* assumption that there may be intercommunicating tubules, although there is no experimental evidence for this assumption; they cannot be seen in the electron microscope. The assumption of invisible channels is useless because invisibility cannot be checked experimentally. In a more recent review Hall (1972) discusses the present data on the role of the vesicles. He concludes that the hypothesis of the role of the vesicles in the transmission by AcCh, dominant for almost 20 years, has no convincing experimental basis.

The author believes that the alternative assumption of AcCh being attached to the storage protein of the assemblies surrounding the gateway accounts in a much more satisfactory way for the action of AcCh in the excitable membrane. It is more in line with the biochemical thinking of the efficiency of cellular mechanisms, such as electron transfer, fatty acid synthesis, oxidative phosphorylation, than is the idea of little glands secreting the transmitter, especially in a cell function which is one of the fastest known and requires a high degree of precision.

VI. Concluding Remarks: Concepts and Axioms in Science

The momentous developments in biological sciences of the last 15 years have profoundly altered our views on cellular mechanisms, e.g., on some molecular events in cell membranes in general. These advances have also revealed many new pertinent aspects of the function of the excitable membrane. Only some highlights of the progress in the analysis of the biochemical basis of nerve excitability have been presented. Even in this restricted area those aspects that have a particular bearing on the function of the proteins processing AcCh have been emphasized. As already mentioned, many important details of the present knowledge of the molecular properties of these proteins will be evaluated in the new monograph in preparation. The use of refined instruments and methods, unavailable at the time of the first printing, required modifications and corrections of some of the data I presented in 1959.

However, the basic concept, developed several decades ago, of the essential role of the AcCh cycle for the control of ion permeability in excitable membranes, in their conducting and synaptic parts, during electrical activity has found an ever increasing amount of experimental support. The chemical concept proved extremely fruitful for devising new experiments in the analysis of the molecular events in excitable membranes and has stimulated many research efforts which in turn have greatly enhanced our knowledge of this vital problem of nerve function. The most exciting recent development resulting from the vast amount of biochemical data accumulated is, in the view of the author, the attempt of integration of the many biochemical and electrophysiological data, as outlined in the Preface. The suggestion of specific molecular events underlying nerve excitability, no matter how incomplete, was the foundation on which Eberhard Neumann has tried to interpret many functional aspects of electrical activity in physicochemical terms (Supplement II). It is fortunate that Dubois and Schoffeniels (1974) have now started efforts in the same direction.

In view of the two opposing theories on the role of AcCh in nerve activity, that of the author and that postulating AcCh to be a neurohumoral transmitter, it appears appropriate to make a few comments on the decisive importance of concepts and basic notions for the efforts of scientists, who determine the thinking, the methods of approach, the design of experiments (see Einstein's remark to Heisenberg in the Preface), and the interpretations. As history has shown time and again, concepts once widely accepted prevail for a long time and it usually takes several decades until a new concept, no matter how sound and convincing, is accepted and the preceding one is abandoned. One famous example, e.g., is the *phlogiston* theory of Stahl. This hypothesis had so strongly penetrated the thinking of scientists of the eighteenth century that many brilliant scientists (Cavendish, Scheele, Fourcroy, and many others) refused to accept Lavoisier's view. Lavoisier had used the balance and had explained the weight increase of metals on heating by the uptake of oxygen. It is informative that Priestly, the discoverer of oxygen, died in 1804 still firmly believing in phlogiston. In his article, "Theoretical Concepts in Biological Sciences," Krebs (1966) quotes several interesting and illuminating examples for the rejection of new concepts—the hostility, vituperation, and vitriolic comments with which other colleagues met the new ideas. This response seems to be common and may, at least in part, reflect a natural human reaction as the author has discussed in some length in his Prefatory Chapter to the *Annual Review of Biochemistry* (1972).

The objections to new concepts may not be as surprising as it may seem to those who think of science in terms of a human endeavor essentially based on objective data and experimental observations. When one realizes, however, to what extent basic notions and methods of approach affect our reasoning, conflicts and opposition to new ideas appear almost unavoidable. The biochemist (or molecular biologist), analyzing a problem on cellular, subcellular, and molecular levels, will use criteria, methods, and notions different from those used by a physiologist who studies overall manifestations of complex preparations or of organs characterized by a nearly infinite diversity of structure and organization. As repeatedly emphasized, even a subcellular structure, such as a cell membrane, is extremely complex. Biochemists have become increasingly aware during the last decade of the many big loopholes in our knowledge of membranes, their chemical composition, ultrastructure, and the impact of their complex organization on chemical reactions. Investigators using the oscilloscope in combination with pharmacological tests, will of necessity arrive at conclusions frequently in contradiction to those based on biochemical analysis. The author, versed in the history of science, was not surprised about the vigorous opposition to his views. He found it more difficult to understand the vigorous rejection prevailing for two decades, even to consider the possibility of chemical reactions controlling electrical activity and nerve excitability. Nobody after all can question in our age that many theoretical concepts of biology have been profoundly deepened and clarified by the biochemical approach. The author is amused to occasionally read statements by authorities, that his work is apt to create confusion among young scientists. This reminds him of the story in Plato, which he read (in the original text) in his high school years, about the trial of Socrates: Socrates was accused of misguiding youth and teaching them of gods in which the authorities of the state did not believe. The author feels fortunate that although his opponents may accuse him of the same crime, they have not the power to inflict upon him the fate of Socrates.

In biochemistry, as in all sciences including physics, there are certain basic notions and concepts—fundamental assumptions which cannot be further reduced, but have to be accepted as an act of faith (Born, 1949). Max Planck (1922) expressed it in the following words: "*Auch in der Physik gilt der Satz, dass man nicht selig werden kann ohne den Glauben.*" Axioms applied in the biochemical approach differ in many respects from those of the physiological one. Two fundamental notions of biochemical thinking may be mentioned as an illustration. The one is the biochemical unity of life, a notion which has played an essential role in the development of biochemistry since the time of Pasteur. Nature has shown little imagination in modifying chemical mechanisms associated with given functions. The specific chemical processes associated with a cellular mechanism, are remarkably similar from the most primitive cells to those in the most highly developed organisms. In hundreds of millions of years of evolution the cytochromes have changed their essential amino acid composition only to a limited degree (Smith, 1968; Margoliash, 1972). Thus, it appears reasonable to assume that such a specific system as the AcCh cycle associated with ion permeability changes during electrical activity, is ubiquitously required for all forms of bioexcitability, one of the universal and most vital characteristics of living organisms. The adaptation of the same specific system to the great diversity of the needs in a variety of functions may be achieved by many additional factors, such as changes of form and shape, of structure and organization, and a variety of additional chemical factors—altogether they may modify the effects of the specific chemical systems in any way required.

A second important notion may be briefly mentioned. Typical for chemical processes in cell is their cyclic character. This important principle was first recognized for the process of glycolysis by Otto Meyerhof half a century ago. He found that only about one-sixth of the lactic acid formed anerobically from glycogen during muscular contraction by a series of intermediary steps is oxidized, as was later established, via the citric acid cycle, but five-sixths of the lactic acid formed is resynthesized during recovery to glycogen by the energy derived from the oxidation. His observations provided the basis for the explanation of the well-known Pasteur effect. Meyerhof was the first to recognize the far-reaching general implications of these findings. The series of cycles associated with most cellular functions apparently prevent unnecessary waste by minimum energy dissipation using the excess energy for reversing most of the processes and thereby restoring the initial state with maximum efficiency. The AcCh cycle may well have a similar bioenergetic function.

Neither the views on the biochemical foundation of nerve excitability and the role of the AcCh cycle described so far nor the attempt of integrating these data with the biophysical events in the model discussed in Supplement II claim to provide final answers; *they are working hypotheses*. Many questions remain open. J. J. Thompson once said "a theory is a tool and not a creed." An hypothesis derives its value from its ability to stimulate new experiments. The results in turn may be used for correcting, modifying, or even discarding a working hypothesis. The elements on which the biochemical data and the model are based may be subdivided into three categories: (i) well-established biophysical and biochemical data; (ii) indirect evidence which requires further experimental support; and (iii) postulates which, however sound the reasoning may be, present a challenge for experimental tests.

But in contrast to the theories and hypothesis developed, the work presented is essentially based on three axioms, on fundamental notions which have guided the author in all his endeavors. According to the first axiom no manifestation of life is conceivable without molecular changes, i.e., without chemical processes (see also Liebig's statement on p. 10). The second axiom is the basic similarity of the elementary chemical forces, involved in a specific cellular mechanism, throughout all living organisms. The third axiom is the paramount role of proteins, including enzymes, in all mechanisms of living cells.

However, no matter what our present concepts and theories are, we should be aware that in spite of all the miraculous achievements in the physical and biological sciences during this century, we are still far from understanding the universe and man. It is interesting to read the statement of Albert Einstein quoted at the beginning of his biography ["Albert Einstein Creator and Rebel" by Banesh Hoffmann, with the collaboration of Helen Dukas (1972)]:

> One thing I have learned in a long life: that all our science, measured against reality, is primitive and childlike—and yet it is the most precious thing we have.

Acknowledgements

The author gratefully acknowledges the encouragement and help of Professor Paul A. Marks, Vice President of Health Sciences, and of Professor Lewis P. Rowland, Chairman of the Department of Neurology. They have made possible the writing of this revised monograph and the continuation of the work, carried out at Columbia University for more than three decades, by a vigorous, young, and competent group.

He would like to express his thanks to Dr. George B. Koelle for the permission to reproduce the still unpublished electron micrograph of the electroplax, to Dr. Virginia Tennyson for permission to reproduce the electron micrograph of the squid fibers, and also to Dr. Terrone L. Rosenberry for the permission to include his still unpublished diagrammatic scheme of the hydrolysis of AcCh by AcCh-esterase. The efficiency and devotion of Ms. Eva-Renate Busse was invaluable in preparing this monograph.

The author is greatly indebted to the National Science Foundation, the National Institutes of Health, and the New York Heart Association for their generous and unfailing financial support over decades.

Reviews by the Author Since 1959*

1963 In "Handbuch der experimentellen Pharmakologie," (G. B. Koelle, ed.), Vol. 15, pp. 40–45 and 701–740. Springer-Verlag, Berlin and New York.

1963 *Biochem. Z.* **338**, 454–473.

1964 *New Perspect. Biol., Proc. Symp., 1963* (M. Sela, ed.) BBA (Biochim. Biophys. Acta) Libr., Vol. 4, pp. 176–204.

1964 *In* "Tribute to V. A. Engelhardt, Molecular Biology: Problems and Perspectives," pp. 282–303. Acad. Sci. U.S.S.R., Moscow.

1964 *J. Mt. Sinai Hosp., New York* **31**, 549–583.

1965 *Nova Acta Leopold.* [N.S.] **30**, 207–233.

1965 *Isr. J. Med. Sci.* **1**, 1201–1219.

1966 *In* "Nerve as a Tissue" (K. Rodahl, ed.), pp. 141–161. McGraw-Hill, New York.

1966 *In* "Current Aspects of Biochemical Energetics: Fritz Lipmann Dedicatory Volume" (N. O. Kaplan and E. P. Kennedy, eds.), pp. 145–172. Academic Press, New York.

1966 *Ann. N.Y. Acad. Sci.* **137**, 877–900.

1968 *Proc. Nat. Acad. Sci. U.S.* **61**, 1034–1041.

1969 *J. Gen. Physiol.* **54**, 187S–224S.

1970 *Science* **168**, 1059–1066.

1971 *In* "Handbook of Sensory Physiology" (W. R. Loewenstein, ed.), Vol. I. pp. 18–102. Springer-Verlag, Berlin and New York.

1972 *Annu. Rev. Biochem.* Vol. 41, Prefatory Chapter, pp. 1–28.

1972 *Proc. Nat. Acad. Sci. U.S.* **68**, 3170–3174.

1973 *Biochim.* **55**, 365–376.

1973 *In* "The Structure and Function of Muscle" (G. H. Bourne, ed.), 2nd ed., Vol. 3, pp. 32–117. Academic Press, New York.

1974 *In* "Central Nervous System: Studies on Metabolic Regulation and Function" (E. Genazzini and H. Herken, eds.), pp. 121–137. Springer-Verlag, Berlin and New York.

* These summaries written by the author since 1959 reflect the continuous adjustments of his views to the new information becoming available at an ever increasing rate. Some aspects are discussed in several reviews in more details than in this Supplement. Moreover, since they have been published in several languages and countries, their quotation may be a convenience for many readers.

SUPPLEMENT II

Toward a Molecular Model of Bioelectricity

I. Introduction

Bioelectrical excitability is a universal property of all higher organisms. Although numerous experimental data have been accumulated in the various disciplines working on bioelectricity, the mechanism of nerve excitability and synaptic transmission, however, is still an unsolved problem. Many mechanistic interpretations of nerve behavior and the various molecular and mathematical models generally cover only a part of the known facts and are thus selective and only of limited value. According to Agin, there is no need for further (physically) unspecific mathematical models (such as for instance the Hodgkin–Huxley scheme for squid giant axons). "What is needed is a quantitative theory based on elementary physicochemical assumptions and which is detailed enough to produce calculated membrane behavior identical with that observed experimentally. At the present time such a theory does not exist" (Agin, 1972).

The first proposal of a *physically specific* mechanism for the control of bioelectricity is due to Nachmansohn (1955a, 1971). The chemical hypothesis of Nachmansohn remained, however, on a qualitative level and until recently there was no detailed link between the numerous biochemical data on excitability and the various electrophysiological observations. A first attempt at integrating some basic data of biochemical and pharmaco-electrophysiological studies had been initiated by the late Aharon Katchalsky and a first report has been recently given (Neumann *et al.*, 1973).

The present account is a further step toward a specific quantitative physicochemical model for the control of ion flows during excitation. After an introductory survey on fundamental electrophysiological and biochemical observations our integral model is developed, and some basic electrophysiological excitation parameters are formulated in terms of specific membrane processes.

II. Biomembrane Electrochemistry

It is well known that bioelectricity and nerve excitability are electrically manifested in stationary membrane potentials and the various forms of transient potential changes such as, e.g., the action potential (Hodgkin, 1964; Tasaki, 1968). Although these electrical properties reflect membrane processes, bioelectricity and excitability are intrinsically coupled to the metabolic activity of the excitable cells. Now, inherent to all living cells is a high degree of coupling between energy and material flows. But already on the subcellular level of membranes intensive chemodiffusional flow coupling occurs, and apparently time-independent properties reflect balance between active and passive flows of cell components.

The specific function of the excitable membrane requires the maintenance of nonequilibrium states (Katchalsky, 1967). The intrinsic nonequilibrium nature of membrane excitability is most obviously reflected in the unsymmetric ion distributions across excitable membranes. These nonequilibrium distributions are metabolically mediated and maintained by "active transport" (e.g., Na^+/K^+ exchange pump). A fundamental requirement for such an active chemodiffusional flow coupling is spatial anisotropy of the coupling medium (Prigogine, 1968), including the membrane structure (Katchalsky, 1967).

Anisotropy is apparently a characteristic property of biological membranes. Furthermore, cellular membranes including nerve membranes may be physicochemically described in terms of a layer structure. For instance it was found that the internal layer of the axonal membrane contains proteins required for excitability. Proteolytic action on this layer causes irreversible loss of this property (Tasaki and Singer, 1966).

The membrane components of excitable membranes include ionic constituents: fixed charges (some of them may serve to facilitate locally perm-selective ion diffusion), mono- and divalent metal ions, especially Ca^{2+} ions, and in particular ionic side groups of membrane proteins directly involved in the control of electric properties. Local differences in the distribution of fixed charges may create membrane regions with different permeability characteristics, and (externally induced) ionic currents may amplify either the accumulation or the depletion of ions within the membrane. Structural inhomogeneity of excitable membranes may thus lead to the observed nonlinear dependencies between certain physical parameters, e.g., between current intensity and potential (Cole, 1968). A well-known example of this nonlinearity is the current rectification in resting stationary states of excitable membranes. Correspondent to current rectification, there is a straightforward dependence of the membrane potential on the logarithm of ion activities only in limited concentration ranges (Hodgkin and Keynes, 1955; Tasaki and Takenada, 1964). Thus, various chemical and physical membrane parameters show the intrinsic structural and functional anisotropy of excitable membranes. This recognition reveals the approximate nature of any classical equilibrium and nonequilibrium approach to electrochemical membrane parameters on the basis of linearity and involving the assumption of a homogeneous isotropic membrane. Since, however, details of the membrane anisotropy are not known, any exact quantitative nonequilibrium analysis of membrane processes and ion exchange currents across excitable membranes faces great difficulties for the time being.

III. Stationary Membrane Potentials

It is only in the frame of a number of partially unrealistic, simplifying assumptions that we may approximately describe the electrochemical behavior of excitable membranes, within a limited range of physicochemical state variables (Agin, 1967). Explicitly, we use the formalism of classical irreversible thermodynamics restricted to linearity between flows and driving forces. The application of the Nernst–Planck equation to the ion flows across excitable membranes represents such a linear approximation widely used to calculate stationary membrane potentials.

There is experimental evidence that large contributions to the resting stationary membrane potential $\Delta\psi_r$ are attributable to ion selectivities. Indeed, it appears that excitable membranes have developed dynamic structures which are characterized by permselectivity for certain ion types and ion radii. This property may be described by "Nernst terms" which, however, are strictly valid only for electrochemical equilibria. Nevertheless, the Nernst approximation for the resting stationary potential is useful for a number of practical cases. The Nernst terms may be calculated by application of the Nernst-Planck equation relating the ion flow J_i to the gradient of the electrochemical potential $\nabla\tilde{\mu}_i$ of the ion type i,

$$J_i = U_i C_i \nabla(-\tilde{\mu}_i) \tag{1}$$

where U_i is the ionic mobility and C_i is the molar concentration. The electrochemical potential of ion i of valency Z_i is defined by

$$\tilde{\mu}_i = \mu_i^0 + RT \ln a_i + Z_i F \psi \tag{2}$$

In Eq. (2), μ_i^0 is the standard chemical potential and a_i the thermodynamic activity which in dilute solution may be approximated by $a_i \simeq C_i$; RT is the (molar) thermal energy, F is the Faraday constant, and ψ is the electrical potential.

In the frame of the linear model the flux component $J_i(x)$ perpendicular to the membrane surfaces, at the point x within the membrane is given by

$$J_i(x) = U_i(x) C_i(x) \frac{d}{dx}(-\tilde{\mu}_i) \tag{3}$$

Integration of Eq. (3) within the membrane boundaries ($x = 0$ and $x = d$, the membrane "thickness") in a closed form requires restrictive assumptions. Despite the experimentally suggested inhomogeneity of excitable membranes, the following approximations are used: (i) the ion mobilities are the same throughout the membrane, and (ii) space charge effects are negligible, so that we may assume approximate microscopic electroneutrality. Thus, $\nabla^2\psi \simeq 0$ within the membrane. The Laplace equation $\nabla^2\psi = 0$ is equivalent to the constant field condition; that usually is not an appropriate physical assumption (Agin, 1967; Zelman and Shih, 1972).

For the condition of stationarity, i.e., at zero net membrane current ($I_m = 0$) the uniform model treatment yields the Nernst contributions

$$(\Delta\psi_N)_{I_m=0} = \frac{RT}{F}\sum_i \frac{t_i}{Z_i} \ln \frac{a_i^{(o)}}{a_i^{(i)}} \tag{4}$$

where t_i is the transference number representing the fraction of membrane current carried by ion type i. (The fraction t_i may vary between 0 and 1.)

Although stationary states of excitable membranes always reflect balances between active transport processes and passive "leakage" fluxes, we may safely neglect flow contributions associated with active transport as long as the time scale considered does not exceed the millisecond range.

The electrical potential difference $\Delta\psi_r$ of an excitable membrane under "resting" conditions (stationary case of $I_m = 0$) may thus be approximated by

$$\Delta\psi_r \simeq (\Delta\psi)_{I_m=0} = \frac{RT}{F}\sum_i \frac{t_i}{Z_i} \ln \frac{a_i^{(o)}}{a_i^{(i)}} \tag{5}$$

It should be realized that the measured potential difference (measured for instance with calomel electrodes in connection with salt bridges) cannot be used to calculate the exact value for the average electric field $\bar{E}$ across the permeation barrier of thickness d. $\bar{E}$ may be defined by

$$\bar{E} = -(1/d)\,\Delta\psi' \tag{6}$$

The electrical potential difference $\Delta\psi'$ associated with the permeation barrier results from different sources. Although these sources are mutually coupled we may formally separate the various contributions. Owing to the presence of fixed charges at membrane surfaces (Segal, 1968) there are Donnan potentials $\Delta\psi_D^{(o)}$ and $\Delta\psi_D^{(i)}$ from the inside (i) and outside (o) interfaces between membrane and environment. Furthermore, there are interdiffusion potentials $\Delta\psi_U$ arising from differences in ionic mobilities within ion exchange domains of intramembraneous fixed charges (Tasaki, 1968). We may sum up these terms to $\Delta\psi_{DU} = \Delta\psi_D^{(o)} + \Delta\psi_D^{(i)} + \Delta\psi_U$. Thus,

$$\Delta\psi' = (\Delta\psi_N)_{I_m=0} + \Delta\psi_{DU} \tag{7}$$

Compared to $\Delta\psi_{DU}$ the "Nernst" contributions to $\Delta\psi'$ appear relatively large. Inserting Eqs. (4) and (7) in Eq. (6) we obtain for the resting state:

$$(\bar{E})_{I_m=0} = -\frac{1}{d}\left(\Delta\psi_{DU} + \frac{RT}{F}\sum_i \frac{t_i}{Z_i} \ln \frac{a_i^{(o)}}{a_i^{(i)}}\right) \tag{8}$$

Among the various ions known to contribute to $\Delta\psi$ (*in vivo*) are the metal ions Na^+, K^+, Ca^{2+} and protons and Cl^- ions. In the resting state the value of the stationary membrane potential, $\Delta\psi_r$, is different for different cells and tissues. Distribution and density of fixed charges and ion gradients and the extent to which they contribute to $\Delta\psi'$ differ for different excitable membranes.

Variations of the natural ionic environment lead also to alterations in the Donnan and interdiffusion terms. Since ions are an integral part of the membrane structure, for instance as counterions of fixed charges, any change in ion type and salt concentration, but also variations in temperature and pressure, may cause changes in the lipid phase (Traeuble and Eibl, 1974) and in the conformation of membrane proteins. As a consequence of such structural changes, membrane "fluidity" and permeability properties may be considerably altered. It is known that, for instance, an increase in the external Ca^{2+} ion concentration increases the electrical resistance of excitable membranes (cf., e.g., Cole, 1968).

IV. Transient Changes of Membrane Potentials

A central problem in bioelectricity is the question: what is the mechanism of the various transient changes of membrane parameters associated with nerve activity? Before discussing a specific physicochemical model for the control of electrical activity, it appears necessary to recall a few fundamental electrophysiological and chemical observations on excitable membranes.

The various transient expressions of nerve activity such as depolarizations and hyperpolarizations of the resting stationary potential level reflect changes of a dielectric capacitive nature, interdiffusional ion redistributions, and "activations" or "inactivations" of different gradients or changes of the extent with which these gradients contribute to the measured parameters. At constant pressure and temperature, the ion gradient contributions to $\Delta\psi$ may be modeled as variations of the t_i parameters; see Eq. (4).

A. Threshold Behavior

Transient changes in $\Delta\psi$ may be caused by various physical and chemical perturbations orginating from adjacent membrane regions, from adjacent excitable cells, or from external stimuli. Phenomenologically we differentiate sub- and suprathreshold responses of the excitable membrane to stimulation. Subthreshold changes are, for instance, potential changes which attenuate with time and distance from the site of perturbation; this phenomenon has been called electrotonic spread. If, however, the intensity of depolarizing stimulation exceeds a certain threshold value, a potential change is triggered that does not attenuate but propagates as such along the entire excitable membrane. This suprathreshold response is called "regenerative" and has been christened action potential or nerve impulse.

B. Stimulus Characteristics

In order to evoke an action potential the intensity of the stimulation has to reach the threshold with a certain minimum velocity (Cole, 1968). If the external stimulus for instance is a current the intensity I of which gradually increases with time t the minimum slope condition for the action potential may be written

$$dI/dt \geq (dI/dt)^{\min} \tag{9}$$

The observation of a minimum slope condition for the generation of action potentials is of crucial importance for any mechanistic approach to excitability.

On the other hand, when reactangular "current pulses" are applied, there is an (absolute) minimum intensity, I_{th}, called rheobase, and a minimum time interval, Δt, of stimulus duration. It is found that the product of suprathreshold intensity and minimum duration is approximately constant. Thus, for $I \geq I_{\text{th}}$,

$$I\,\Delta t = \text{const} \tag{10}$$

describing the well-known strength-duration curve. Whereas the strength-duration product does not depend on temperature, the temperature coefficient of the rheobase (dI_{th}/dT related to a temperature increase of 10 °C) Q_{10} is generally about 2 (see, e.g., Cole, 1968).

It is recalled that the value of I_{th} is history dependent. The threshold changes as a function of previous stimulus intensity and duration. Hyperpolarizing and depolarizing subthreshold prepulses also change the threshold intensity of the stimulating current. See Section IV,D.

It appears that, in general, the induction of the action potential requires the *reduction* of the intrinsic membrane potential $\Delta\psi_r$ below a certain threshold value $\Delta\psi_{th}$. This potential decrease $\Delta(\Delta\psi) = \Delta\psi_r - \Delta\psi_{th}$ is usually about 15 to 20 mV and has to occur in the *form of an impulse,* $\int \Delta(\Delta\psi)\, dt$, in which a minimum condition between membrane potential and time must be fulfilled.

The condition for the initiation of an action potential may be written

$$\Delta\psi_r - \int \left(\frac{d\,\Delta\psi}{dt}\right)^{\min} dt = \Delta\psi_{th} \tag{11}$$

or in the more general form

$$\Delta\psi_r - \frac{1}{\Delta t}\int \Delta(\Delta\psi)\, dt \leq \Delta\psi_{th} \tag{12}$$

The potential change $\Delta(\Delta\psi)$ is equivalent to a change in the intrinsic field, ΔE, across the membrane of the thickness d. Thus $\Delta E = -\Delta(\Delta\psi)/d$, and with Ohm's law we have $d\,\Delta E = -R_m I$, where R_m is the membrane resistance.

Including Eq. (4) we may write

$$\begin{aligned}\int \Delta(\Delta\psi)\, dt &= -d\int \Delta E\, dt = \int R_m I\, dt \\ &= \frac{RT}{Z_i F}\int \Delta \ln\left(\frac{a_i^{(o)}}{a_i^{(i)}}\right) dt\end{aligned} \tag{13}$$

With Eqs. (12) and (13), we see how the intrinsic membrane potential may be changed: by an (external) field (voltage or current) pulse or by an ion pulse involving those ions that determine the membrane potential. Such an ion pulse may be produced, if the concentration of K^+ ions on the outside of excitable membranes is sufficiently increased within a sufficiently short time interval (impulse condition). Similarly, a pH change can cause action potentials (Tasaki *et al.*, 1965).

As seen in Eq. (4) the membrane potential is dependent on temperature. Therefore, temperature (and also pressure) changes alter the membrane potential in excitable membranes. Moreover, action potentials can be evoked by thermal and mechanical shocks (see, e.g., Julian and Goldman, 1962).

It is furthermore remarked that under natural conditions electrical depolarization produces an action potential only if the excitable membrane was kept above a certain (negative) level of polarization; in squid giant axons the minimum membrane potential is about —30 to —40 mV. This directionality of membrane polarization and the requirement of field reduction for regenerative excitation are a further manifestation of the intrinsic anisotropy of excitable membranes.

C. Propagation of Local Activity

In many excitable cells $\Delta\psi_r$ is about —60 to —90 mV, where the potential of the cell interior is negative. Assuming an average membrane thickness $d = 100$ Å, we calculate with Eq. (8) an average field intensity of appproximately 60 to 90 kV/cm (cf., however, Carnay and Tasaki, 1971). The field vector in the resting stationary state is directed from the outside to the inside across the membrane of the excitable cell. (It is this electric field which partially compensates the chemical potential gradient of the K^+ ions.) Any perturbation which is able to change locally the intrinsic membrane potential, i.e., the membrane field, will also affect adjacent parts of the membrane or even adjacent cells. If the field change remains below threshold or does not fulfill the minimum slope condition there will be only subthreshold attenuation of this field change. A nerve impulse in its rising phase may, however, reduce the membrane field in adjacent parts to such an extent and within the required minimum time interval, that suprathreshold responses are triggered. It is recalled that electric fields represent long-range forces (decaying with distance).

D. Refractory Phases

An important observation suggestive for the nature of bioelectricity is the refractory phenomena. If a nerve fiber is stimulated a second time shortly after a first impulse is induced, there is either an impulse that propagates much slower or only a subthreshold change. Immediately after stimulation a fiber is absolutely refractory, no further impulse can be evoked. This period is followed by a relatively refractory phase during which the threshold potential is more positive and only a slowly propagated impulse can be induced. The various forms of history-dependent behavior are also called *accommodation* or *adaption* of the fiber to preceding manipulations (see, e.g., Tasaki, 1968).

E. Time Constants

If rectangular current pulses of fixed duration but of variable amplitude and amplitude directions are locally applied to an excitable membrane, a series of local responses are obtained. Figure 1 shows schematically the time course of these local responses upon hyper- and depolarizing stimulations: subthreshold potential changes and action potential (Katz, 1966).

It appears that the threshold level reflects a kind of instability. When the stimulus is switched off, fluctuations either result in the action potential (path c → c′) or in a return to the resting level (path c → c″). If, for phase c the stimulus of the same intensity would last longer, an action potential would develop immediately for the hump visible in the late phase of stimulation.

It has been found that large sections of the time course of subthreshold responses and the first phase of the action potential (also called latency; see Fig. 1) can be described with linear differential equations of first order. Furthermore, the time constants τ_m associated with these sections are nearly equal.

The actual values for τ_m are different for different excitable membranes and are dependent on temperature. In squid giant axons the value

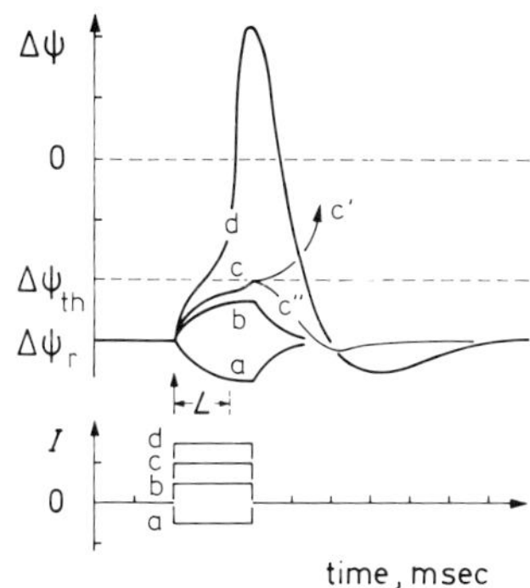

FIG. 1. Sub- and suprathreshold responses of the membrane potential $\Delta\psi$ to stimulating rectangular current pulses of fixed duration but variable intensity I. a is the inward current pulse causing hyperpolarization; b, subthreshold outward current ($I < I_{th}$) causing depolarization; c, threshold outward current (I_{th}) causing either an action potential (path c′) or return to resting stationary potential $\Delta\psi_r$ (path c″); d, suprathreshold outward pulse ($I > I_{th}$) causing action potential. L, latency phase of the action potential; $\Delta\psi_{th}$, threshold potential.

of τ_m at about 20°C is about 1 msec; the corresponding value for lobster giant axons is about 3 msec. The temperature coefficient ($d\tau_m^{-1}/dT$ related to a temperature increase of 10°C) Q_{10} is about 2 (Cole, 1968).

As outlined by Cole, a time constant in the order of milliseconds can hardly be modeled by simple electrodiffusion: an ion redistribution time of 1 msec requires the assumption of extremely low ion mobilities (about 10^{-8} cm sec^{-1}/V cm^{-1}). However, even in "sticky" ion exchange membranes ionic mobilities are not below 10^{-6} cm sec^{-1}/V cm^{-1}. Furthermore, the temperature coefficients of simple electrodiffusion are only about 1.2 to 1.5.

Thus, the magnitude and temperature coefficient of τ_m suggest that electrodiffusion is not rate limiting for a large part of the ion redistributions following perturbations of the membrane field.

This conclusion is supported by various other observations. The time course of the action potential can be formally associated with a series of time constants, all of which have temperature coefficients of about 2 to 3 (see, e.g., Cole, 1968). The various phases of the action potential are prolonged with decreasing temperature. In the framework of the classical Hodgkin–Huxley phenomenology, prolonged action potentials should correspond to larger ion movements. However, it has been recently found that in contrast to this prediction, the amount of ions actually transported during excitation decreases with decreasing temperature (Landowne, 1973).

It thus appears that ion movements (caused by perturbations of stationary membrane states) are largely rate limited by membrane processes. Such processes may comprise phase changes of lipid domains or conformational changes of membrane proteins. Changes of these types usually are cooperative, and temperature dependencies are particularly pronounced within the cooperative transition ranges.

Prolonged potential changes and smaller ion transport at decreased temperature may be readily modeled, if at lower temperature, configurational rearrangements of membrane components involve smaller fractions of the membrane than at higher temperature.

There is a formal resemblance between the time constant τ of a phase change or a chemical reaction and the *RC* term of an electrical circuit with resistance R and capacitance C (Oster *et al.* 1973). If a membrane process is associated with the thermodynamic affinity A and a rate $J_r = d\xi/dt$, where ξ is the fractional advancement of this process, we may formally define a reaction resistance $R_r = (\partial A/\partial J_r)$ and an average reaction capacitance $\bar{C}_r = -(\partial A/\partial \xi)^{-1}$. The time constant is then given by

$$\tau = R_r \bar{C}_r \tag{14}$$

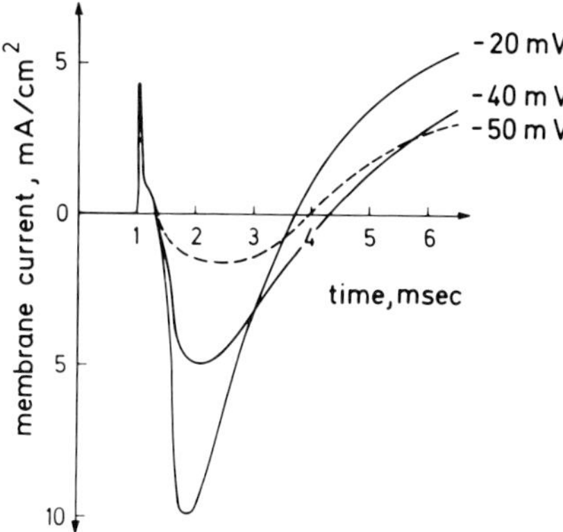

FIG. 2. Time course of membrane currents during voltage clamp experiments of internally perfused giant axons of squid *Loligo pealii;* resting potential $\Delta\psi_r \simeq -65$ mV. Dashed line: response to a clamp of -50 mV, i.e., a depolarization of 15 mV that, under unclamped conditions, does not evoke an action potential. The solid lines represent suprathreshold membrane currents. (Scheme of data by Narahashi *et al.,* 1969).

The time course of electrical parameters may thus be formally modeled in terms of reaction time constants for membrane processes of the type discussed above. For the chemical contribution of the observed subthreshold change in the membrane potential.

$$\Delta\psi = I_m R_m[1 - \exp(-t/\tau_m)] \tag{15}$$

caused by the rectangular current pulse of the intensity I_m across the membrane resistance R_m, we have $\tau_m = R_r\bar{C}_r$.

As already mentioned, the time constants for subthreshold changes and for the first rising phase of the action potential are almost the same. Furthermore, the so-called "Na^+ ion activation–inactivation curves (in voltage clamp experiments, see Fig. 2) for both sub- and suprathreshold conditions are similar in shape, although much different in amplitude" (Plonsey, 1969). These data strongly suggest that sub- and suprathreshold responses are essentially based on the same control mechanism; it is then most likely the molecular organization of the control system that accounts for the various types of response.

V. Proteins Involved in Excitation

Evidence is accumulating that proteins and protein reactions are involved in transient changes of electrical parameters during excitation. As briefly mentioned, the action of proteases finally leads to inexcitability. Many membrane parameters such as ionic permeabilities are pH dependent. In many examples this dependency is associated with a p*K* value of about 5.5 suggestive for the participation of carboxylate groups of proteins.

Sulfhydryl reagents and oxidizing agents interfere with the excitation mechanism, e.g., prolonging the duration and finally blocking the action potential (Huneeus-Cox *et al.*, 1966). The membrane of squid giant axons stain for SH groups provided the fibers had been stimulated (Robertson, 1970). These findings strongly suggest the participation of protein specific redox reactions in the excitation process (see also Marquis and Mautner, 1974).

Of particular interest appear the effects of ultraviolet radiation on nodes of Ranvier. The spectral radiation sensitivity of the rheobase and of the fast transient inward current (normally due to Na^+ ions) in voltage clamp is very similar to the ultraviolet absorbance spectrum for proteins (von Muralt and Stämpfli, 1953; Fox, 1972). The results of Fox also indicate that the fast transient inward component of the action current is associated with only a small fraction of the node membrane.

The conclusions on locally limited excitation sites are supported by the results obtained with certain nerve poisons. Extremely low concentrations of tetrodotoxin reduce and finally abolish the fast inward component of voltage clamp currents (for review, see Evans, 1972). The block action of this toxin is pH dependent and is associated with a p*K* value of about 5.3.

It thus appears that the ionic gateways responsible for the transient inward current involve protein organizations comprising only a small membrane fraction.

VI. Impedance and Heat Changes

A. *Impedance Changes*

The impedance change accompanying the action potential is one of the basic observations in electrophysiological studies on excitability (Cole, 1968). It is generally assumed that the impedance change reflects changes in ionic conductivities (ionic permeabilities). Figure 3 shows schematically that the conductance first rises steeply and in a second phase decays gradually toward the stationary level. In contrast to the pronounced resistance change, the membrane capacity apparently does not change during excitation (Cole, 1970). This result, too, suggests that only a small fraction of the membrane is involved in the rather drastic permeability changes during excitation.

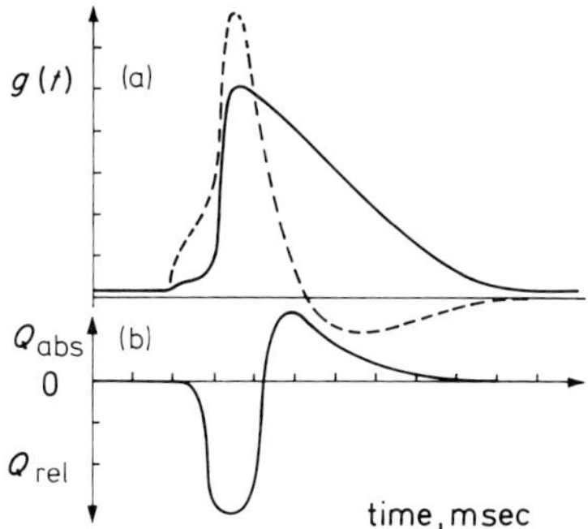

FIG. 3. Schematic representation of (a) conductivity change $g(t)$ accompanying the action potential (dashed line) and (b) heat exchange cycle. Q_{rel}, heat released during rising phase; Q_{abs}, heat absorbed during falling phase of the action potential.

B. Heat Changes

The action potential is accompanied by relatively large heat changes (Abbott *et al.*, 1958; Howarth *et al.*, 1968) (see Fig. 3). These heat changes may be thermodynamically modeled in terms of a cyclic variation of membrane states. The complex chain of molecular events during a spike may be simplified by the sequence of state changes $A \rightarrow B \rightarrow A$, where A represents the resting stationary state and B symbolizes the transiently excited state of higher ionic permeability.

The heat changes occur under practically isothermal–isobaric conditions. If we now associate the Gibbs free energy change ΔG, with the (overall) excitation process $A \rightleftharpoons B$, we may write

$$\Delta G = \Delta H - T\,\Delta S \tag{16}$$

where ΔH is the reaction enthalpy (as heat exchangeable with the environment) and ΔS the reaction entropy.

More recently, it has been confirmed that the rising phase of the action potential is accompanied by heat release Q_{rel}, while during the falling phase the heat Q_{abs} is reabsorbed. (Howarth *et al.*, 1968) (see Fig. 3). In our simple *A–B* model, the first phase is associated with

$$\Delta G_{A\rightarrow B} = \Delta H_{A\rightarrow B} - T\,\Delta S_{A\rightarrow B}$$

and the second one with

$$\Delta G_{B\rightarrow A} = \Delta H_{B\rightarrow A} - T\,\Delta S_{B\rightarrow A}$$

In general, for a cyclic process (where the original state is restored),

$$\oint dG = 0 \qquad \text{or} \qquad \Delta G_{A\rightarrow B} + \Delta G_{B\rightarrow A} = 0$$

In the hypothetical case of ideality (reversibility),

$$\Delta H = Q \qquad \text{and} \qquad Q_{\text{rel}} + Q_{\text{abs}} = 0$$

Since no *natural* process occurs ideally (i.e., completely reversible) there are always irreversible contributions. This means a part of $\Delta G_{A\to B}$ and of $\Delta G_{B\to A}$ will dissipate into heat. We may formally split ΔG in a reversible (exchangeable) contribution ΔG^{rev} and an irreversible contribution ΔG^{irr} (Neumann, 1973). Thus,

$$\Delta G_{A\to B} = \Delta G^{\text{rev}}_{A\to B} + \Delta G^{\text{irr}}_{A\to B}$$

$$\Delta G_{B\to A} = \Delta G^{\text{rev}}_{B\to A} + \Delta G^{\text{irr}}_{B\to A}$$

By definition $\Delta G^{\text{irr}} = -T\,\Delta S^{\text{irr}} \leq 0$, since the change in the inner entropy ΔS^{irr} is always larger or equal to zero (see, e.g., Prigogine, 1968). The measured heats are then given by

$$Q_{\text{rel}} = \Delta H^{\text{rev}}_{A\to B} + \Delta G^{\text{irr}}_{A\to B}$$

$$Q_{\text{abs}} = \Delta H^{\text{rev}}_{B\to A} + \Delta G^{\text{irr}}_{B\to A}$$

Since in a cycle $\Delta H^{\text{rev}}_{A\to B} + \Delta H^{\text{rev}}_{B\to A} = 0$, we find for the "difference" between heat released and heat absorbed,

$$\Delta Q = Q_{\text{rel}} + Q_{\text{abs}} = (\Delta G^{\text{irr}}_{A\to B} + \Delta G^{\text{irr}}_{B\to A}) = -T\,\Delta S^{\text{irr}} \leq 0$$

Owing to irreversible contributions we have for the absolute values

$$|Q_{\text{rel}}| > |Q_{\text{abs}}| \qquad (Q_{\text{rel}} \text{ counting negative!})$$

It is found experimentally, that $|Q_{\text{abs}}| \simeq 0.9\,|Q_{\text{rel}}|$. As stressed by Guggenheim (1949), $\Delta G^{\text{irr}} < 0$ or $\Delta S^{\text{irr}} > 0$, *only if phase changes and/or chemical reactions are involved.* Then, for our case we may write

$$\Delta Q = \Delta G^{\text{irr}} = -\sum_j A_j \xi_j < 0$$

where A is the affinity and ξ is the extent of membrane processes j involved (Neumann, 1973).

Since the action potential most likely involves only a small fraction of the excitable membrane, the measured heat changes Q_{rel} and Q_{abs} appear to be very large. Since on the other hand the mutual transition $A \to B \to A$ "readily" occurs, the value of ΔG ($= \Delta G_{A\to B} = -\Delta G_{B\to A}$) cannot be very large. In order to compensate a large ΔH (here $\simeq Q$), there must be a large value for ΔS; see Eq. (16). This means that the *entropy change associated with the membrane permeability change during excitation is also very large.*

It is, in principle, not possible to deduce from heat changes, the nature of the processes involved. However, large configurational changes —equivalent to a large overall ΔS—in biological systems frequently arise from conformational changes of macromolecules or macromolecular organizations such as membranes or from chemical reactions. In certain polyelectrolytic systems such changes even involve metastable states and irreversible transitions of domain structures (Neumann, 1973).

The large absolute values of Q (and the irreversible contribution ΔQ) suggest structural changes and/or chemical reactions to be associated with the action potential. Furthermore, there are observations indicating the occurrence of metastable states and nonequilibrium transitions in excitable membranes at least for certain perfusion conditions (Tasaki, 1968).

VII. The Cholinergic System and Excitability

Among the oldest known macromolecules associated with excitable membranes are some proteins of the cholinergic system. The cholinergic apparatus comprises acetylcholine (AcCh), the synthesis enzyme choline *O*-acetyltransferase (Ch-T), acetylcholinesterase (AcCh-E), acetylcholine-receptor (AcCh-R) and a storage site (S) for AcCh. Since details of this system are outlined elsewhere (Neumann and Nachmansohn, 1975), only a few aspects essential for our integral model of excitability are discussed here.

A. *Localization of the AcCh System*

Chemical analysis has revealed that the concentration of the cholinergic system is very different for various excitable cells. For instance, squid giant axons contain much less AcCh-E (and Ch-T) than axons of lobster walking legs (see, e.g., Brzin *et al.*, 1965); motor nerves are generally richer in cholinergic enzymes than sensory fibers (Gruber and Zenker, 1973).

The search for cholinergic enzymes in nerves was greatly stimulated by the observation that AcCh is released from isolated axons of various excitable cells, provided inhibitors of AcCh-E such as physostigmine (eserine) are present. This release is appreciably increased upon electric stimulation (Calabro, 1933; Lissak, 1939) or when axons are exposed to higher external K^+ ion concentrations (Dettbarn and Rosenberg, 1966). Since larger increases of external K^+ concentration depolarize excitable membranes, reduction of membrane potential appears to be a prerequisite for AcCh liberation from the storage site.

Evidence for the presence of axonal AcCh storage sites and receptors are still mainly indirect. However, direct evidence continuously accumulates from extrajunctional AcCh receptors (Porter *et al.*, 1973); binding studies with α-bungarotoxin (a nerve poison with a high affinity to AcCh-R) indicate the presence of receptor like proteins in axonal membrane fragments (Denburg *et al.*, 1972).

Recently, another protein which also binds AcCh and α-bungarotoxin, has been isolated by Chang from the electric organ of *Electrophorus electricus* (Chang, 1974). We recall that the receptor protein (R) may be operationally defined as the membrane component which upon binding of AcCh and other agonists causes a membrane permeability change and which upon binding of α-bungarotoxin and other antagonists prevents AcCh- (or electrically) induced permeability changes. To account for specificity and efficiency, the receptor protein should bind AcCh with a high binding constant and with a binding stoichiometry of not more than one or a few AcCh molecules per receptor protein. The other protein (called AcCh-R II by Chang) would not match this operational definition, because it has a high binding capacity for AcCh associated with a relatively low binding constant. This protein would, however, meet the properties required for a membrane storage site for AcCh and may thus be called S protein.

There are still discrepancies as to the presence and localization of the cholinergic system. However, the differences in the interpretations of chemico-analytical data and the results of histochemical light- and electron-microscopic investigations gradually begin to resolve; there appears progressive confirmation for the early chemical data of an ubiquitous cholinergic system. For instance, AcCh-E reaction products can be made visible in the excitable membranes of more and more nerves formerly called noncholinergic (see, e.g., Koelle, 1971). Catalysis products of AcCh-E are demonstrated in pre- and postsynaptic parts of excitable membranes (see, e.g., Barrnett, 1962; Koelle, 1971; Lewis and Shute, 1966). In a recent study, stain for the α-bungarotoxin-receptor complex is visible also in presynaptic parts of axonal membranes (see Fig. 1 in Porter *et al.*, 1973). These findings suggest the presence of the cholinergic system in both junctional membranes and thus render morphological support for the results of previous studies on the pre- and postsynaptic actions of AcCh and inhibitors and activators of the cholinergic system (Masland and Wigton, 1940; Riker *et al.*, 1959).

B. The Barrier Problem

Histochemical, biophysical, biochemical, and particularly, pharmacological studies on nerve tissue face the great difficulty of an enormous morphological and chemical complexity. It is now recognized that owing to various structural features not all types of nerve tissue are equally suited for certain investigations.

The excitable membranes are generally not easily accessible. The great majority of nerve membranes is covered with protective tissue layers of myelin, of Schwann, or glia cells. These protective layers insulating the excitable membrane frequently comprise structural and chemical barriers that impede the access of test compounds to the excitable membrane. In particular, the lipid-rich myelin sheaths are impervious to many quaternary ammonium compounds such as AcCh and *d*-tubocurarine (curare).

In some examples such as the frog neuromuscular junction, externally applied AcCh or the receptor inhibitor curare have relatively easy access to the synaptic gap whereas the excitable membranes of the motor nerve and the muscle fiber appear to be largely inaccessible. On the other hand, the cholinergic system of neuromuscular junctions of lobsters are protected against external action of these compounds whereas the axons of the walking legs of lobster react to AcCh and curare (cf., e.g., Dettbarn, 1967).

Penetration barriers also comprise absorption of test compounds within the protective layers. Furthermore, chemical barriers in the form of hydrolytic enzymes frequently cause decomposition of test compounds before they can reach the nerve membrane. For instance, phosphoryl phosphatases in the Schwann cell layer of squid giant axons cause hydrolysis of organophosphates such as the AcCh-E inhibitor diisopropyl fluorophosphate (DFP), and impulse conduction is blocked only at very high concentrations of DFP (Hoskin *et al.*, 1966).

A very serious source of error in concentration estimates and in interpretations of pharmacological data resides in procedures that involve homogenization of lipid-rich nerve tissue (see, e.g., Nachmansohn, 1969). For instance, homogenization liberates traces of inhibitors (previously applied) which despite intensive washing still adhered to the tissue. Even when the excitable membrane was not reached by the inhibitor, membrane components react with the inhibitor during homogenization (Hoskin *et al.*, 1969). Thus blockage of enzyme activity observed after homogenization is not necessarily an indicator for blockage during electrical activity (cf. e.g., Katz, 1966, p. 90).

As demonstrated in radiotracer studies, failure to interfere with bioelectricity is often concomitant with the failure of test compounds to reach the excitable membrane. Compounds such as AcCh or curare act on squid giant axons only after (enzymatic) reduction of structural barriers (Rosenberg and Hoskin, 1963). Diffusion barriers, even after partial reduction, are often the reason for longer incubation times and higher concentration of test compounds as compared to less protected membrane sites.

In this context it should be mentioned that the enzyme choline *O*-acetyltransferase, sometimes considered to be a more specific indicator of the cholinergic system, is frequently difficult to identify in tissue and is *in vitro* extremely unstable (Nachmansohn, 1963).

In the light of barrier and homogenization problems it appears obvious that any statement on the absence of the cholinergic system or on the failure of blocking compounds to interfer with excitability are *only useful*, if they are based on evidence that the test compound had actually reached the nerve membrane.

C. Electrogenic Aspects of the AcCh System

Particularly suggestive for the bioelectric function of the cholinergic system in axons are electrical changes resulting from eserine and curare application to nodes of Ranvier, where permeability barriers are less pronounced (Dettbarn, 1960a,b). Similar to the responses of certain neuromuscular junctions, eserine prolongs potential changes also at nodes; curare first reduces the amplitude of the nodal action potential and then also decreases the intensity of subthreshold potential changes in a similar way as known for frog junctions (see Fig. 4).

In all nerves the generation of action potentials is readily blocked by (easily permeating) local anesthetics such as procaine or tetracaine. Owing to structural and certain functional resemblance to AcCh which

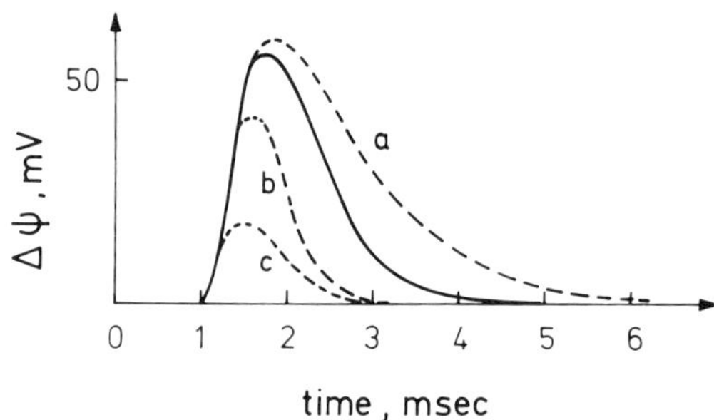

FIG. 4. Action potentials measured at a node of Ranvier of a single fiber of frog sciatic nerve, $T = 23°C$; solid line: reference; Curve a, in the presence of 5×10^{-6} *M* eserine at pH 7.7 (10^{-3} *M* eserine blocks the action potential within 30 sec). Curves b and c, gradual decrease of the amplitude with time in the presence of 10^{-3} *M* curare. The inhibitor solution has been applied externally on top of the Schwann cell layers covering the excitable membrane at the nodes. (Scheme from data by Dettbarn, 1960a).

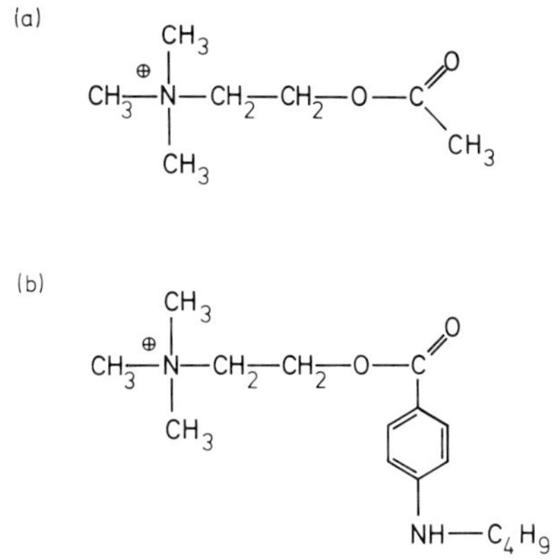

FIG. 5. Chemical structure of (a) the acetylcholine ion, and (b) the tetracaine ion. Note that the structural difference is restricted to the acid residue: (a) CH_3 and (b) the aminobenzoic acid residue rendering tetracaine lipid premeable.

is particularly pronounced for tetracaine (see Fig. 5), these compounds may be considered as analogs of AcCh. In a recent study it is convincingly demonstrated how (by chemical substitution at the ester group) AcCh is successively transformed from a receptor activator (reaching junctional parts only) to the receptor inhibitor tetracaine reaching readily junctional and axonal parts of excitable membranes (Bartels, 1965). Local anesthetics are also readily absorbed in lipid bilayer domains of biomembranes (for review, see Seeman, 1972).

External application of AcCh without esterase inhibitors faces not only diffusion barriers but also esterase activity which increases the local proton concentration (Dettbarn, 1967); the resulting changes in pH may contribute to changes in membrane potential.

Only very few nerve preparations appear to be suited to demonstrate a direct electrogenic action of externally applied AcCh. Diffusion barriers and differences in local concentrations of the cholinergic system may be the reason that the impulse condition for the generation of action potentials cannot be everywhere fulfilled (see Section IV,B). Some neuroblastoma cells produce subthreshold potential changes and action potentials upon electrical stimulation as well as upon AcCh application (Harris and Dennis, 1970; Nelson *et al.*, 1971; Hamprecht, 1974).

There are thus, without any doubt, many pharmacological and chemical similarities between synaptic and axonal parts of excitable membranes *as far as the cholinergic system is concerned.* On the other hand, there are various differences. But it seems that these differences can be accounted for by structural and chemical factors.

As to this problem two extreme positions of interpretation may be distinguished. On the one side more emphasis is put to the differences between axonal and junctional membranes. An extreme view considers the responses of axons to AcCh and structural analogs as a pharmacological curiosity (Ritchie, 1963); and, in general, the cholinergic nature of excitable membrane is not recognized and acknowledged. However, the presence of the cholinergic system in axons and the various similarities to synaptic behavior suggest the same basic mechanism for the cholinergic system in axonal and synaptic parts of excitable membranes.

Since there is until now *no direct experimental evidence for AcCh to cross the synaptic gap*, the action of AcCh may be alternatively assumed to be restricted to the interior of the excitable membrane of junctions and axons. This assumption is based on the fact that no trace of AcCh is detectable outside the nerve unless AcCh-E inhibitors are present. According to this alternative hypothesis, *intramembraneous* AcCh combines with the receptor and causes permeability changes mediated by conformational changes of the AcCh-receptor.

This is the basic postulate of the chemical theory of bioelectricity, attributing the primary events of all forms of excitability in biological organisms to the cholinergic system: in axonal conduction, for subthreshold changes (electronic spread) in axons and pre- and postsynaptic parts of excitable membranes.

In the framework of the chemical model, the various types of responses, excitatory and inhibitory synaptic properties are associated with structural and chemical modifications of the *same basic mechanism involving the cholinergic system.* Participation of neuroeffectors like the catecholamines or GABA and other additional reactions within the synapse possibly give rise to the various forms of depolarizing and hyperpolarizing potential changes in postsynaptic parts of excitable membranes. The question of coupling between pre- and postsynaptic events during signal transmission cannot be answered for the time being. It is, however, suggestive to incorporate in transmission models the transient increase of the K^+ ion concentration in the synaptic gap after a presynaptic impulse.

D. Control Function of AcCh

It is recalled that transient potential changes such as the action potential result from permeability changes caused by proper stimulation. However, a (normally proper) stimulus does not cause an action potential, if (among others) certain inhibitory analogs of the cholinergic agents such as, e.g., tetracaine are present. Tetracaine also reduces the amplitudes of subthreshold potential changes; in the presence of local anesthetics, for instance procaine, mechanical compression does not evoke action potentials (Julian and Goldman, 1962). It thus appears that the (electrical and mechanical) stimulus does *not directly* effect sub- and suprathreshold permeability changes, suggesting preceding events involving the cholinergic system.

If the AcCh-E is inhibited or the amount of this enzyme is reduced by protease action, subthreshold potential changes and (postsynaptic) current flows, as well as the action potential, are prolonged (see, e.g., Dettbarn, 1960a; Takeuchi and Takeuchi, 1972). Thus AcCh-E activity appears to play an essential role in terminating the transient permeability changes. The extremely high turnover number of this enzyme (about 1.4×10^4 AcCh molecules per second, i.e., a turnover time of 70 μsec) is compatible with a rapid removal of AcCh (Nachmansohn, 1955a).

In summary, the various studies using activators and inhibitors of the cholinergic system indicate that both initiation and termination of the permeability changes during nerve activity are (active) processes associated with AcCh. It seems, however, possible to decouple the cholinergic control system from the ionic permeation sites or gateways. Reduction of the external Ca^{2+} ion concentration appears to cause such a decoupling; the result is an increase in potential fluctuations or even random, i.e. uncontrolled firing of action potentials (see, e.g., Cole, 1968).

VIII. The Integral Model

In the previous sections some basic electrophysiological observations and biochemical data are discussed that any adequate model for bioelectricity has to integrate. It is stressed that among the features excitability models have to reproduce, are the threshold behavior, the various similarities of sub- and suprathreshold responses, stimulus characteristics, and the various forms of conditioning and history-dependent behavior.

In the present account we explore some previously introduced concepts for the control of electrical membrane properties by the cholinergic system (Neumann *et al.*, 1973). Among these fundamental concepts are the notion of a basic excitation unit, the assumption of an AcCh storage site particularly sensitive to the electric field of the excitable membrane, and the idea of a continuous sequential translocation of AcCh through the cholinergic proteins (AcCh cycle). Finally we proceed towards a formulation of various excitation parameters in terms of nonequilibrium thermodynamics.

A. *Key Processes*

In order to account for the various interdependencies between electrical and chemical parameters it is necessary to distinguish between a minimum number of single reactions associated with excitation.

A possible formulation of some of these processes in terms of chemical reactions has been previously given (Neumann *et al.*, 1973). The reaction scheme is briefly summarized.

a. Supply of AcCh to the membrane storage site (S), following synthesis (formally from the hydrolysis products choline and acetate):

$$S + AcCh \rightleftharpoons S(AcCh) \quad (17)$$

For the uptake reaction two assumptions are made: (*i*) the degree of AcCh association with the binding configuration S increases with increasing membrane potential (cell interior negative), (*ii*) the uptake rate is limited by the conformational transition from state S to S(AcCh), and is slow as compared to the following translocation steps (see also Section VIII,E,3).

Vesicular storage of AcCh as indicated by Whittaker and co-workers (see, e.g., Whittaker, 1973) is considered as additional storage for membrane sites of high AcCh turnover, for instance at synapses.

b. Release of AcCh from the storage form S(AcCh), for instance by depolarizing stimulation:

$$S(AcCh) \rightleftharpoons S' + AcCh \quad (18)$$

Whereas S(AcCh) is stabilized at large (negative) membrane fields, S′ is more stable at small intensities of the membrane field. The field-dependent conformational changes of S are assumed to gate the path of AcCh to the AcCh-receptors.

The assumptions for the dynamic behavior of the storage translocation sequence

$$S + AcCh \underset{k_{-1}}{\overset{k_1}{\rightleftharpoons}} S(AcCh) \underset{k_{-2}}{\overset{k_2}{\rightleftharpoons}} S' + AcCh \quad (19)$$

may be summarized as follows: the rate constant k_2 for the release step is larger than the rate constant k_1 for the uptake, and also $k_2 \gg k_{-2}$ (see also Section VIII,E).

c. Translocation of released AcCh to the AcCh-receptor (R) and association with the Ca^{2+}-binding conformation $R(Ca^{2+})$. This association is assumed to induce a conformational change to R′ that, in turn, releases Ca^{+} ions:

$$R(Ca^{2+}) + AcCh \rightleftharpoons R'(AcCh) + Ca^{2+} \quad (20)$$

d. Release of Ca^{+} ions is assumed to change structure and organization of gateway components, G. The structural change from a closed configuration, G, to an open state, G′, increases the permeability for passive ion fluxes.

e. AcCh hydrolysis: Translocation of AcCh from R′ (AcCh) to the AcCh-E, E, involving a conformational transition from E to E′:

$$R'(AcCh) + E \rightleftharpoons E'(Ch^{+}, Ac^{-}, H^{+}) + R' \tag{21}$$

The hydrolysis reaction causes the termination of the permeability change by reuptake of Ca^{2+},

$$R' + Ca^{2+} \rightleftharpoons R(Ca^{2+}) \tag{22}$$

concomitant with the relaxation of the gateway to the closed configuration, G.

Thus, the reactions (21) and (22) "close" a reaction cycle which is formally "opened" with reactions (17) and (19).

Since under physiological conditions (i.e., without esterase inhibitor) no trace of AcCh is detectable outside the excitable membrane (axonal and synaptic parts), the sequence of events modeled in the above reaction scheme, is suggested to occur in a specifically organized structure of the cholinergic proteins; a structure that is intimately associated with the excitable membrane.

B. Basic Excitation Unit

Before proceeding toward a model for the organization of the cholinergic system, it is instructive to consider the following well-known electrophysiological observations.

In a great variety of excitable cells the threshold potential change to trigger the action potential is about 20 mV. This voltage change corresponds to an energy input per charge or charged group within the membrane field of only about 1 kT unit of thermal energy (k, Boltzmann constant; T, absolute temperature) at body temperature. If only one charge or charged group would be involved, thermal motion should be able to initiate the impulse. Since random "firing" is very seldom, we have to conclude that several ions and ionic groups have to "cooperate" in a concerted way in order to cause a suprathreshold permeability change.

Furthermore, there are various electrophysiological data which suggest at least two types of gateways for ion permeation in excitable membranes (for summary, see Cole, 1968): a rapidly operating ion passage normally gating the passive flow of Na^+ ions (into the cell interior) and permeation sites that normally limit passive K^+ ion flow.

There are various indications, such as the direction of potential change and of current flow, suggesting that the rising phase of the action potential has contributions predominantly from the "rapid gateway"; the falling phase of the overall permeability change involves larger contributions of the K ion gateways (see also Neher and Lux, 1973). There is certainly coupling between the two gateway types: electrically through field changes and possibly also through Ca^{2+} ions transiently liberated from the "rapid gateways." As explicitly indicated in Eqs. (20) and (22), Ca^{2+} ion movement precedes and follows the gateway transitions.

Recent electrophysiological studies on neuroblastoma cells confirm the essential role of Ca^{2+} ions in subthreshold potential changes and the gating phase of the action potential (Spector *et al.*, 1973).

At the present stage of our model development we associate the direct cholinergic control of permeability changes only to the rapidly operating gateway, G. As seen in Fig. 3 the rising phase of the conductance change caused by the permeability increase is rather steep. This observation, too, supports a cooperative model for the mechanism of the action potential.

The experimentally indicated functional cooperativity together with the (experimentally suggested) locally limited excitation sites, suggests a structural anchorage in a cooperatively stabilized membrane domain.

In order to account for the various boundary conditions discussed above we have introduced the *notion of a basic excitation unit* (*BEU*). Such a unit is suggested to consist of a gateway G, that is surrounded by the cholinergic control system. The control elements are interlocked complexes of storage (S), receptor (R), and esterase (E), and are called SRE-assemblies. These assemblies may be organized in different ways and, for various membrane types, the BEU's may comprise different numbers of SRE assemblies. As an example, the BEU schematically represented in Fig. 6* contain 6 SRE units controlling the permeation site G.

The core of the BEU is a region of dynamically coupled membrane components with fixed charges and counterions such as Ca^{2+} ions. Figure 6 shows that the receptors of the SRE assemblies form a ringlike array. We assume that this structure is cooperatively stabilized and, through Ca^{2+} ions, intimately associated with the gateway components. In this

* Figure 6 is identical with Fig. 18 in Supplement I. It is reproduced there for the convenience of the reader.

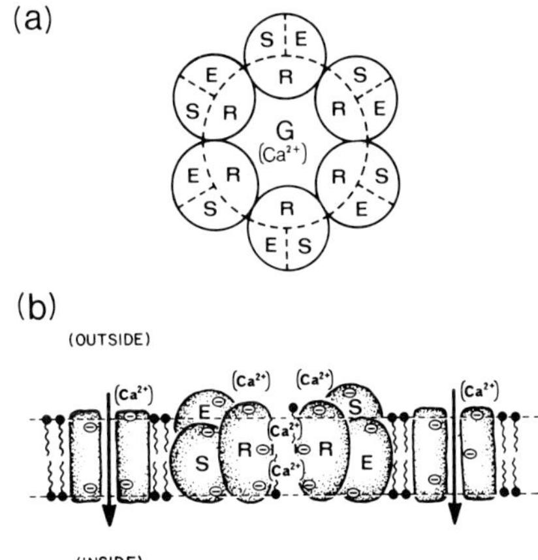

FIG. 6. Scheme of the AcCh-controlled gateway G. (a) Basic excitation unit (BEU) containing in this example 6 SRE-assemblies, viewed perpendicular to the membrane surface. S, AcCh-storage site; R, AcCh-receptor protein; E, AcCh-esterase. (b) Cross section through a BEU flanked by two units which model ion passages for K^+ ions; the arrows represent the local electrical field vectors due to partial permeselectivity to K^+ ions in the resting stationary state. The minus signs ⊖ symbolize negatively charged groups of membrane components. (Neumann, 1974.)

way the Ca^{2+}-dependent conformational dynamics of the receptors is coupled to the transition behavior of the gateway.

The receptor ring of a BEU is surrounded by the "ring" of the storage sites and (spatially separated) by the "ring" of the AcCh-esterases. The interfaces between the different rings define local reaction spaces through which AcCh is exchanged and translocated.

The BEU's are assumed to be distributed over the entire excitable membrane; the BEU density may vary for different membrane parts. The high density of cholinergic proteins found in some examples may be due to clustering of BEU's.

Different membrane types may not only vary in the number of SRE assemblies per BEU but also in the type and organization of the gateway components, thus assuring permselectivities for various ion types, particularly in synaptic parts of excitable membranes. It is only the cholinergic control system, the SRE-element, which is assumed to be same for all types of rapidly controlled gateways for passive ion flows.

1. Action Potential

In the framework of the integral model, the induction of an action potential is based on cooperativity between several SRE assemblies per

BEU. In order to initiate an action potential, a certain critical number of receptors, $\bar{m}^c$ per BEU, has on the average to be activated within a certain critical time interval Δt^c (impulse condition). During this time interval at least, say 4 out of 6 SRE assemblies have to process AcCh in a concerted manner. Under physiological conditions only a small fraction of BEU's is required to generate and propagate the nerve impulse.

2. Subthreshold Responses

Subthreshold changes of the membrane are seen to involve only a few single SRE assemblies of a BEU. On average not more than one or two SRE elements per BEU are assumed to contribute to the measured responses (within time intervals of the duration of Δt^c).

The (small) permeability change caused by Ca^+ release from the receptor thus results from only a small part of the interface between receptor and gateway components of a BEU: the ion exchanges $AcCh^+/Ca^{2+}$ and Na^+ are locally limited.

In the framework of this model, spatially and temporally attenuating electrical activity such as subthreshold axonal, postsynaptic, and dendritic potentials are the sum of spatially and temporally additive contributions resulting from the local subthreshold activity of many BEU's.

Although the permeability changes accompanying local activity are very small (as compared to those causing the action potentials) the summation over many contributions may result in large overall conductivity changes. Such changes may even occur (to a perhaps smaller extent) when the core of the gateway is blocked. It is suggested that compounds like tetrodotoxin and saxitoxin interact with the gateway core only, thus essentially not impeding subthreshold changes at the interface between receptor ring and gateway.

Influx of Ca^{2+}, particularly through pre- and postsynaptic membranes, may affect various intracellular processes leading, e.g., to release of hormones, catecholamines, etc.

C. Translocation Flux of AcCh

As discussed before, the excitable membrane as a part of a living cell is a nonequilibrium system characterized by complex chemodiffusional flow coupling. Although modern theoretical biology tends to regard living organisms only as quasistationary, with oscillations around a steady average, our integral model for the subthreshold behavior of excitable membranes is restricted to stationarity. We assume that the "living" excitable membrane (even under resting conditions) is in a state of continuous subthreshold activity (maintained either aerobically or anaerobically). However, the nonequilibrium formalism developed later on in this section, can also be extended to cover nonlinear behavior such as oscillations in membrane parameters.

In the framework of the integral model, continuous subthreshold activity is also reflected in a continuous sequential translocation of AcCh through the cholinergic system. The SRE elements comprise reaction spaces with continuous input by synthesis (Ch-T) and output by the virtually irreversible hydrolysis of AcCh. Input and output of the control system are thus controlled by enzyme catalysis.

1. Reaction Schemes

Since AcCh is a cation, translocation may most readily occur along negatively fixed charges, and may involve concomitant anion transport or cation exchange. The reaction scheme formulated in Section VIII,A gives therefore only a rough picture. Storage, receptor, and esterase represent macromolecular subunit complexes with probably several binding sites, and the exact stoichiometry of the AcCh reactions is not known.

The conformationally mediated translocation of the AcCh ion, A^+, may then be reformulated by the following sequence:

a. Storage reaction

$$S(A^+) + C^+ \rightleftharpoons S'(C^+) + A^+ \tag{23}$$

(C^+ symbolizes a cation, 2 C^+ may be replaced by Ca^{2+}.)

b. Receptor reaction

$$A^+ + R(Ca^{2+}) \rightleftharpoons R'(A^+) + Ca^{2+} \tag{24}$$

c. Hydrolysis reaction

$$R'(A^+) + E \rightleftharpoons E'(A^+) + R' \rightarrow (Ch^+, Ac^-, H^+) \tag{25}$$

As already mentioned, the nucleation of the gateway transition (causing the action potential) requires the association of a critical number of A^+, $\bar{n}^c$, with the cooperative number of receptors, $\bar{m}^c$, in the Ca^{2+}-binding form $R(Ca^{2+})$, within a critical time interval Δt^c. This time interval is determined by the lifetime of a single receptor–acetylcholine association.

Using formally $\bar{n}^c$ and $\bar{m}^c$ as stoichiometric coefficients, the concerted reaction inducing gateway transition may be written:

$$\bar{n}^c A^+ + \bar{m}^c R(Ca^{2+}) = \bar{m}^c R'(\bar{n}^c A^+) + \bar{m}^c Ca^{2+} \tag{26}$$

Storage and receptor reactions, Eqs. (23) and (24), represent gating processes preceding gateway opening ("Na activation") and causing the latency phase of the action potential. The hydrolysis process causes closure of the cholinergically controlled gateway ("Na inactivation"). In the course of these processes the electric field across the membrane changes, affecting all charged, dipolar, and polarizable components within the field. These field changes particularly influence the storage site and the membrane components controlling the K permeation regions (see Adam, 1970). Figure 7 shows a scheme modeling the "resting" stationary state and a transient phase of the excited membrane.

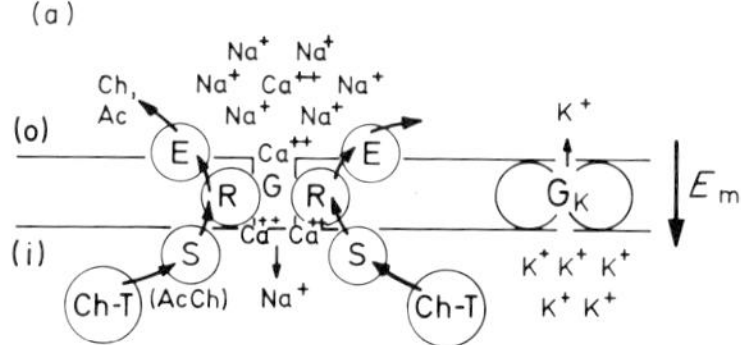

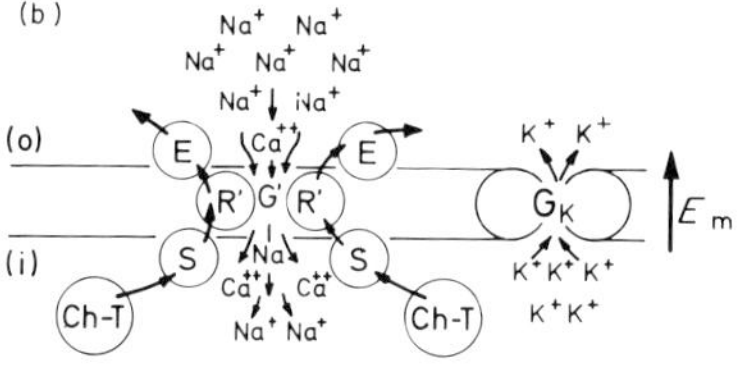

FIG. 7. Schematic representation of a membrane section (a) in the "resting" stationary state and (b) in a transient phase of excitation. In (a), the majority of the acetylcholine-receptors (R) is in the Ca^{2+}-binding conformation R; the cholinergically controlled rapidly operating gateway (G) is in the closed state G and the permeability for Na^+ (and Ca^{2+}) ions is very small as compared to the permeability for K^+ ions through the slow gateway, G_K. The electric field vector E_m pointing from the outside boundary (o) to the inside boundary (i) of the membrane is largely due to the K^+ ion gradient. In (b), most of the receptors are in the acetylcholine-binding conformation R and the rapid gateway is in its open configuration G (Na-activation phase). The change in the electric field (directed outward during the peak phase of the action potential) accompanying the transient Na^+ (and Ca^{2+}) influx causes a transient (slower) increase in the permeability of G_K thus inducing a (delayed) transient efflux of K^+ ions. Hydrolysis of acetylcholine (AcCh) leads to relaxation of R′ and G′ to R and G restoring the resting stationary state. Translocation of AcCh, occasionally in the resting stationary state and in a cooperatively increased manner after suprathreshold stimulation, through a storage site (S) of relatively large capacity, receptor, and AcCh-esterase (E) is indicated by the curved arrows. The hydrolysis products choline (Ch) and acetate (Ac) are transported through the membrane where intracellular choline *O*-acetyltransferase (Ch-T) may resynthetize AcCh (with increased rate in the refractory phase). (Neumann, 1974.)

The complexity of the nonlinear flow coupling underlying suprathreshold potential changes may be tractable in terms of the recently developed network thermodynamics covering inhomogeneity of the reaction space and nonlinearity (Oster *et al.*, 1973). An attempt at such an approach, which formally includes conformational metastability and hysteretic flow characteristics (Katchalsky and Spangler, 1968; see also Blumenthal *et al.*, 1970) is in preparation (Rawlings and Neumann, 1975).

2. Reaction Fluxes

For the nonequilibrium description of the translocation dynamics we may associate reaction fluxes with the translocation sequence, Eqs. (23) to (25).

a. The release flux is defined by

$$J(\mathrm{S}) = d[\bar{n}_{\mathrm{r}}]/dt \tag{27}$$

where $\bar{n}_{\mathrm{r}}$ is the average number of A^+ released into the reaction space between storage and receptor ring.

b. The receptor flux including association of A^+ and conformation change of R is given by

$$J(\mathrm{R}) = d[\bar{n}]/dt \tag{28}$$

where $\bar{n}$ is the average number of A^+ associated with R.

c. The esterase (or decomposition flux) is defined by

$$J(\mathrm{E}) = d[\bar{n}_{\mathrm{e}}]/dt \tag{29}$$

where $\bar{n}_{\mathrm{e}}$ is the average number of A^+ processed through AcCh-E.

Stationary states of the cholinergic activity are characterized by constant overall flow of AcCh; neither accumulations nor depletions of locally processed AcCh occur outside the limit of fluctuations. Thus, for stationary states,

$$J(\mathrm{S}) = J(\mathrm{R}) = J(\mathrm{E}) = \mathrm{const} \tag{30}$$

Statistically occurring small changes in membrane properties such as the so-called miniature end-plate potentials are interpreted to reflect amplified fluctuations in the subthreshold activity of the cholinergic system.

Oscillatory excitation behavior observed under certain conditions (see, e.g., Cole, 1968) may be modeled by periodic accumulation and depletion of AcCh in the reaction spaces of the BEU's.

D. Field Dependence of AcCh Storage

In the simplest case, a change of the membrane potential affects the chain of translocation events already at the beginning, i.e., at the storage site. Indeed, the observation of AcCh release by electrical stimulation or in response to K^+ ion-induced depolarization support the assumption of a field-dependent storage site for AcCh.

Denoting by $\bar{n}_b$ the amount of AcCh bound on the average to S we may define a distribution constant for the stationary state of the storage translocation by $K = \bar{n}_b/\bar{n}_r$. This constant (similar to an equilibrium constant) is a function of temperature T, pressure p, ionic strength I, and of the electric field E. A field dependence of K requires that the storage translocation reaction involve ionic, dipolar, or polarizable groups.

The isothermal–isobaric field dependence of K at constant ionic strength may be expressed by the familiar relation:

$$\left(\frac{\partial \ln K}{\partial E}\right)_{p,T,I} = \frac{\Delta M}{RT} \tag{31}$$

where ΔM is the reaction moment; ΔM is (proportional to) the difference in the permanent (or induced) dipole moments of reaction products and reactants. If a polarization process is associated with a finite value of ΔM, K should be proportional to E^2 (for relatively small field intensities up to 100 kV/cm). Furthermore, a small perturbation of the field causes major changes in K only on the level of higher fields (see, e.g., Eigen, 1967a). It is therefore of interest to recall that, under physiological conditions, excitable membranes generate action potentials only above a certain (negative) potential difference.

The suggestion of a field-induced conformational change in a storage protein to release AcCh derives from recent studies on field effects in macromolecular complexes and biomembranes. It has been found that electric impulses in the intensity similar to the depolarization voltage changes for the induction of action potentials are able to cause structural changes in biopolyelectrolytes (Neumann and Katchalsky, 1972; Revzin and Neumann, 1974) and permeability changes in vesicular membranes (Neumann and Rosenheck, 1972). In order to explain the results, a polarization mechanism has been proposed that is based on the displacement of the counterion atmosphere of polyelectrolytes or of oligoelectrolytic domains in membrane organizations.

If the conformational dynamics of the storage site indeed involves a polarization mechanism, we may represent the dependence of bound AcCh, $\bar{n}_b$, on the electric field of the membrane as shown in Fig. 8. Increasing membrane potential increases the amount of bound AcCh

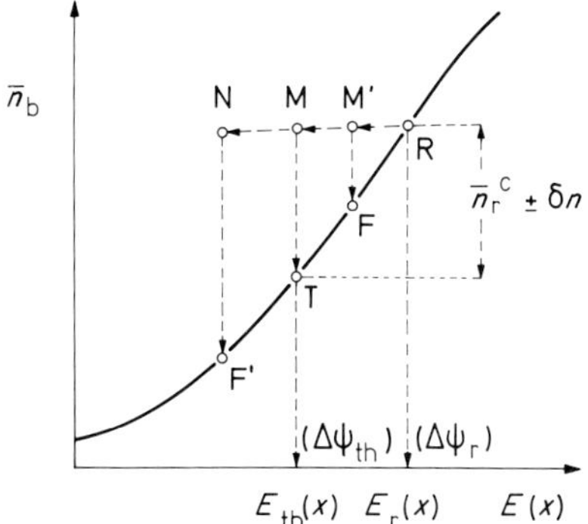

FIG. 8. Model representation of the field-dependent stationary states for AcCh storage. The mean number, $\bar{n}_b$, of AcCh ions bound to the storage site at the membrane site x of the release reaction, as a function of the electric field $E(x)$ (at constant pressure, temperature and ionic strength). The intervals M′F, MT, and NF′ correspond to the maximum number of AcCh ions released, $\bar{n}_r$, for 3 different depolarization steps: a subthreshold change from the resting state R to F, a threshold step R to T releasing the threshold or critical number $\bar{n}_r^c$ ($\pm\delta n$, fluctuation), and a suprathreshold step R to F′ with $\bar{n}_r > \bar{n}_r^c$. (Neumann, 1974.)

and thus also the number of AcCh ions that, after fast reduction of the membrane potential, is translocatable to the receptor.

E. Relaxation of AcCh Translocation Fluxes

It is recalled that the receptor reaction [cf. Eqs. (24) and (26)] plays a key role in coupling the control function of AcCh with the permeability change of the gateway. Uptake of AcCh from the storage ring, conformational transition, and Ca^{2+} release comprise a sequence of three single events. It is, therefore, assumed that the processing of AcCh through the receptor is slower than the preceding step of AcCh release from the storage. The receptor reaction is thus considered to be rate limiting. Therefore, any (fast) change in the membrane field will lead to either a transient accumulation or a depletion of AcCh in the reaction space between S-ring and R-ring of a BEU.

In the case of a fast depolarization there is first a transient increase in the storage flux $J(S)$ causing transient accumulation of AcCh in the S—R reaction space.

Since flux intensities increase with increasing driving forces (see, e.g., Katchalsky, 1967), $J(S)$ will increase with increasing perturbation intensity, thus causing an increase in the rate of all following processes.

It is recalled that in the framework of our integral model the time course of changes in electrical membrane parameters, such as the membrane potential, is controlled by the cholinergic system and the gateway dynamics.

1. Subthreshold Relaxations

Subthreshold perturbations do not induce the gateway transitions and are considered to cause membrane changes of small extent only. The time constant, τ_m, for subthreshold relaxations of chemical contributions to membrane potential changes [see Eq. (15)] is thus equal to the time constant τ_R of the rate-limiting receptor flux. For squid giant axons $\tau_R = \tau_m \simeq 1$ msec, at 20°C.

The relaxation of $J(R)$ to a lasting subthreshold perturbation (e.g., current stimulation) is given by

$$\frac{dJ(R)}{dt} = -\frac{1}{\tau_R}[J(R) - J'(R)] \tag{32}$$

where $J'(R)$ is the stationary value of the new flux. Equivalent to Eq. (32), we have for $\bar{n}$,

$$\frac{d[\bar{n}]}{dt} = -\frac{1}{\tau_R}([\bar{n}] - [\bar{n}]') \tag{33}$$

describing an exponential "annealing" to a new level of AcCh, $\bar{n}'$, processed through the receptor ring.

It is evident from the reaction scheme, Eqs. (23) to (25), that the time constant τ_R is the relaxation time of a coupled reaction system.

Applying a few simplifying assumptions and using normal mode analysis (see Eigen and DeMaeyer, 1963), we may readily calculate τ_R as a function of the various reaction parameters (Neumann, 1975).

2. Parameters of Suprathreshold Changes

It is recalled that the induction of an action potential is associated with three critical parameters:

$$\bar{n}^c, \bar{m}^c, \Delta t^c = \tau_R$$

For a perturbation, the intensity of which increases gradually with time, the condition $\bar{n} \geq \bar{n}_r^c$ corresponding to $\bar{n}_r \geq \bar{n}_r^c$ (within a BEU) can only be realized, if the minimum slope condition leading to

$$\frac{dJ(S)}{dt} \geq \frac{dJ(S)^{min}}{dt} \tag{34a}$$

and to [the equivalent expression for $J(\mathrm{R})$]

$$\frac{dJ(\mathrm{R})}{dt} \geq \frac{dJ(\mathrm{R})^{\min}}{dt} = \frac{d}{dt} J(\mathrm{R})_{\mathrm{th}} \tag{34b}$$

is fulfilled.

Equations (34a) and (34b) represent the "chemical minimum slope condition," and the threshold receptor flux $J(\mathrm{R})_{\mathrm{th}}$ may be considered as the equivalent to the rheobase (see Section IV,B).

For the fraction of BEU's necessary to evoke an action potential, the "chemical rheobase" may be specified by

$$J(\mathrm{R})_{\mathrm{th}} = [\bar{n}^{\mathrm{c}}]/\tau_{\mathrm{R}} \tag{35}$$

For rectangular (step) perturbations the threshold condition is

$$J(\mathrm{R}) \geq [\bar{n}^{\mathrm{c}}]/\tau_{\mathrm{R}} \tag{36}$$

Since $J(\mathrm{R})$ increases with the intensity of the (step) perturbation, the time intervals Δt ($<\tau_{\mathrm{R}}$) in which $\bar{n}^{\mathrm{c}}$ AcCh ions start to associate with the receptor, become smaller with larger stimulus intensities.

We may write this "strength-duration" relationship for suprathreshold perturbations in the form

$$J(\mathrm{R})\, \Delta t = [\bar{n}^{\mathrm{c}}] \tag{37}$$

Compare Eq. (10). The time intervals Δt of the receptor activation correspond to the observed latency phases.

Since $\bar{n}_{\mathrm{r}}^{\mathrm{c}}$ and $\bar{n}^{\mathrm{c}}$ are numbers describing functional cooperativity, the strength-duration product Eq. (37) does not depend on temperature. The fluctuations $\pm\delta n$ for n_{r}, however, increase with increasing temperature (and may finally lead to thermal triggering of action potentials). The chemical rheobase, Eq. (35) is a reaction rate which, in general, is temperature dependent (normally with a Q_{10} coefficient of about 2).

There are further aspects of electrophysiological observations which the integral model at the present stage of development may (at least qualitatively) reproduce.

If the membrane potential is slowly reduced, subthreshold flux relaxation of the ratio $\bar{n}_{\mathrm{b}}/\bar{n}_{\mathrm{r}}$ may keep $\bar{n}_{\mathrm{r}}$ always smaller than $\bar{n}_{\mathrm{r}}^{\mathrm{c}}$. Thus, corresponding to experience, slow depolarization does not (or only occasionally) evoke action potentials.

In order to match the condition $\bar{n}_{\mathrm{r}} \geq \bar{n}_{\mathrm{r}}^{\mathrm{c}}$, starting from the resting potential, the depolarization has in any case to go beyond the threshold potential, where

$$\bar{n}_{\mathrm{b}}(M) - \bar{n}_{\mathrm{b}}(T) = \bar{n}_{\mathrm{r}}^{\mathrm{c}}$$

(see Fig. 8).

For stationary membrane potentials $\Delta\psi < \Delta\psi_{th}$, the maximum number of AcCh ions that can be released by fast depolarization within Δt^c is smaller than $\bar{n}_r^c$. Thus, corresponding to experience, below a certain membrane potential, near below $\Delta\psi_{th}$, no nerve impulse can be generated.

3. Refractory Phenomena

After the gateway transition to the open state, the receptors of the BEU's have to return to the Ca^{2+}-binding conformation $R(Ca^{2+})$ before a second impulse can be evoked. Even if $\bar{n}^c$ AcCh ions would already be available, the time interval for the transition of $\bar{m}^c$ receptors is finite and causes the observed absolutely refractory phase.

Hyperpolarizing prepulses shift the stationary concentration of $\bar{n}_b$ to higher values (see Fig. 8). Owing to increased "filling degree" the storage site appears to be more sensitive to potential changes (leading, among others, to the so-called "off-responses"). On the other hand, depolarizing prepulses and preceding action potentials temporarily decrease the actual value of $\bar{n}_b$, thus requiring increased stimulus intensities for the induction of action potentials.

The assumptions for the kinetic properties of the storage site mentioned in the discussion of Eq. (19) are motivated by the accommodation behavior of excitable membranes. The observation of a relatively refractory phase suggests that the uptake of AcCh into the storage from $S(A^+)$ is slow compared to the release reaction. Therefore, after several impulses there is partial "exhaustion" of the storage site. If during the (slow) refilling phase there is a new stimulation, $\bar{n}_b$ may still be lower than the stationary level. Therefore, the membrane has to be depolarized to a larger extent in order to fulfill the action potential condition $\bar{n}_r \geq \bar{n}_r^c$.

F. The AcCh Control Cycle

The cyclic nature of a cholinergic permeability control in excitable membranes by processing AcCh through storage, receptor, esterase, and synthetase is already indicated in a reaction scheme developed 20 years ago (see Fig. 11 of Nachmansohn, 1955a). The complexity of mutual coupling between the various cycles directly or indirectly involved in the permeability control of the cholinergic gateway is schematically represented in Fig. 9. In this representation it may be readily seen that manipulations such as external application of AcCh and its inhibitory or activating congeners may interfere at *several sites* of the AcCh cycle.

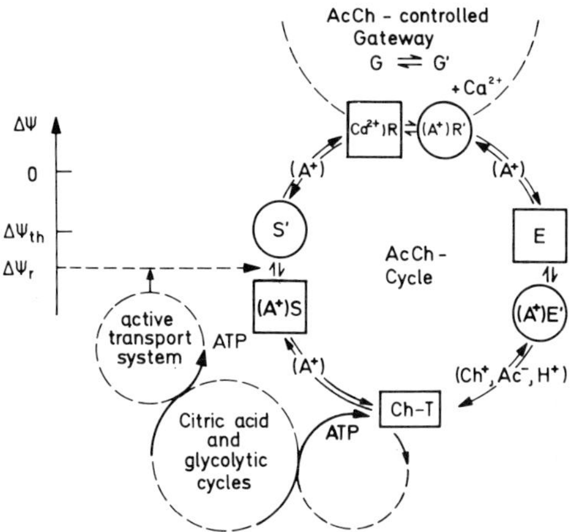

FIG. 9. Acetylcholine (AcCh) cycle for the cyclic chemical control of stationary membrane potentials $\Delta\psi$ and transient potential changes. The binding capacity of the storage site for AcCh is assumed to be dependent on the membrane potential $\Delta\psi$ and is thereby coupled to the active transport system (and the citric acid and glycolytic cycles). The control cycle for the gateway G (Ca^{2+} binding and closed) and G′ (open) comprises the SRE-assemblies (see Fig. 4) and the choline *O*-acetyltransferase (Ch-T); Ch-T couples the AcCh synthesis cycle to the translocation pathway of AcCh through the SRE assemblies. The continuous subthreshold flux of AcCh through such a subunit is maintained by the virtually irreversible hydrolysis of AcCh to choline (Ch^+), acetate (Ac^-), and protons (H^+) and by steady supply flux of AcCh to the storage from the synthesis cycle. In the resting stationary state, the membrane potential ($\Delta\psi_r$) reflects dynamic balance between active transport (and AcCh synthesis) and the flux of AcCh (through the control cycles surrounding the gateway) and of the various ions unsymmetrically distributed across the membrane. Fluctuations in membrane potential (and exchange currents) are presumably amplified by fluctuations in the local AcCh concentrations maintained at a stationary level during the continuous translocation of AcCh through the cycle (Neumann and Nachmansohn, 1975).

In particular, the analysis of pharmacological and chemical experiments faces the difficulty of this complexity.

In the previous sections it has been shown that basic parameters of electrophysiological phenomenology may be modeled in the framework of a nonequilibrium treatment of the cholinergic reaction system. The various assumptions and their motivations by experimental observations are discussed and the cholinergic reaction cycle is formulated in a chemical reaction scheme.

In conclusion, the integral model at the present level of development appears to cover all essential pharmaco-electrophysiological and biochemical data on excitable membranes. The model is expressed in specific reactions subject to further experimental investigations involving the reaction behavior of isolated membrane components as well as of membrane fragments containing these components in structure and organization.

IX. Summary

The mechanism of nerve excitability is still an unsolved problem. There are various mechanistic interpretations of nerve behavior, but these approaches cover only a part of the known facts and are thus selective and unsatisfactory.

An attempt at an integral interpretation of basic data well-established by electrophysiological, biochemical, and biophysical investigations was inspired by the late Aharon Katchalsky and a first essay has been given (Neumann *et al.,* 1973). The present account on nerve excitability is a further step towards a specific physicochemical theory of bioelectricity. In order to account for the various pharmacoelectrophysiological and biochemical observations on excitable membranes, the notion of a basic excitation unit is introduced. This notion is of fundamental importance for modeling details of sub- and suprathreshold responses such as threshold behavior and strength–duration curve, in terms of kinetic parameters for specific membrane processes.

Our integral model of excitability is based on the original chemical hypothesis for the control of bioelectricity (Nachmansohn, 1955a, 1971). This specific approach includes some frequently ignored experimental facts on AcCh-processing proteins in excitable membranes. According to the integral model, acetylcholine ions are continuously processed through the basic excitation units within excitable membranes; axonal, pre-, and postsynaptic parts. Excitability, i.e., the generation and propagation of nerve impulses, is due to an cooperative increase in the rate of AcCh translocation through the cholinergic control system.

At the present stage of the model, the cholinergic control is restricted to the rapidly operating ion (normally Na^+) carrying permeation sites. The variations in the electric field of the membrane, caused by the cholinergically controlled rapid gateway, in turn, affects the permeability of the slower ion (normally K^+) carrying permeation sites in the excitable membrane.

The basic biochemical data suggesting a cyclic cholinergic control (AcCh cycle) of the ion movements are presented in a concise form and some of the controversial interpretations of biochemical and electrophysiological data on excitability are discussed.

Note Added in Manuscript. Recently, it has been found that acetylcholine induces a conformational change in the isolated acetylcholine-receptor protein (from *Electrophorus electricus*). This configurational change alters the binding of calcium ions to the polyelectrolytic macromolecule. The kinetic analysis of this fundamentally important biochemical reaction [see Eq. (16)] results in number values for apparent rate constants and equilibrium parameters of the participating elementary processes, but also reveals the stoichiometry of the interactions between receptor, acetylcholine, and calcium ions (Chang and Neumann, 1975).

Acknowledgments

This study is based on numerous discussions with Professor David Nachmansohn whom I thank for the many efforts to reduce my ignorance in the biochemistry of excitable membranes. The critical comments of Dr. E. Neher on the electrophysiological part are gratefully acknowledged. Thanks are also due to Professor Manfred Eigen for his critical interest and generous support of this work. Finally, I would like to thank the Stiftung Volkswagenwerk for a grant.

References to Supplements I and II

Abbott, B. C., Hill, A. V., and Howarth, J. V. (1958). *Proc. Roy. Soc., Ser B.* **148,** 149.

Adam, G. (1970). *In* "Physical Principles of Biological Membranes" (F. Snell *et al.,* eds.), pp. 35–67. Gordon & Breach, New York.

Agin, D. (1967). *Proc. Nat. Acad. Sci. U.S.* **57,** 1232.

Agin, D. (1972). *In* "Foundations of Mathematical Biology" (R. Rosen, ed.), Vol. 1, pp. 253–277. Academic Press, New York.

Ariëns, E. J., ed. (1964). "Molecular Pharmacology," Vols. 1 and 2. Academic Press, New York.

Armett, J., and Ritchie, J. M. (1960). *J. Physiol. (London)* **152,** 141.

Ascher, L. (1925). *Pfluegers Arch. Gesamte Physiol. Menschen Tiere* **210,** 689.

Auger, D., and Fessard, A. (1939). *In* "Livro Homnagem aos Professores Alvaro e Miguel Ozorio de Almeida," p. 25. Rio de Janeiro, Brazil.

Augustinsson, K.-B., and Heimburger, G. (1955). *Acta Chem. Scand.* **8,** 310.

Baker, P. F., Hodgkin, A. L., and Shaw, T. I. (1962). *J. Physiol. (London)* **164,** 355.

Barrnett, R. J. (1962). *J. Cell Biol.* **12,** 247.

Bartels, E. (1962). *Biochem. Biophys. Acta* **63,** 365.

Bartels, E. (1965). *Biochim. Biophys. Acta* **109,** 194.

Bartels, E., and Nachmansohn, D. (1965). *Biochem. Z.* **342,** 359.

Bartels, E., Deal, W., Karlin, A., and Mautner, H. G. (1970). *Biochim. Biophys. Acta* **203,** 568.

Bartels, E., Wassermann, N. H., and Erlanger, B. F. (1971). *Proc. Nat. Acad. Sci. U.S.* **68,** 1820.

Bender, M. L. (1962). *J. Amer. Chem. Soc.* **84,** 2582.

Bender, M. L., and Kezdy, F. J. (1964). *J. Amer. Chem. Soc.* **86,** 3704.

Bender, M. L., and Kezdy, F. J. (1965). *Annu. Rev. Biochem.* **34,** 49.

Bender, M. L., Kezdy, F. J., and Gunter, C. R. (1964). *J. Amer. Chem. Soc.* **86,** 3714.

Benson, A. A. (1968). *In* "Membrane Models and the Formation of Biological Membranes" (L. Bolis and P. A. Pethica, eds.), p. 190. North-Holland Publ., Amsterdam.

Berg, P. (1956). *J. Biol. Chem.* **222,** 991 and 1015.

Bergmann, F., and Segal, R. (1954). *Biochem. J.* **58,** 692.

Bergmann, F., Riman, S., and Segal, R. (1958). *Biochem. J.* **68,** 493.

Berman, J. D., and Young, M. (1971). *Proc. Nat. Acad. Sci. U.S.* **68,** 395.

Berman, R., Wilson, I. B., and Nachmansohn, D. (1953). *Biochim. Biophys. Acta* **12,** 315.

Berman-Reisberg, R. (1954). *Biochim. Biophys. Acta* **14,** 442.

Berman-Reisberg, R. (1957). *Yale J. Biol. Med.* **29,** 403.

Betz, W., and Sakmann, B. (1971). *Nature New Biol. (London)* **232,** 94.

Beychok, S. (1965). *Biochem. Pharmacol.* **14,** 1249.

Biesecker, G. (1973). *Biochemistry* **12,** 4403.

Bieth, J., Vratsanos, S. M., Wassermann, N., and Erlanger, B. F. (1969). *Proc. Nat. Acad. Sci. U.S.* **64,** 1103.

Blasie, J. K., and Worthington, C. R. (1969). *J. Mol. Biol.* **39,** 417.

Blow, D. M. (1969). *Biochem. J.* **112**, 261.

Blow, D. M. (1971). *In* "The Enzymes" (P. D. Boyer, ed.), 3rd ed., Vol. 3, p. 185. Academic Press, New York.

Blow, D. M., and Steitz, T. A. (1970). *Annu. Rev. Biochem.* **39**, 63.

Blow, D. M., Birktoft, J. J., and Hartley, B. S. (1969) *Nature (London)* **221**, 337.

Blumenthal, R., Changeux, J.-P., and Lefevre, R. (1970). *J. Membrane Biol.* **2**, 351.

Bon, S., Rieger, F., and Massoulié, J. (1973). *Eur. J. Biochem.* **35**, 372.

Born, M. (1949). "Natural Philosophy of Cause and Chance." Oxford Univ. Press, London and New York.

Bourgeois, J. P., Popot, J. L., Ryter, A., and Changeux, J.-P. (1973). *Brain Res.* **62**, 557.

Brown, G. L., Dale, H. H., and Feldberg (1936). *J. Physiol. (London)* **87**, 394.

Brink, F. (1954). *Pharmacol. Rev.* **6**, 243.

Bruice, T. C. (1969). *Proc. Nat. Acad. Sci. U.S.* **47**, 1924.

Brzin, M. (1966). *Proc. Nat. Acad. Sci. U.S.* **56**, 1560.

Brzin, M., and Dettbarn, W.-D. (1967). *J. Cell Biol.* **32**, 577.

Brzin, M., Dettbarn, W.-D., Rosenberg, P., and Nachmansohn, D. (1965). *J. Cell Biol.* **26**, 353.

Burt, A. M., and Silver, A. (1973). *Nature (London)* **243**, 157.

Bullock, T. H., Nachmansohn, D., and Rothenberg, M. A. (1946). *J. Neurophysiol.* **9**, 9.

Calabro, Q. (1933). *Riv. Biol.* **15**, 299.

Canepa, F. G., Pauling, P., and Sörum, H. (1966). *Nature (London)* **210**, 907.

Carnay, L. D., and Tasaki, I. (1971). *In* "Biophysics and Physiology of Excitable Membranes" (W. J. Adelman, Jr., ed.), pp. 379–422. Van Nostrand-Reinhold, Princeton, New Jersey.

Carraway, K. L., Spoerl, P., and Koshland, D. E. (1969). *J. Mol. Biol.* **42**, 133.

Cecil, R. (1963). *In* "The Proteins" (H. Neurath, ed.), 2nd ed., Vol. 1, p. 456. Academic Press, New York.

Chang, C. C., and Lee, C. Y. (1963). *Arch. Int. Pharmacodyn. Ther.* **144**, 241.

Chang, H. W. (1974). *Proc. Nat. Acad. Sci. U.S.* **71**, 2113.

Chang, H. W., and Neumann, E. (1975). *Proc. Nat. Acad. Sci. U.S.* (in preparation).

Changeux, J.-P., and Podleski, T. R. (1968). *Proc. Nat. Acad. Sci. U.S.* **59**, 944.

Changeux, J.-P., and Thiéry, J. (1968). *BBA (Biochim. Biophys. Acta) Libr.* **11**, 116.

Changeux, J. P., Tung, Y., and Kittell, C. (1967). *Proc. Nat. Acad. Sci. U.S.* **57**, 335.

Changeux, J.-P., Gerhart, J. C., and Schachman, H. K. (1968a). *Biochemistry* **7**, 531.

Changeux, J.-P., Podleski, T. R., and Wofsy, L. (1968b). *Proc. Nat. Acad. Sci. U.S.* **58**, 2063.

Changeux, J.-P., Gautron, J., Israel, M., and Podleski, T. (1969). *C.R. Acad. Sci., Ser. D* **269**, 1788.

Changeux, J.-P., Kasai, M., and Lee, C. Y. (1970). *Proc. Nat. Acad. Sci. U.S.* **67**, 1241.

Chen, Y. T., Rosenberry, T. L., and Chang, H. W. (1974). *Arch. Biochem. Biophys.* **161**, 479.

Chu, S. H., and Mautner, H. G. (1966). *J. Org. Chem.* **31**, 308.

Chu, S. H., Hillman, G. R., and Mautner, H. G. (1972). *J. Med. Chem.* **15**, 760.

Clark, A. J. (1937). *In* "Handbuch der experimentellen Pharmakologie" (W. Heubner

and J. Schueller, eds.), Vol. IV, p. 63. Springer-Verlag, Berlin and New York.
Cleland, W. W. (1964). *Biochemistry* **3**, 480.
Cole, K. S. (1965). *Physiol. Rev.* **45**, 340.
Cole, K. S. (1968). *In* "Membranes, Ions and Impulses" (C. A. Tobias, ed.). Univ. of California Press, Berkeley.
Cole, K. S. (1970). *In* "Physical Principles of Biological Membranes" (F. Snell *et al.*, eds.), pp. 1–15. Gordon & Breach, New York.
Condrea, E., and Rosenberg, P. (1968). *Biochim. Biophys. Acta* **150**, 271.
Cowan, S. L. (1934). *Proc. Roy. Soc., Ser. B* **115**, 216.
Cuatrecasas, P. (1970). *J. Biol. Chem.* **245**, 3059.
Cuatrecasas, P. (1974). *Annu. Rev. Biochem.* **43**, 169.
Cuatrecasas, P., and Anfinsen, C. B. (1971a). *Annu. Rev. Biochem.* **40**, 259.
Cuatrecasas, P., and Anfinsen, C. B. (1971b). *In* "Methods in Enzymology" (W. B. Jakoby, ed.), Vol. 22, p. 345. Academic Press, New York.
Cuatrecasas, P., Wilcheck, M., and Anfinsen, C. B. (1968). *Proc. Nat. Acad. Sci. U.S.* **61**, 636.
Culvenor, C. C. J., and Ham, N. S. (1966). *Chem. Commun.* **15**, 537.
Currier, S. F., and Mautner, H. G. (1974). *Proc. Nat. Acad. Sci. U.S.* **71**, 3355.
Cushley, R. J., and Mautner, H. G. (1970). *Tetrahedron* **26**, 2151.
Dale, H. H., Feldberg, W., and Vogt, M. (1936). *J. Physiol. (London)* **86**, 353.
Davis, F. A., and Nachmansohn, D. (1964). *Biochim. Biophys. Acta* **88**, 384.
Deal, W. J., Erlanger, B. F., and Nachmansohn, D. (1969). *Proc. Nat. Acad. Sci. U.S.* **64**, 1230.
Del Castillo, J., Escobar, I., and Gijon, E. (1971). *Int. J. Neurosci.* **1**, 199.
Del Castillo, J., Bartels, E., and Sobrino, J. A. (1972). *Proc. Nat. Acad. Sci. U.S.* **69**, 2081.
De Lorenzo, A. J. D., Dettbarn, W.-D., and Brzin, M. (1968). *J. Ultrastruct. Res.* **24**, 367.
Denburg, J. L., Eldefrawi, M. E., and O'Brien, R. D. (1972). *Proc. Nat. Acad. Sci. U.S.* **69**, 177.
De Robertis, E., and Bennett, H. S. (1955). *J. Biophys. Biochem. Cytol.* **1**, 47.
De Robertis, E. D. P., De Iraldi, A. P., del Arnaiz, G. R., and Salganicoff, L. (1962). *J. Neurochem.* **9**, 23.
Dettbarn, W.-D. (1960a). *Biochim. Biophys. Acta* **41**, 377.
Dettbarn, W.-D. (1960b). *Nature (London)* **186**, 891.
Dettbarn, W.-D. (1961). *In* "Bioelectrogenesis" (C. Chagas and A. Paes de Carvalho, eds.), pp. 237–261. Elsevier (*London*) Amsterdam.
Dettbarn, W.-D. (1962). *Nature (London)* **194**, 1175.
Dettbarn, W.-D. (1963). *Life Sci.* **12**, 910.
Dettbarn, W.-D. (1967). *Ann. N.Y. Acad. Sci.* **144**, 483.
Dettbarn, W.-D., and Davis, F. A. (1963). *Biochim. Biophys. Acta* **66**, 397.
Dettbarn, W.-D., and Rosenberg, P. (1962). *Biochem. Pharmacol.* **11**, 1025.
Dettbarn, W.-D., and Rosenberg, P. (1966). *J. Gen. Physiol.* **50**, 447.
Dettbarn, W.-D., Rosenberg, P., and Nachmansohn, D. (1964). *Life Sci.* **3**, 55.
Dixon, G. H., Kauffman, D. L., and Neurath, H. (1958). *J. Biol. Chem.* **233**, 1373.
Dohan, J. S., and Woodward, G. E. (1939). *J. Biol. Chem.* **129**, 393.
Dubois, D. M., and Schoffeniels, E. (1974). *Proc. Nat. Acad. Sci. U.S.* **71**, 2858.
Dudai, Y., and Silman, I. (1971). *FEBS Lett.* **16**, 324.
Dudai, Y., and Silman, I. (1973). *FEBS Lett.* **30**, 49.
Dudai, Y., and Silman, I. (1974a). *J. Neurochem.* **22**, 19.

Dudai, Y., and Silman, I. (1974b). *Biochem. Biophys. Res. Commun.* **59**, 117.

Dudai, Y., Silman, I., Shinitzky, M., and Blumberg, S. (1972). *Proc. Nat. Acad. Sci. U.S.* **69**, 2400.

Dudai, Y., Herzberg, M., and Silman, I. (1973). *Proc. Nat. Acad. Sci. U.S.* **70**, 2473.

Eccles, J. C. (1935). *J. Physiol. (London)* **84**, 50p.

Edwards, G. A., Ruska, H., and de Harven, E. (1958). *J. Biophys. Biochem. Cytol.* **4**, 107.

Ehrenpreis, S. (1960). *Biochim. Biophys. Acta* **44**, 561.

Ehrlich, P. (1900). *Proc. Roy. Soc., Ser. B* **66**, 424.

Ehrlich, P. (1909). *Ber.,* **42**, 17.

Eigen, M. (1967a). *In* "The Neurosciences" (G. C. Quarton, T. Melnechuk, and F. O. Schmitt, eds.), pp. 130–142. Rockefeller Univ. Press, New York.

Eigen, M. (1967b). *In* "Fast Reactions and Primary Processes in Chemical Kinetics" (S. Claesson, ed.), p. 333. Wiley, New York.

Eigen, M., and DeMayer, L. (1963). *Tech. Org. Chem.* **8**, Part II, 895.

Eigen, M., and Hammes, G. G. (1963). *Advan. Enzymol.* **25**, 1.

Elbers, P. F. (1964). *Recent Progr. Surface Sci.* **3**, 443.

Eldefrawi, M. E., and Eldefrawi, A. T. (1973). *Arch. Biochem. Biophys.* **159**, 362.

Ellman, G. L. (1959). *Arch. Biochem.* **82**, 70.

Ernest, M. J., and Kim, K. H. (1973). *J. Biol. Chem.* **248**, 1550.

Evans, M. H. (1972). *Int. Rev. Neurobiol.* **15**, 83.

Feldberg, W., and Vartiainen, A. (1934). *J. Physiol. (London)* **83**, 103.

Fessard, A. (1946). *Ann. N.Y. Acad. Sci.* **47**, 501.

Fessard, A. (1958). *In* "Traité de Zoologie" (P.-P. Grassé, ed.), Vol. 13, pp. 1143–1238. Masson, Paris.

Fox, J. M. (1972). Ph.D. Dissertation, Univ. of Homburg-Saarbruecken.

Franklin, G. I., and Potter, L. T. (1972). *FEBS Lett.* **28**, 101.

Froede, H. C., and Wilson, I. B. (1970). *Isr. J. Med. Sci.* **6**, 179.

Fulpius, B., Cha, S., Klett, R., and Reich, E. (1972). *FEBS Lett.* **24**, 323.

Furchgott, R. F. (1964). *Annu. Rev. Pharmacol.* **4**, 21.

Gerhart, J. C., and Schachman, H. K. (1965). *Biochemistry* **4**, 1054.

Glover, V. A. S., and Potter, L. T. (1971). *J. Neurochem.* **18**, 571.

Gray, E. G., and Whittaker, V. P. (1962). *J. Anat.* **96**, 79.

Green, D. E., and MacLennan, D. H. (1969). *BioScience* **19**, 213.

Green, D. E., and Perdue, J. F. (1966). *Proc. Nat. Acad. Sci. U.S.* **55**, 1295.

Gruber, H., and Zenker, W. (1973). *Brain Res.* **51**, 207.

Guggenheim, E. A. (1949). "Thermodynamics." Interscience, New York.

Hall, Z. W. (1972). *Annu. Rev. Biochem.* **41**, 925.

Hall, Z. W., and Kelly, R. B. (1971). *Nature, New Biol.* **232**, 62.

Hamprecht, B. (1974). *Hoppe-Seyler's Z. Physiol. Chem.* **355**, 109.

Harris, A. J., and Dennis, M. J. (1970). *Science* **167**, 1253.

Hartley, B. S. (1964). *Nature (London)* **201**, 1284.

Hartley, B. S. (1970). *Phil. Trans. Roy. Soc. London, Ser. B* **257**, 77.

Hartley, B. S. (1974). *Ann. N.Y. Acad. Sci.* **227**, 438.

Hartley, B. S., and Kauffman, D. L. (1966). *Biochem. J.* **101**, 229.

Hartley, B. S., and Shotton, D. M. (1971). *In* "The Enzymes" (P. D. Boyer, ed.), 3rd ed., Vol. 3, p. 323. Academic Press, New York.

Hazelbauer, G. L., and Changeux, J.-P. (1974). *Proc. Nat. Acad. Sci. U.S.* **71**, 1479.

Hebb, C. O. (1957). *Physiol. Rev.* **37**, 196.

Hecht, S., Shlaer, S., and Pirenne, M. H. (1941). *J. Gen. Physiol.* **25**, 819.

Hendricks, S. B., and Borthwick, H. A. (1967). *Proc. Nat. Acad. Sci. U.S.* **58**, 2125.
Hess, G. H. (1971). *In* "The Enzymes" (P. D. Boyer, ed.), 3rd ed., Vol. 3, p. 213. Academic Press, New York.
Hestrin, S. (1950). *Biochim. Biophys. Acta* **4**, 310.
Higman, H. B., and Bartels, E. (1961). *Biochim. Biophys. Acta* **54**, 543.
Higman, H. B., and Bartels, E. (1962). *Biochim. Biophys. Acta* **57**, 77.
Higman, H. B., Podleski, T. R., and Bartels, E. (1963). *Biochim. Biophys. Acta* **75**, 187.
Higman, H. B., Podleski, T. R., and Bartels, E. (1964). *Biochim. Biophys. Acta* **79**, 138.
Hill, A. V. (1960). *In* "Molecular Biology" (D. Nachmansohn, ed.), pp. 17–24. Academic Press, New York.
Hinterbuchner, L. P., and Nachmansohn, D. (1960). *Biochim. Biophys. Acta* **44**, 554.
Hobbiger, F. W. (1963). *In* "Handbuch der experimentellen Pharmakologie" (G. B. Koelle, ed.), Vol. 15, p. 921. Springer-Verlag, Berlin and New York.
Hodgkin, A. L. (1951). *Biol. Rev. Cambridge Phil. Soc.* **26**, 338.
Hodgkin, A. L. (1964). "The Conduction of the Nervous Impulse." Thomas, Springfield, Illinois.
Hodgkin, A. L., and Huxley, A. F. (1952). *J. Physiol. (London)* **117**, 500.
Hodgkin, A. L., and Keynes, R. D. (1955). *J. Physiol. (London)* **128**, 61.
Hoffmann, B., with the collaboration of H. Dukas (1972). "Albert Einstein Creator and Rebel." Viking Press, New York.
Holzer, A. (1969). *Advan. Enzymol.* **32**, 297.
Hoskin, F. C. G., and Rosenberg, P. (1964). *J. Gen. Physiol.* **46**, 1117.
Hoskin, F. C. G., Rosenberg, P., and Brzin, M. (1966). *Proc. Nat. Acad. Sci. U.S* **55**, 1231.
Hoskin, F. C. G., Kremzner, L. T., and Rosenberg, P. (1969). *Biochem. Pharmacol.* **18**, 1727.
Howarth, J. V., Keynes, R. D., and Ritchie, J. M. (1968). *J. Physiol. (London)* **194**, 745.
Hubbard, J. I., and Schmidt, R. F. (1963). *J. Physiol. (London)* **166**, 145.
Hucho, F., and Changeux, J.-P. (1973). *FEBS Lett.* **38**, 11.
Huneeus-Cox, F., and Smith, F. H. (1965). *Biol. Bull.* **129**, 408.
Huneeus-Cox, F., Frenandez, H. L., and Smith, B. H. (1966). *Biophys. J.* **6**, 675.
Husain, S. S., and Mautner, H. G. (1973). *Proc. Nat. Acad. Sci. U.S.* **70**, 3749.
Jaffe, M. J. (1970). *Plant Physiol.* **46**, 768.
Jencks, W. P. (1963). *Annu. Rev. Biochem.* **32**, 639.
Julian, F. J., and Goldman, D. E. (1962). *J. Gen. Physiol.* **46**, 197.
Kalderon, N., Silman, I., Blumberg, S., and Dudai, Y. (1970). *Biochim. Biophys. Acta* **207**, 560.
Karlin, A. (1967). *J. Theor. Biol.* **16**, 306.
Karlin, A. (1969). *J. Gen. Physiol.* **54**, 245s.
Karlin, A., and Bartels, E. (1966). *Biochim. Biophys. Acta* **126**, 525.
Karlin, A., and Cowburn, D. (1973). *Proc. Nat. Acad. Sci. U.S.* **70**, 3636.
Karlin, A., and Cowburn, D. A. (1974). *In* "Neurochemistry of Cholinergic Receptors" (E. De Robertis and J. Schacht, eds.), pp. 37–48. Raven, New York.
Karlin, A., and Winnik, M. (1968). *Proc. Nat. Acad. U.S.* **60**, 668.
Karlsson, E., Heilbronn, E., and Widlund, L. (1972). *FEBS Lett.* **28**, 107.
Kasai, M., and Changeux, J.-P. (1970). *C.R. Acad. Sci., Ser. D* **270**, 1400.

Kasai, M., and Changeux, J.-P. (1971). *J. Membrane Biol.* **6**, 73.
Katchalsky, A. (1964). *In* "Connective Tissue: Intercellular Macromolecules," pp. 9–41. Little, Brown, Boston, Massachusetts.
Katchalsky, A. (1967). *In* "The Neurosciences" (G. C. Quarton, T. Melnechuk, a F. O. Schmitt, eds.), pp. 326–343. Rockefeller Univ. Press, New York.
Katchalsky, A., and Spangler, R. (1968). *Quart. Rev. Biophys.* **1**, 127.
Katz, B. (1966). "Nerve, Muscle and Synapse," McGraw Hill, New York.
Katz, B., and Miledi, R. (1965). *Proc. Roy. Soc., Ser. B* **161**, 453.
Kaufman, H., Vratsanos, S. M., and Erlanger, B. F. (1968). *Science* **162**, 1487.
Kewitz, H., Wilson, I. B., and Nachmansohn, D. (1956). *Arch. Biochem. Biophys.* **64**, 456.
Keynes, R. D., and Aubert, X. (1964). *Nature (London)* **203**, 261.
Kirtley, M. E., and Koshland, D. E., Jr. (1967). *J. Biol. Chem.* **242**, 4192.
Kitz, R. J., and Wilson, I. B. (1963). *J. Biol. Chem.* **238**, 745.
Kitz, R. J., Ginsburg, S., and Wilson, I. B. (1965). *Biochem. Pharmacol.* **14**, 1471.
Klett, R. P., Fulpius, B. W., Cooper, D., Smith, M., Reich, E., and Possani, L. D. (1973). *J. Biol. Chem.* **248**, 6841.
Koelle, G. B., ed. (1963a). "Handbuch der experimentellen Pharmakologie," Vol. 15. Springer-Verlag, Berlin and New York.
Koelle, G. B. (1963b). *In* "Handbuch der experimentellen Pharmakologie" (G. B. Koelle, ed.), Vol. 15, pp. 187–298. Springer-Verlag, Berlin and New York.
Koelle, G. B. (1971). *Ann. N.Y. Acad. Sci.* **183**, 5.
Koelle, G. B., and Friedenwald, J. S. (1949). *Proc. Soc. Exp. Biol. Med.* **70**, 617.
Koelle, G. B., Davis, R., Smyrl, E. G., and Fine, A. Z. (1974). *J. Histochem. Cytochem.* **22**, 252.
Korn, E. D. (1966). *Science* **153**, 1491.
Koshland, D. E., Jr. (1959). *In* "The Enzymes" (P. D. Boyer, H. Lardy, and K. Myrbäck, eds.), 2nd ed., Vol. 1, pp. 317–318. Academic Press, New York.
Koshland, D. E., Jr. (1969). *Harvey Lect.* **65**, 33.
Koshland, D. E., Jr. (1970). *In* "The Enzymes" (P. D. Boyer, ed.), 3rd ed., Vol. 1, p. 341. Academic Press, New York.
Koshland, D. E., Jr., and Neet, K. E. (1968). *Annu. Rev. Biochem.* **37**, 359.
Koshland, D. E., Jr., Strumeyer, D. H., and Ray, W. J., Jr. (1962). *Brookhaven Symp. Biol.* **15**, 101.
Kraut, J. (1971). *In* "The Enzymes" (P. D. Boyer, ed.), 3rd ed., Vol. 3, p. 547. Academic Press, New York.
Krebs, H. A. (1966). *In* "Current Aspects of Biochemical Energetics" (N. O. Kaplan and E. Kennedy, eds.), p. 83. Academic Press, New York.
Kremzner, L. T., and Rosenberg, P. (1971). *Biochem. Pharmacol.* **20**, 2953.
Kremzner, L. T., and Wilson, I. B. (1963). *J. Biol. Chem.* **238**, 1714.
Krupka, R. M. (1966a). *Biochemistry* **5**, 1983.
Krupka, R. M. (1966b). *Biochemistry* **5**, 1988.
Landowne, D. (1973). *Nature (London)* **242**, 457.
Lawler, H. C. (1961). *J. Biol. Chem.* **236**, 2296.
Lawler, H. C. (1963). *J. Biol. Chem.* **238**, 132.
Leach, S. L., Meschers, A., and Swanepoel, O. A. (1965). *Biochemistry* **4**, 23.
Lee, C. Y., and Chang, C. C. (1966). *Mem. Inst. Butantan, Sao Paulo* **33**, 555.
Lee, C. Y., Tseng, L. F., and Chin, T. H. (1967). *Nature (London)* **215**, 1177.
Lennard, J., and Singer, S. J. (1966). *Proc. Nat. Acad. Sci. U.S.* **56**, 1828.
Leuzinger, W., and Baker, A. L. (1967). *Proc. Nat. Acad. Sci. U.S.* **57**, 446.

Leuzinger, W., Baker, A. L., and Cauvin, E. (1968). *Proc. Nat. Acad. Sci. U.S.* **59**, 620.
Leuzinger, W., Goldberg, M., and Cauvin, E. (1969). *J. Mol. Biol.* **40**, 217.
Levitzki, A., and Koshlond, D. E., Jr. (1969). *Proc. Nat. Acad. Sci. U.S.* **62**, 1121.
Lewis, P. R., and Shute, C. C. D. (1966). *J. Cell Sci.* **1**, 381.
Lindstrom, J., and Patrick, J. (1973). *Proc. Nat. Acad. Sci. U.S.* **70**, 3334.
Lissak, K. (1939). *Amer. J. Physiol.* **127**, 263.
Loewenstein, W. R., ed. (1966). *Ann. N.Y. Acad. Sci.* **137**, 403.
Loewi, O. (1921). *Pfluegers Arch. Gesamte Physiol. Menschen Tiere* **189**, 239.
Lorente de Nó, R. (1938). *Amer. J. Physiol.* **121**, 331.
Lowry, O. H., Rosebrough, N. J., Farr, A. L., and Randall, R. J. (1951). *J. Biol. Chem.* **193**, 265.
Lundin, S. J., and Hellström, B. (1968). *Z. Zellforsch. Mikrosk. Anat.* **85**, 264.
McIntyre, A. R. (1959). *In* "Curare and Curare-like Agents" (D Bovet, F. Bovet-Nitti, and G. B. Marini-Bettolo, eds.), p. 211. Elsevier, Amsterdam.
Manson, L. A., ed. (1971). "Biomembranes," Vol. I, Plenum, New York.
Marchbanks, R. M. (1968). *Biochem. J.* **106**, 110 and 533.
Marchbanks, R. M. (1970). *FEBS Symp.* **21**, 285.
Margoliash, E. (1972). *Harvey Lect.* **66**, 171–247.
Marquis, J. K., and Mautner, H. G. (1974). *J. Membrane Biol.* **15**, 249.
Martin, R., and Rosenberg, P. (1968). *J. Cell Biol.* **36**, 341.
Masland, R. L., and Wigton, R. S. (1940). *J. Neurophysiol.* **3**, 269.
Massoulié, J., and Rieger, F. (1969). *Eur. J. Biochem.* **11**, 441.
Massoulié, J., Rieger, F., and Bon, S. (1971). *Eur. J. Biochem.* **21**, 542.
Mautner, H. G. (1967). *Pharmacol. Rev.* **19**, 107.
Mautner, H. G., and Bartels, E. (1970). *Proc. Nat. Acad. Sci. U.S.* **67**, 74.
Mautner, H. G., and Günther, W. H. H. (1961). *J. Amer. Chem. Soc.* **83**, 3342.
Mautner, H. G., Bartels, E., and Webb, G. D. (1966). *Biochem. Pharmacol.* **15**, 187.
Mazur, A. (1946). *J. Biol. Chem.* **164**, 271.
Mehrotra, K. N. (1961). *J. Insect Physiol.* **6**, 215.
Meister, A. (1968). *Harvey Lect.* **63**, 159.
Meunier, J.-C., Huchet, M., Boquet, P., and Changeux, J.-P. (1971). *C.R. Acad. Sci., Ser.* D **272**, 117.
Meunier, J.-C., Olsen, R. W., Menez, A., Fromageot, P., Boquet, P., and Changeux, J.-P. (1972). *Biochemistry* **11**, 1200.
Meunier, J.-C., Sugiyama, H., Cartaud, J., Sealock, R., and Changeux, J.-P. (1973). *Brain Res.* **62**, 307.
Monod, J., Changeux, J.-P., and Jacob, F. (1963). *J. Mol. Biol.* **6**, 306.
Monod, J., Wyman, J., and Changeux, J.-P. (1965). *J. Mol. Biol.* **12**, 88.
Morris, D. (1966). *Biochem. J.* **98**, 745.
Morris, D., Maneckjee, A., and Hebb, C. (1971). *Biochem. J.* **125**, 857.
Mounter, L. A. (1963). *In* "Handbuch der experimentellen Pharmakologie" (G. B. Koelle, ed.), Vol. 15, p. 486. Springer-Verlag, Berlin and New York.
Moyed, H. S., and Umbarger, H. E. (1962). *Physiol. Rev.* **42**, 444.
Myers, D. K., and Kemp, A., Jr. (1954). *Nature (London)* **173**, 33.
Nachmansohn, D. (1952). *In* "Modern Trends in Physiology and Biochemistry" (E. S. G. Barrón, ed.), p. 230. Academic Press, New York.
Nachmansohn, D. (1955a). *Harvey Lect.* **49**, 57.
Nachmansohn, D. (1955b). *Ergeb. Physiol., Biol. Chem. Exp. Pharmakol.* **48**, 575.

Nachmanson, D. (1955c). *In* "A Textbook of Physiology" (J. F. Fulton, ed.), 17th ed., p. 192. Saunders, Philadelphia, Pennsylvania.
Nachmansohn, D. (1963). *In* "Handbuch der experimentellen Pharmakologie" (G. B. Koelle, ed.), Vol. 15, pp. 40–45 and 701–740. Springer-Verlag, Berlin and New York.
Nachmansohn, D. (1966). *In* "Current Aspects of Biochemical Energetics" (N. O. Kaplan and E. P. Kennedy, eds.), pp. 145–192. Academic Press, New York.
Nachmansohn, D. (1969). *J. Gen. Physiol.* **54**, 187S–224S.
Nachmansohn, D. (1971). *In* "Handbook of Sensory Physiology" (W. R. Loewenstein, ed.), Vol. I, pp. 42–48. Springer-Verlag, Berlin and New York.
Nachmansohn, D. (1972a). *Annu. Rev. Biochem.* **41**, 1–28.
Nachmansohn, D. (1972b). *Proc. Nat. Acad. Sci. U.S.* **68**, 3170–3174.
Nachmansohn, D. (1973). *In* "The Structure and Function of Muscle" (G. H. Bourne, ed.), Vol. 3, pp. 103–104. Academic Press, New York.
Nachmansohn, D. (1975). *In* "Biochemistry of Sensory Functions" (L. Jaenicke, ed.), p. 43. Springer Verlag, Berlin and New York.
Nachmansohn, D., and Meyerhof, B. (1941). *J. Neurophysiol.* **4**, 348.
Nachmansohn, D., and Wilson, I. B. (1951). *Advan. Enzymol.* **12**, 259.
Narahashi, T., Moore, J. W., and Poston, R. N. (1969). *J. Neurobiol.* **1**, 3.
Nastuk, W. L. (1954). *Fed. Proc., Fed. Amer. Soc. Exp. Biol.* **13**, 104.
Neher, E., and Lux, H. D. (1973). *J. Gen. Physiol.* **61**, 385.
Nelson, P. G., Peacock, J. H., and Amano, T. (1971). *J. Cell. Physiol.* **77**, 353.
Nernst, W. (1888). *Z. Phys. Chem.* **2**, 613.
Nernst, W. (1889). *Z. Phys. Chem.* **4**, 129.
Neumann, E. (1973). *Angew. Chem. Int. Ed. Engl.* **12**, 356.
Neumann, E. (1974). *In* "Physics and Mathematics of the Nervous System" (M. Coward *et al.*, eds.), Vol. 4, pp. 42–81. Springer Verlag, Berlin and New York.
Neumann, E. (1975). In "Biochemistry of Sensory Functions" (L. Jaenicke, ed.), p. 465. Springer-Verlag, Berlin and New York.
Neumann, E., and Katchalsky, A. (1972). *Proc. Nat. Acad. Sci. U.S.* **69**, 993.
Neumann, E., and Nachmansohn, D. (1975). *In* "Biomembranes" (L. A. Manson, ed.) Vol. 7, p. 99. Plenum, New York.
Neumann, E., and Rosenheck, K. (1972). *J. Membrane Biol.* **10**, 279.
Neumann, E., Nachmansohn, D., and Katchalsky, A. (1973). *Proc. Nat. Acad. Sci. U.S.* **70**, 727.
Neurath, H. (1964). *Abstr., Int. Congr. Biochem., 6th, 1964* Vol. 4, p. 249.
Neurath, H., and Dixon, G. H. (1957). *Fed. Proc., Fed. Amer. Soc. Exp. Biol.* **16**, 792.
Nistratova, S. H., and Turpaev, T. M. (1959). *Biochemistry* (U.S.S.R.) **24**, 155.
O'Brien, R. D. (1960). "Toxic Phosphorus Esters." Academic Press, New York.
Olsen, R. W., Meunier, J.-C., and Changeux, J.-P. (1972). *FEBS Lett.* **28**, 96.
Oppenheimer, H. L., Labouesse, B., and Hess, G. P. (1966). *J. Biol. Chem.* **241**, 2720.
Oster, G. F., Perelson, A. S., and Katchalsky, A. (1973). *Quart. Rev. Biophys.* **6**, 1.
Planck, M. (1890a). *Ann. Phys. Chem.* [3] **39**, 161.
Planck, M. (1890b). *Ann. Phys. Chem.* [3] **40**, 561.
Planck, M. (1922). "Gesammelte Reden und Aufsaetze." Hirzel, Stuttgart.
Plonsey, R. (1969). "Bioelectric Phenomena." McGraw-Hill, New York.
Podleski, T. R. (1966). Ph.D. Dissertation, Faculty of Pure Science, Columbia University, New York.

Podleski, T. R. (1967). *Proc. Nat. Acad. Sci. U.S.* **58**, 268.

Podleski, T. R. (1969). *Biochem. Pharmacol.* **18**, 211.

Podleski, T. R. and Bartels, E. (1963). *Biochim. Biophys. Acta* **75**, 387.

Podleski, T. R., and Nachmansohn, D. (1966). *Proc. Nat. Acad. Sci. U.S.* **56**, 1034.

Podleski, T., Meunier, J.-C., and Changeux, J.-P. (1969). *Proc. Nat. Acad. Sci. U.S.* **63**, 1239.

Pohle, W., and Matthies, H. (1959). *Naunyn-Schmiedebergs Arch. Exp. Pathol. Pharmakol.* **236**, 253.

Pontremoli, S., and Horecker, B. L. (1970). *Curr. Top. Cell. Regul.* **2**, 173.

Porter, C. W., Chiu, T. H., Wieckowski, J., and Barnard, E. A. (1973). *Nature (London), New Biol.* **241**, 3.

Porter, K. R., and Bonneville, M. A. (1964). "An Introduction to the Fine Structure of Cells and Tissues." Lea & Febiger, Philadelphia, Pennsylvania.

Potter, L. T., Glover, V. A. S., and Saelens, J. K. (1968). *J. Biol. Chem.* **243**, 3864.

Poziomek, E. J., Kramer, D. D. N., Fromm, B. W., and Mosher, W. A. (1961a). *J. Org. Chem.* **26**, 423.

Poziomek, E. J., Mosher, W. A., and Michel, H. O. (1961b). *J. Amer. Chem. Soc.* **83**, 3916.

Prigogine, I. (1968). "Thermodynamics of Irreversible Processes," 3rd ed. Thomas, Springfield, Illinois.

Prince, A. K. (1966). *Biochem. Pharmacol.* **15**, 411.

Prince, A. K. (1967). *Proc. Nat. Acad. Sci. U.S.* **57**, 1117.

Racker, E. (1970). "Membranes of Mitochondria and Chloroplasts." Van Nostrand-Reinhold, New York.

Raftery, M. A., Schmidt, J., and Clark, D. G. (1972). *Arch. Biochem. Biophys.* **152**, 882.

Rama Sastry, B. V., and Henderson, G. I. (1972). *Biochem. Pharmacol.* **21**, 787.

Rawlings, P. K., and Neumann, E. (1975). *Proc. Nat. Acad. Sci. U.S.* (in press).

Revzin, A., and Neumann, E. (1974). *Biophys. Chem.* **2**, 144.

Rieger, F., Bon, S., and Massoulie, J. (1972a). *C.R. Acad. Sci., Ser. D.* **274**, 1753.

Rieger, F., Tsuji, S., and Massoulie, J. (1972b). *Eur. J. Biochem.* **30**, 73.

Riker, W. F. (1953). *Pharmacol. Rev.* **5**, 1.

Riker, W. F., Jr., Werner, G., Roberts, J. and Kuperman, A. (1959). *Ann. N.Y. Acad. Sci.* **81**, 328.

Ritchie, J. M. (1963). *Biochem. Pharmacol.* **12**(S), 3.

Robertson, J. D. (1960a). *In* "Progress in Biophysics" (B. Katz and J. A. V. Butler, eds.), pp. 343–418. Pergamon, Oxford.

Robertson, J. D. (1960b). *In* "Molecular Biology" (D. Nachmansohn, ed.), p. 137. Academic Press, New York.

Robertson, J. D. (1970). *In* "The Neurosciences" (F. O. Schmitt, ed.), Vol. 2, pp. 715–728. Rockefeller Univ. Press, New York.

Rosenberg, P. (1965). *Toxicon* **3**, 125.

Rosenberg, P. (1966). *Mem. Inst. Butantan, Sao Paulo* **33**, 477.

Rosenberg, P., and Condrea, E. (1968). *Biochem. Pharmacol.* **17**, 2033.

Rosenberg, P., and Dettbarn, W.-D. (1967). *Toxicon* **4**, 296.

Rosenberg, P., and Hoskin, F. C. G. (1963). *J. Gen. Physiol.* **46**, 1065.

Rosenberg, P., and Mautner, H. G. (1967). *Science* **155**, 1569.

Rosenberg, P., Mautner, H. G., and Nachmansohn, D. (1966). *Proc. Nat. Acad. Sci. U.S.* **55**, 835.

Rosenberry, T. L. (1975). *Advan. Enzymol.* (in press).

Rosenberry, T. L., and Bernhard, S. A. (1971). *Biochemistry* **10**, 4114.
Rosenberry, T. L., Chang, H. W., and Chen, Y. T. (1972). *J. Biol. Chem.* **247**, 1555.
Rosenberry, T. L., Chen, Y. T., and Bock, E. (1974). *Biochemistry* **13**, 3068.
Rossi-Fanelli, A., Antonini, E., and Caputo, A. (1964). *Advan. Protein Chem.* **19**, 73.
Rothenberg, M. A., and Nachmansohn, D. (1947). *J. Biol. Chem.* **168**, 223.
Rothenberg, M. A., Sprinson, D. B., and Nachmansohn, D. (1948). *J. Neurophysiol.* **11**, 111.
Rothfield, L. I., ed. (1971). "Structure and Function of Biological Membranes." Academic Press, New York.
Schachman, H. K. (1974). *Harvey Lect.* **68**, 67–113.
Schmidt, J., and Raftery, M. A. (1973). *Biochemistry* **12**, 852.
Schoellmann, G., and Shaw, E. (1963). *Biochemistry* **2**, 252.
Schoffeniels, E. (1957). *Biochim. Biophys. Acta* **26**, 585.
Schoffeniels, E. (1958). *Science* **127**, 1117.
Schoffeniels, E. (1959). Thèse d'agrégation, Université de Liège, Liège.
Schoffeniels, E., and Nachmansohn, D. (1957). *Biochim. Biophys. Acta* **26**, 1.
Schrodinger, E. (1944). "What is Life?" Cambridge Univ. Press, London and New York.
Seeman, P. (1972). *Pharmacol. Rev.* **24**, 585.
Segal, J. R. (1968). *Biophys. J.* **8**, 470.
Shefter, E., and Kennard, O. (1966). *Science* **153**, 1389.
Shefter, E., and Mautner, H. G. (1967). *J. Amer. Chem. Soc.* **89**, 1249.
Shefter, E., and Mautner, H. G. (1969). *Proc. Nat. Acad. Sci. U.S.* **63**, 1253.
Shintzky, M., Dudai, Y., and Silman, I. (1973). *FEBS Lett.* **30**, 125.
Silman, I. (1969). *J. Gen. Physiol.*, **54S**, 50 and 265.
Silman, I. (1970). *FEBS Symp.* **21**, 337.
Silman, I., and Karlin, A. (1967). *Proc. Nat. Acad. Sci. U.S.* **58**, 1664.
Silman, I., and Karlin, A. (1969). *Science* **164**, 1420.
Singer, S. J. (1971). *In* "Structure and Function of Biological Membranes" (L. I. Rothfield, ed.), p. 145. Academic Press, New York.
Sjöstrand, F. S. (1962). *Symp. Int. Soc. Cell Biol.* **1**, 47.
Sjöstrand, F. S. (1963). *J. Ultrastruct. Res.* **9**, 561.
Sjöstrand, F. S., and Barajas, L. (1968). *J. Ultrastruct. Res.* **25**, 121.
Sjöstrand, F. S., and Barajas, L. (1970). *J. Ultrastruct. Res.* **32**, 293.
Smith, E. L. (1968). *Harvey Lect.* **62**, 231–256.
Smith, H. M. (1958). *J. Cell. Comp. Physiol.* **51**, 161.
Spector, I., Kimhi, Y., and Nelson, P. G. (1973). *Nature* (*London, New Biol.* **246**, 124.
Spencer, T., and Sturtevant, J. M. (1959). *J. Amer. Chem. Soc.* **81**, 1874.
Stadtman, E. R. (1966). *Advan. Enzymol.* **28**, 41.
Stadtman, E. R. (1970a). *Harvey Lect.* **65**, 97.
Stadtman, E. R. (1970b). *In* "The Enzymes" (P. D. Boyer, ed.), 3rd ed., Vol. 1, p. 397. Academic Press, New York.
Stämpfli, R. (1956). *J. Physiol.* (*Paris*) **48**, 710.
Takeuchi, A., and Takeuchi, N. (1972). *Advan. Biophys.* **3**, 45–95.
Tasaki, I. (1968). "Nerve Excitation." Thomas, Springfield, Illinois.
Tasaki, I., and Singer, I. (1966). *Ann. N.Y. Acad. Sci.* **137**, 793.
Tasaki, I., and Takenada, T. (1964). *In* "The Cellular Functions of Membrane Transport" (J. F. Hoffman, ed.), p. 95. Prentice-Hall, Englewood Cliffs, New Jersey.

Tasaki, I., Singer, I., and Takenaka, T. (1965). *J. Gen. Physiol.* **48**, 1095.
Taylor, P., Jones, J. W., and Jacobs, N. M. (1974). *Mol. Pharmacol.* **10**, 78.
Tomas, N., Davis, R., and Koelle, G. B. (1975). In press.
Träuble, H., and Eibl, H. (1974). *Proc. Nat. Acad. Sci. U.S.* **71**, 214.
Umbarger, H. E. (1964). *Science* **145**, 674.
Van Asperen, K. (1958). *Nature (London)* **181**, 355.
Vanderkooi, G., and Green, D. E. (1971). *BioScience* **21**, 409.
Vincent, P., and Lazdunski, M. (1972). *Biochemistry* **11**, 2967.
von Muralt, A., and Staempfli, R. (1953). *Helv. Physiol. Pharmacol. Acta* **11**, 182.
Wald, G. (1968). *Science* **162**, 230.
Walsh, K. A., and Neurath, H. (1964). *Proc. Nat. Acad. Sci. U.S.* **52**, 884.
Webb, G. D. (1965). *Biochim. Biophys. Acta* **102**, 172.
Weber, M., and Changeux, J.-P. (1973). *Mol. Pharmacol.* **10**, 15 and 35.
Werner, G., and Kuperman, A. S. (1963). *In* "Handbuch der experimentellen Pharmakologie" (G. B. Koelle, ed.), Vol. 15, pp. 570–678. Springer-Verlag, Berlin and New York.
Wescoe, W. C., and Riker, W. F., Jr. (1951). *Ann. N.Y. Acad. Sci.* **54**, 438.
Westheimer, F. H. (1957). *Proc. Nat. Acad. Sci. U.S.* **43**, 969.
White, H. L., and Cavallito, C. J. (1970). *Biochim. Biophys. Acta* **206**, 343.
Whittaker, V. P. (1968). *Proc. Nat. Acad. Sci. U.S.* **60**, 1081.
Whittaker, V. P. (1973). *Naturwissenschaften* **60**, 281.
Wilson, I. B. (1951). *J. Biol. Chem.* **190**, 111.
Wilson, I. B. (1952). *J. Biol. Chem.* **197**, 215.
Wilson, I. B., and Meislich, E. K. (1953). *J. Am. Chem. Soc.* **75**, 4628.
Wilson, I. B., Ginsburg, S., and Meislich, E. K. (1955). *J. Amer. Chem. Soc.* **77**, 4286.
Wilson, I. B., and Ginsburg, S. (1955a). *Arch. Biochem. Biophys.* **54**, 569.
Wilson, I. B., and Ginsburg, S. (1955b). *Biochim. Biophys. Acta* **18**, 168.
Wilson, I. B., and Cabib, E. (1956). *J. Amer. Chem. Soc.* **78**, 202.
Wilson, I. B., and Harrison, M. A. (1961). *J. Biol. Chem.* **236**, 2292.
Wilson, I. B., and Quan, C. (1958). *Arch. Biochem. Biophys.* **73**, 131.
Wilson, I. B., Bergmann, F., and Nachmansohn, D. (1950). *J. Biol. Chem.* **186**, 781.
Wofsy, L., Metzger, H., and Singer, S. J. (1962). *Biochemistry* **1**, 1031.
Wright, C. S., Alden, R. A., and Kraut, J. (1969). *Nature (London)* **221**, 235.
Wyman, J., Jr. (1964). *Advan. Protein Chem.* **19**, 233.
Yeh, J. Z., and Narahashi, T. (1974). *J. Pharmacol. Exp. Ther.* **189**, 697.
Zelman, A., and Shih, H. H. (1972). *J. Theor. Biol.* **37**, 373.
Zenker, W., and Gruber, H. (1973). *Brain Res.* **51**, 207.

Index

B

C

D

E

F

G

H

I

K

O

P